21st Century Surface Science - a Handbook

Edited by Phuong Pham, Pratibha Goel,
Samir Kumar and Kavita Yadav

Published in London, United Kingdom

IntechOpen

Supporting open minds since 2005

21st Century Surface Science - a Handbook
http://dx.doi.org/10.5772/intechopen.87891
Edited by Phuong Pham, Pratibha Goel, Samir Kumar and Kavita Yadav

Contributors
Jie Zhang, Liu Hong, Orkut Sancakoğlu, Krishnacharya Khare, Meenaxi Sharma, Yeeli Kelvii Kwok, Rongfu Wen, Xuehu Ma, Rakesh Joshi, Shivanjali Saxena, Pedro J. Rivero, Rafael J. Rodriguez, Adrian Vicente, Aravind Kumar, Krithiga Thangavelu, Venkatesan Dhanancheyan, Azusa N Hattori, Ken Hattori, Xoan Xosé Fernández Sánchez-Romate, Silvia G Prolongo, Alberto Jiménez-Suárez, Phuong Viet Pham, Imtisal Akhtar, Malik Abdul Rehman, Yongho Seo, Mujtaba Ikram, Bilal Tariq, Rayha Khan, Husnain Ahmad, Abdullah Khan Durrani, Muhammad Ikram, Asghari Maqsood, Sana Arbab, Muhammad Aamir Iqbal, Xu-Yang Yao, Bao-Jun Bai and Wang Ren

First published in London, United Kingdom, 2020 by IntechOpen
IntechOpen is the global imprint of INTECHOPEN LIMITED, registered in England and Wales, registration number: 11086078, 5 Princes Gate Court, London, SW7 2QJ, United Kingdom
Printed in Croatia

British Library Cataloguing-in-Publication Data
A catalogue record for this book is available from the British Library

Additional hard and PDF copies can be obtained from orders@intechopen.com

21st Century Surface Science - a Handbook
Edited by Phuong Pham, Pratibha Goel, Samir Kumar and Kavita Yadav
p. cm.
Print ISBN 978-1-78985-199-1
Online ISBN 978-1-78985-200-4
eBook (PDF) ISBN 978-1-83962-640-1

For EU product safety concerns:
IN TECH d.o.o., Prolaz Marije Krucifikse Kozulić 3, 51000 Rijeka, Croatia, info@intechopen.com or visit our website at intechopen.com.

Meet the editors

Phuong Viet Pham is Distinguished Research Fellow at College of Information Science and Electronic Engineering and Zhejiang University-University of Illinois at Urbana-Champaign Joint Institute (ZJU-UIUC), Zhejiang University, China. He earned a PhD from SKKU Advanced Institute of Nanotechnology (SAINT), Sungkyunkwan University (SKKU), South Korea (2016). After obtaining his degree, Dr. Pham spent a few years as a postdoctoral researcher and research fellow at the School of Advanced Materials Science and Engineering, SKKU, and the Center for Multidimensional Carbon Materials (CMCM), Institute for Basic Science (IBS), South Korea, respectively. He is the recipient of the NSF Career Award of China for Excellent Young Scientist (2019). His research interest focuses on low-dimensional materials, 2D material synthesis, new doping technique development, nanocomposites, block copolymers, plasma engineering for organic light-emitting diodes (OLEDs), transistors, sensors, photodetectors, flexible displays, and wearable electronics.

Dr. Pratibha Goel received her PhD from the Indian Institute of Technology Delhi (IITD) in 2016. Since then, she has been a postdoctoral fellow at Peking University (China), CSIR-NPL (India), and Imperial College London (UK). Currently, she is working as a postdoctoral research associate at the University of Notre Dame, USA. Her research interests include nanostructured surfaces for tunable wetting properties and Surface-enhanced Raman Spectroscopy (SERS). Currently, she is working on nanopipette-based, single-molecule-based biosensors.

Dr. Kumar is an experimental physicist with expertise in the development and study of sculptured thin films and interfaces. Dr. Kumar received his PhD in Physics from the Indian Institute of Technology Delhi, India, in 2017. He has been working at Kyoto University since February 2018. His primary area of research is the synthesis of novel nanostructures by glancing angle deposition for plasmonics, surface-enhanced spectroscopy, photocatalysis, water repellent surfaces, and bio/chemical sensing applications.

Dr Kavita Yadav obtained her PhD from the Indian Institute of Technology Delhi (IITD) in 2016 and received the SJSS Sodha research award for best PhD publication. She has also worked as Senior Project Scientist at Nanoscale Research Facility, IIT Delhi. She won the DST INSPIRE faculty award from the Department of Science and Technology, Govt. of India, New Delhi, and joined the Department of Physics at the Central University of Haryana as DST Inspire Faculty in November 2017. Currently, she is working in the Department of Higher Education Haryana as Assistant Professor in Physics. Her research interests include formation of solid surfaces with tunable wetting properties, oil-water filters, CO2 hydrogenation, and gas sensing.

Contents

Preface

Surface sciences elucidate the fundamental aspects of physics and chemistry at a wide range of surfaces/interfaces of arbitrary objects. Nowadays, one of the emerging edges of surface sciences lies in micro-nano surface/interface structures of low-dimensional materials (0D, 1D, 2D) and three-dimensional (3D) materials, which are attracting great interest owing to their breakthroughs in high-performance applications. Among them, silicon, CVD graphene, graphene oxide, transition metal dichalcogenides, carbon nanotubes, carbon nanoparticles, transparent conducting oxide, metal oxides, and so on emerge as representative materials for "the nano era of the twenty-first century" with intriguing characteristics in electronics and optoelectronics. On another edge of surface science, the wetting phenomenon is also a representative behavior that controls the equilibrium of the surface energy of a liquid deposited on a surface. The wettability of solid surfaces is raising considerable interest because of its novel applications in various fields, from microfluidics to chemistry. This book provides a comprehensive overview of the important achievements of surface science from the aspects of high-quality synthesis, surface modifications, smart coating based on nanoparticles, the wettability of various surfaces, and physics/chemistry characterizations, as well as theoretical growth kinetics of thin films.

This book is divided into five sections. The first section describes the synthesis processes and characteristics of thin films, and the second section discusses the etching and lithography techniques of thin films such as graphene and silicon. The third section explains the wetting and permeability behavior of materials. This section introduces readers to the wetting phenomenon and describes different types of wetting. It explains the static and dynamic contact angles of liquid, discusses the effect of roughness on the contact angles, and evaluates the impact of roughness on surface wettability. In addition, this section introduces smart surfaces with tunable wettability via external stimuli or suitable coatings. It presents various techniques such as electric field, temperature, light, mechanical strain, pH, and so on for tuning surface wetting properties, which are extremely useful for various commercial applications. The fourth section describes the electrospinning surface engineering technique for the development of surfaces with different wettability and potential industrial applications for the different electrospun fibrous coatings. It reviews the electrospinning process and describes in detail the design of superhydrophobic surfaces obtained by electrospun fibers. Finally, the fifth section shows the fabrication of a scanning probe using multiwalled carbon nanotube (MWCNT) tips as one of the best candidates for imaging material surfaces such as silicon pillars.

Chapter 1 introduces five basic stages of the film deposition process, including vapor adsorption, surface diffusion, the reaction of the adsorbed species with each other and the surface to form the bonds of the film material, nucleation, and microstructure formation. It also analyzes the influence of deposition process parameters on the three basic growth modes of films, focuses on the relationship between the control parameters of homoepitaxy and heteroepitaxy and the film structure, and gives the dynamic characteristics of each growth stage.

Chapter 2 introduces various synthesis approaches for carbon nanotubes (CNTs) such as chemical deposition of vapor, discharge using electric arc, and laser ablation mechanisms, which are driven by functionalization, chemical addition, doping, and filing such that in-depth characterization and manipulation of CNTs is possible. In addition, the chapter discusses the elasticity, electromechanical, chemical, and optical properties of CNTs.

Chapter 3 provides a summary of semiconductors, bandgap theory, thin film applications and traditional thin-film processing methods, and the aerosol deposition technique of several materials for semiconductor devices.

Chapter 4 presents recent advances in new etching technologies for nanomaterials (e.g., graphene) as well as emerging applications based on these advanced technologies.

Chapter 5 is a significant contribution to the graphene industry as it explains the novel and modified fabrication techniques for ceramics–graphene hybrids. The improved physical properties may be used to set ceramics–graphene hybrids as a standard for electrical, mechanical, thermal, and energy applications. Further, silica–rGO hybrids may be used as dielectric materials for high-temperature applications due to improved dielectric properties. The fabricated nano assembly is important from a technological point of view, and may be further applied as electrolytes, catalysts, or conductive, electrochemically active, and dielectric materials for high-temperature applications. Finally, the chapter discusses porous carbon as a massive source of electrochemical energy for supercapacitors and lithium-ion batteries.

Chapter 6 introduces methods of evaluation by reflection high-energy electron diffraction (RHEED) and low-energy electron diffraction (LEED) based on a reciprocal space map, and methods of creating various atomically flat {111} and {100} side-surfaces of three-dimensional Si nano-architectures and tilted {111} facet-surfaces fabricated by lithography dry and wet etching processes followed by annealing treatment in a vacuum.

Chapter 7 presents the wetting phenomenon and describes different types of wetting. It also provides an introduction to static and dynamic contact angles of liquid as well as discusses the effect of roughness on the contact angle and evaluates the impact of roughness on surface wettability.

Chapters 8 and 9 discuss smart surfaces with tunable wettability and in situ surface wettability via external stimuli or suitable coatings. They discuss various techniques such as electric field, temperature, light, mechanical strain, pH, and so on for tuning surface wetting properties, which are extremely useful for various commercial applications.

Chapter 10 deals with the effects of an oligomeric organosilicon surfactant (OSSF) on wettability modification, surface tension reduction, invasion of different fluids, and fluid flow-back. It was found that the amount of spontaneous imbibition and remaining water could be reduced by the surfactant as a result of surface tension reduction and wettability alteration. Besides, the mechanism of OSSF includes the physical obstruction effect, surface tension reduction of external fluids, and wettability alteration of the reservoir generated. Meanwhile, quantum chemical calculations indicate that the adsorbent layer of polydimethylsiloxane (PDMS) could decrease the affinity and adhesion of CH_4 and H_2O on the pore surface.

Chapter 11 examines the electrospinning surface engineering technique for the development of surfaces with different wettability and potential industrial applications for different electrospun fibrous coatings.

Chapter 12 describes smart coatings such as polymer coatings, superhydrophobic and self-cleaning coatings, and nanocomposite coatings that are generally based on a polymer matrix doped with carbon nanoparticles such as carbon nanotubes or graphene. These methods enhance the electrical, thermal, and mechanical properties of and confer new functionalities to materials, turning them into smart materials able to interact with the environment and respond appropriately to external stimuli, making them useful for applications such as health monitoring or resistive heating.

Chapter 13 summarizes the basics of interfacial wetting and droplet dynamics in the condensation process, discusses the underlying mechanisms of droplet manipulation for condensation enhancement, and introduces some emerging works to illustrate the power of surface modification. Moreover, this chapter provides perspectives for future surface design in the field of condensation enhancement.

Chapter 14 explores the scanning algorithm to scan various types of features above (protrusion) or below (holes) silicon pillar surfaces using the MWCNT tips as a scanning probe attached by dielectrophoresis and focused ion beam (FIB) treatment. This study reveals MWCNT as one of the best candidates to image nanomaterials such as silicon pillars.

Finally, the editors of this book wish to acknowledge all of the authors, members of the academic editorial board, service manager, and commissioning editor for their cooperation and contributions. We would also like to thank IntechOpen for giving us the opportunity to complete this exciting book project.

Phuong Pham
Zhejiang University,
Hangzhou, China

Pratibha Goel
University of Notre Dame,
Indiana, USA

Samir Kumar
Kyoto University,
Kyoto, Japan

Kavita Yadav
Department of Higher Education,
Haryana, India

Section 1

Synthesis and Properties of Thin Film

Chapter 1

Growth Kinetics of Thin Film Epitaxy

Hong Liu

Abstract

This chapter mainly introduces five basic stages of the film deposition process (vapor adsorption, surface diffusion, reaction between adsorbed species, reaction of film materials to form bonding surface, and nucleation and microstructure formation), analyzes the influence of deposition process parameters on the three basic growth modes of the film, focuses on the relationship between the control parameters of homoepitaxy and heteroepitaxy and the film structure, gives the dynamic characteristics of each growth stage, and examines the factors determining epitaxy film structure, topography, interfacial properties, and stress. It is shown that two-dimensional nucleation is a key to obtain high-quality epitaxial films.

Keywords: deposition, adsorption, diffusion, nucleation, epitaxy, dynamic characteristics

1. Introduction

Epitaxial thin films and artificial multilayers are grown on solid single-crystal surfaces with atomic monolayer thickness control either by chemical vapor deposition (CVD) [1, 2] or by molecular beam epitaxy (MBE). In CVD, precursor molecules are thermally decomposed in a continuous flow oven in a background atmosphere of clean inert gas, whereas in MBE the surface is held in ultrahigh vacuum (UHV, 10^{-8} Pa). Controlling the growth morphology is a challenge in both fabrication techniques; it requires knowledge of both thermodynamics and of kinetics.

As with other thin films, epitaxial films can provide properties or structures that are difficult or impossible to obtain in bulk materials. Indeed, many materials are easier to grow epitaxially than to grow and shape in bulk form. Compared to polycrystalline films, epitaxial films have at least four advantages, which are elimination of grain boundaries, ability to monitor the growth by surface diffraction, control of crystallographic orientation, and the potential for atomically smooth growth.

Epitaxy is the special type of thin film deposition and is particularly demanding about all aspects of process control. Film quality is readily degraded by small amounts of contamination, nonstoichiometry, and lattice mismatch. On the other hand, when good control is achieved, complex multilayered structures with unique properties can be fabricated with atomic layer precision. Moreover, the precise structural and compositional nature of the epitaxial growth surface allows the use of

IntechOpen

growth monitoring techniques that give detailed information about film growth mechanisms on an atomic scale.

The purpose of this chapter is to guide the new readers who have just entered this field. Based on the in-depth analysis of the main aspects of epitaxy technology by cross-referencing the relevant literature provided by experts, the research and development direction of epitaxy technology are evaluated. Epitaxy refers to the orderly growth of crystal materials on the substrate crystal and the establishment of a clear crystal relationship at the interface between the two crystal lattices. In homoepitaxy, the epitaxial layer and substrate are made of the same material, while in heteroepitaxy, they are made of different materials. If two materials have the same crystal structure, they are called similar, otherwise they are called different. In the epitaxial structure, the same lattice spacing between the epitaxial material and the substrate material in the same direction plane is called lattice matching, otherwise, lattice mismatch. At one growth site, the constituent atoms are bonded to the epitaxial film, in which the bonding leads to the unequal probability of the atoms' attachment and desorption in the equilibrium. Atoms bonded with energy higher than the growth site are considered to be part of the epitaxial film. All atoms bonded with less energy than the growth sites are called adatoms. In the region of relatively high temperature, the mobility of atoms is stronger, and they can aggregate into two-dimensional islands, thus forming a new surface step. The method of epitaxy can be divided into (1) solid phase epitaxy (SPE), (2) liquid phase epitaxy (LPE), and (3) vapor phase epitaxy (VPE). This chapter only discusses the growth kinetics of each stage, including gas adsorption, surface diffusion, interaction of adsorbed species, bonding of surface-forming film materials, and nucleation and microstructure formation of epitaxial growth, rather than specific epitaxial growth methods.

2. General description of epitaxial growth

In the early study of thin films, it was found that the growth process of thin films is a complex process, including atom arrival, atom adsorption, diffusion/migration on the surface, nucleation, and coalescence. It was also found that four parameters influence the film growth: pressure, deposition rate, substrate temperature, and substrate structure. Also, the binding energy of the adsorbent to the substrate is of vital importance, but since this is not a controllable parameter, we will ignore it here. For metals adsorbed on insulator surfaces, we assume that every atom that impinges on the surface stays there. For other systems one may operate with a sticking coefficient, which is the probability of an atom sticking to the surface upon impingement. The adsorbed atoms can exhibit a complicated dynamical behavior at the surface: Atoms can move around on the corresponding surface, and they can diffuse into the substrate or even desorb from the substrate. When two atoms meet, they form metastable nuclei. This is referred to as nucleation. Nuclei can also split up, rotate, or migrate across the surface. At a certain critical size, the nuclei become stable, and this is where actual crystal growth begins. Initial film growth is categorized into three different types of behaviors. The three growth modes are called Volmer-Weber (VW), Stranski-Krastanov (SK), and Frank-van der Merwe (FM) [3]. **Figure 1** illustrates the different growth modes, which can be described as follows. For VW growth the growth is occurring as three-dimensional (3D) nuclei

Figure 1.
Illustration of the three different growth modes. Left: FM growth. Center: SK growth. Right: VW growth.

which later coalesce. SK growth is characterized as the formation of one or more layers upon which nucleation and growth dominate. FM growth or layer-by-layer growth is the growth mode that has our interest because of the well-ordered surfaces produced this way. To achieve layer-by-layer growth of atoms, instead of 3D growth, one must try to reduce the nucleation rate. This can be done by (1) reducing the pressure since it is believed that residual gases can create nucleation sites on the substrate surface, (2) increasing the substrate temperature which promotes the mobility of the atoms on the surface, or (3) reducing the deposition rate. RHEED can be used to verify the growth mode because oscillations of the intensity indicate that layer-by-layer growth is occurring.

Firstly, the heart of the thin film process sequence will be discussed. Deposition may be considered as six sequential substeps, and that will be examined one by one in the next section. The arriving atoms and molecules must first (1) adsorb on the surface, after which they often (2) diffuse some distance before becoming incorporated into the film. Incorporation involves (3) reaction of the adsorbed species with each other and the surface to form the bonds of the film material. The (4) initial aggregation of the film material is called nucleation. As the film grows thicker, it (5) develops a structure, or morphology, which includes both topography (roughness) and crystallography. A film's crystallography may range from amorphous to polycrystalline to single-crystal. The last is obtained by epitaxy—that is, by replicating the crystalline order of a single-crystal substrate. Epitaxy has special techniques and features which are also the focus of this chapter, and (6) diffusional interactions occur within the bulk of the film and with the substrate. These interactions are similar to those of post-deposition annealing, since they occur beneath the surface on which deposition is continuing to occur. Sometimes, after deposition, further heat treatment of a film is carried out to modify its properties. For example, composition can be modified by annealing in a vapor, and crystal growth can be achieved by long annealing or by briefly melting. These post-deposition techniques will not be discussed in this chapter.

The word "epitaxy" comes from the Greek word epi, which means "located on," while "taxis" means "arranged." Epitaxial growth refers to the registration or alignment of the crystal atoms in the single-crystal substrate into the single-crystal film. More precisely, if the atoms of the substrate material at the interface occupy the natural lattice position of the film material, the interface between the film and the substrate crystal is epitaxial, and vice versa. These two materials do not have to be the same crystal, but they are usually like this. When the film material is the same as the substrate material, the crystallographic registration between the film and the substrate is usually called uniform epitaxy. The epitaxial deposition of thin film materials different from substrate materials is called heteroepitaxy.

Epitaxial growth technology has important advantages in material manufacturing of microelectronic and optoelectronic applications. It can be used to prepare films with very good crystal quality. This also makes it possible to fabricate composite films with ideal electronic or optical properties that do not exist in nature. There are many factors that affect the selection of materials and processing methods for epitaxial growth. It includes the chemical compatibility of the film material and the substrate material; the magnitude of the energy band gap of the film material and its relationship with the energy band gap and the edge of the energy band of the substrate material; whether the minimum value of the conduction band energy and the maximum value of the valence band energy are in the same wave vector position is an important factor in optical applications; and the chemical compatibility of the dopant applied to produce the required functional behavior.

In heteroepitaxial film growth, the substrate crystal structure provides a template for locating the atoms of the first arriving film material, and each atomic layer of the film material provides the same function for the next layer formed by FM growth, as described in the previous section. If the substrate is a single crystal with good quality and the vapor supersaturation is moderate, the atoms have a high mobility on the growth surface; this is a common growth mode. If the lattice parameter mismatch is not too large, for example, it is less than 0.5%, the growth tends to plane. If the mismatch is large, the material tends to gather on the surface of the island, but remains epitaxial.

Plane growth is carried out by attaching atoms to the edge of the step, which causes the step to move on the growth surface. Generally speaking, the unstressed lattice size of the thin film material in the direction parallel to the interface, such as a_f, will be different from the lattice size of the substrate, such as a_s; this difference may be as high as a few percent. However, the atomic position of the thin film material is consistent with that of the substrate, and the atomic structure of the thin film material is maintained. The thin film material bears any necessary strain along the interface, which makes it possible. In terms of lattice parameters, this mismatch strain is:

$$\epsilon_m = \frac{a_s - a_f}{a_f} \tag{1}$$

The definition of mismatch strain in Eq. (1) is consistent with the standard definition of tensile elastic strain of material in the state of no stress. Sometimes we use the denominator of a_s instead of a_f to measure the lattice mismatch. In this case, the value of the mismatch strain is slightly different. The definition in Eq. (1) is consistent with the concept of tensile strain, i.e., the change of material element length relative to the initial length, which will be used in all cases here. The process of epitaxial film growth is shown schematically in **Figure 2**. Although this process will lead to the potential large elastic strain in the film, for very thin film, the energy

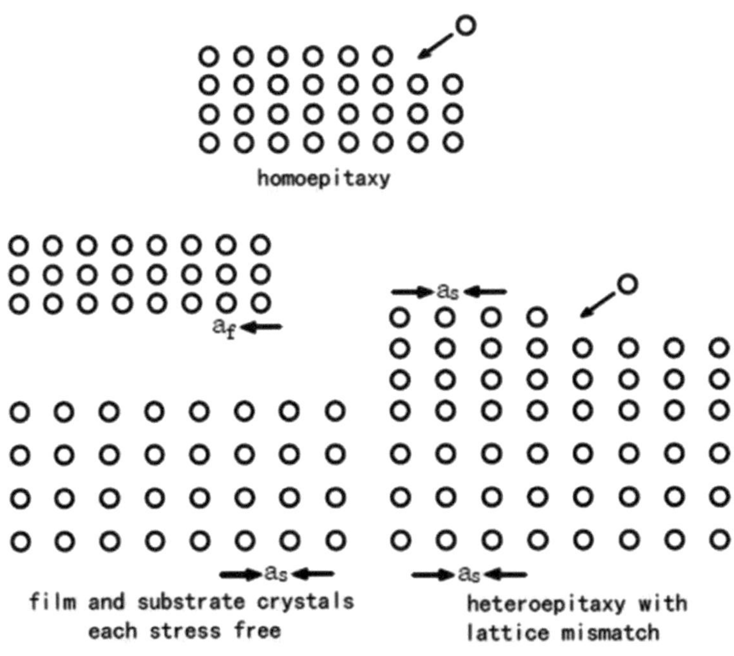

Figure 2.
Schematic illustration of heteroepitaxial film growth with lattice mismatch. The substrate thickness is presumed to be large compared to film thickness, and the structure extends laterally very far compared to any thicknesses. Under these circumstances, the lattice mismatch is accommodated by elastic strain at the deposited film.

cost of adding strain material is easily offset by the energy gain associated with the potential chemical bonding effect in the process. This can be proved by a simple argument, which only depends on the familiar block parameter value.

Take a simple comparison of different forms of energy. Both the elastic energy and the bonding energy can be compared with kT_s (Boltzmann constant times the absolute temperature of the substrate). This quantity is an indicator of the average energy per adatom in the substrate. At a typical deposition temperature of 800 K, the value of kT_s is about 0.07 eV. Interestingly, this value is one order of magnitude smaller than the bond energy estimate, but one order of magnitude larger than the elastic energy estimate. Although these are rough estimates, the calculation is instructive.

3. Dynamic characteristics of each stage of epitaxial growth

3.1 The process from vapor to adatoms

In this section, the factors controlling the early growth of thin films on the substrate are described from the perspective of atomism. This process starts with a clean surface of the substrate, which at a temperature of T_s is exposed to the vapor of the chemically compatible film material, which is at a temperature of T_v. In order to form a single-crystal film, the film material atoms in the vapor must reach the substrate surface, adhere to it, and locate a possible equilibrium position before the structural defects remain in the growth front. On the other hand, in order to form amorphous films, it is necessary to prevent atoms from reaching the growth surface to obtain a stable equilibrium position. In both cases, this must occur in a more or less identical way over a very large area of the substrate surface for structural development. At first glance, the result seems unlikely, but such films are made as usual. The atoms in the vapor touch the surface of the substrate, where they form chemical bonds with the atoms in the substrate. The idea that the temperature of the substrate must be sufficiently low so that the vapor phase is in a sense supersaturated with respect to the substrate will make the following more specific. In the process of adhesion, energy is reduced due to the formation of bond. As shown in the diagram in **Figure 3**, if the energy generated due to thermal fluctuation is enough to overcome the adhesion energy occasionally, some parts of the adhesion atom (called the adsorption atom) can be returned to vapor by evaporation.

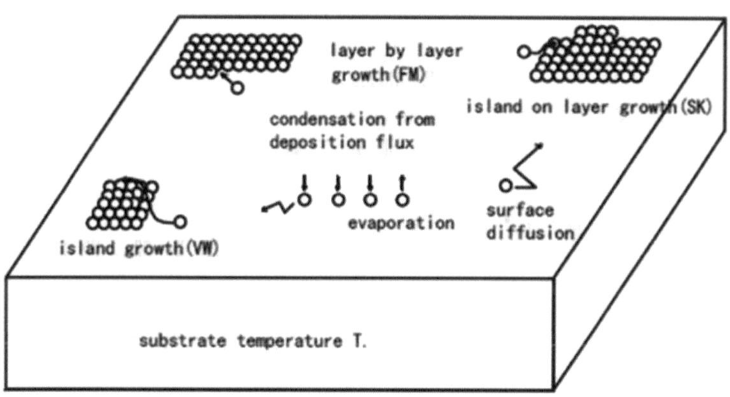

Figure 3.
Schematic showing the atomistics of film formation on substrates.

In order to make the discussion more specific, we assume a simple hexagonal close-packed crystal structure to facilitate the calculation of bonding. It has been recognized that in order to make the film growth possible, the vapor contacting the growth surface must be supersaturated with respect to the substrate at the substrate temperature T_s. For a uniform crystal at a certain temperature in contact with its own vapor at the same temperature, the equilibrium vapor pressure p_e of the system is defined as the pressure under which the vapor atoms condense on the solid surface and the atoms evaporate from the surface at the same rate. In equilibrium, the entropy free energy of each atom in the vapor is equal to the free energy of each atom in the crystal. Due to the effect of chemical bond, the lower internal energy of atoms in the crystal is offset by the lower entropy energy in the crystal. For the net deposition of the substrate surface, the pressure p in the vapor must be higher than the equilibrium vapor pressure p_e at the substrate temperature, that is, the vapor must be supersaturated. For pressure p, the entropy free energy of each atom in the vapor, which is higher than the pressure p_e, is estimated as the work of each atom required to increase the vapor pressure from p_e to p at a constant temperature. According to the ideal gas law, $kT_v lnp/p_e$ where k = 1.38 × 10^{-23} J/K = 8.617 × 10^{-5} eV/K is Boltzmann constant, T_v is the absolute temperature of the vapor. If the vapor becomes supersaturated, there is a difference of free energy between the vapor and the crystal, which provides the chemical potential of the interface driving the interface toward the vapor. As the interface progresses in a self-similar way, some distant layers of vapor of the same mass are transformed into inner layers of the same mass. When these energies are the same, the interface will not move forward or backward.

In thin film deposition, because the vapor phase and the substrate are not the same material phase, and the temperature of the substrate is usually lower than that of the vapor phase, the situation is often complex. In this case, the definition of equilibrium vapor pressure is not clear. However, in most cases, when the vapor pressure is lower than the equilibrium vapor pressure, the film material will not deposit on the growth surface, which is an operational definition of p_e in the film growth process. Once atoms are attached to the substrate, their entropy free energy will be reduced from that of the vapor. These atoms form a two-dimensional vapor distribution on the substrate surface. The deposited material will be rapidly heated to the substrate temperature T_s. On the surface of the crystal, there are some atoms whose density ρ_{ad} is balanced with a straight surface step or wall frame adjacent to some monolayer atoms, that is, the step does not move forward or backward. Step is the boundary between two phases in a homogeneous material system. Therefore, the concept of equilibrium vapor pressure or equilibrium vapor pressure density ρ_{ad}^e can be called in a similar way. In order to realize the film growth by condensation, the free energy of each atom in the two-dimensional gas must exceed the free energy of the fully entrained surface atom, the amount of which is:

$$\varepsilon_{phase} = kTln\left(\rho_{ad}/\rho_{ad}^e\right) \quad (2)$$

Consider a molecule approaching a surface from the vapor phase, as shown in **Figure 4**. Upon arriving within a few atomic distances of the surface, it will begin to feel an attraction due to interaction with the surface molecules. This happens even with symmetrical molecules and with inert gases, neither of which has dipole moments. It happens because even these molecules and atoms act as oscillating dipoles, and this behavior creates the dipole-induced-dipole interaction known as the Van der Waals force or London dispersion force. Polar molecules, having permanent dipoles, are attracted more strongly. The approaching molecule is being attracted into a potential well like the one that was illustrated in **Figure 5** for condensation.

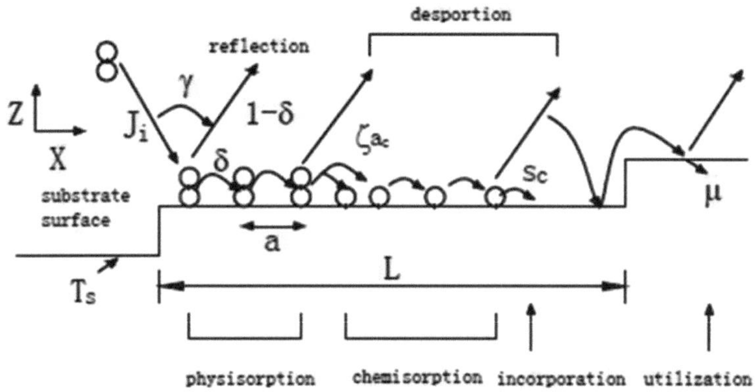

Figure 4.
Adsorption processes and quantities. a, is used only for condensation (adsorption of a material onto itself). A vertical connecting bar denotes a chemical bond.

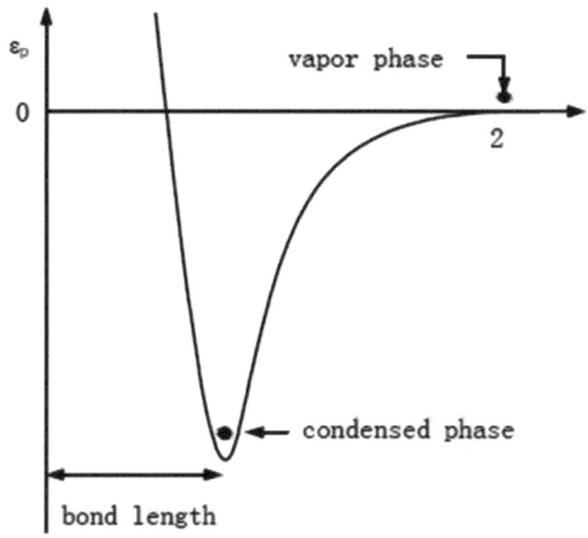

Figure 5.
Molecular potential energy diagram for evaporation and condensation.

Condensation is just a special case of adsorption in which the substrate composition is the same as that of the adsorbate. This is sometimes the case in thin film deposition and sometimes not. In either case, the molecule accelerates down the curve of the potential well until it passes the bottom and is repelled by the steeply rising portion, which is caused by mutual repulsion of the nuclei. If enough of the molecule's perpendicular component of momentum is dissipated into the surface during this interaction, the molecule will not be able to escape the potential well after being repelled, though it will still be able to migrate along the surface. This molecule is trapped in a weakly adsorbed state known as physical adsorption or physisorption. The fraction of approaching molecules so adsorbed is called the trapping probability, δ, and the fraction escaping (reflecting) is (1 − δ) as shown in **Figure 4**. The quantity δ is different from the thermal accommodation coefficient, η, which was defined by Eq. (3):

$$\eta = \frac{T_{rs} - T_{rh}}{T_{rs} - T_h} \tag{3}$$

η represents the degree to which the molecule accommodates itself to the temperature T_h of the surface from which it is reflected. Gas-conductive heat transfer is shown in **Figure 6**.

In general, a molecule is at least partially accommodated thermally to the surface temperature, T_s, even when it is reflected without having been trapped.

In addition to the low temperature T, the physical adsorption molecules are mobile on the surface; as shown in **Figure 4**, the adsorption molecules hop (diffuse) between the surface atomic sites. Adsorption molecules can also be desorbed by obtaining enough energy at the tail of thermal energy distribution, or they can form chemical bonds through further interaction with surface atoms, namely, chemical adsorption. If both adsorption states exist, the physical adsorption state is called precursor state. Chemical adsorption involves the sharing of electrons on the new molecular orbital, which is much stronger than physical adsorption. Physical adsorption only involves dipole interaction. These two types of adsorption can be distinguished in almost all gas phase surface combinations, so they constitute a valuable model for analyzing any surface process. This model [4] has long been applied to heterogeneous catalysis, thin film deposition, and condensation of molecular vapors. Recent theory indicates that even the condensation of a monatomic vapor such as Al can involve both adsorption states, the precursor state in that case being an $Al - Al$ dimer whose bonding to the bulk Al is inhibited by the existence of the dimer bond [5]. In such a case, and in the case of condensing molecular vapors such as As_4, the vapor would not be considered actually condensed until it had become fully incorporated into the solid phase by chemisorption. Thus, the condensation coefficient, α_c, defined by $J_c = \alpha_c J_i$ [J_i, impinging flux; J_c, the portion of J_i that condenses; α_c, corresponding condensation coefficient] is that fraction of the arriving vapor that becomes not only trapped but also chemisorbed, as indicated in **Figure 4**. However, the term α_c is not used in the case of chemisorption on a foreign substrate. Then, the chemisorption reaction probability, ξ, will be derived later. The precursor model may also be applied to cases where both of the adsorption states involve chemical bonding, but where the bonding in one state is weaker than in the other.

These examples will be revisited after a more detailed study of the energetics of the precursor adsorption model.

Consider a hypothetical diatomic gas phase molecule $Y_2(g)$ adsorbing and then dissociatively chemisorbing as two Y atoms. **Figure 7** shows a diagram of the potential energy versus molecular distance, z, from the surface. This is similar to **Figure 5** for condensation except that we have changed from the molecular (ε_p) to the molar (E_p) quantities of potential energy which are more conventional in chemistry. The energy scales shown represent typical bond strengths. Three curves are shown: two alternate ones for the precursor state (a and b) and one for the

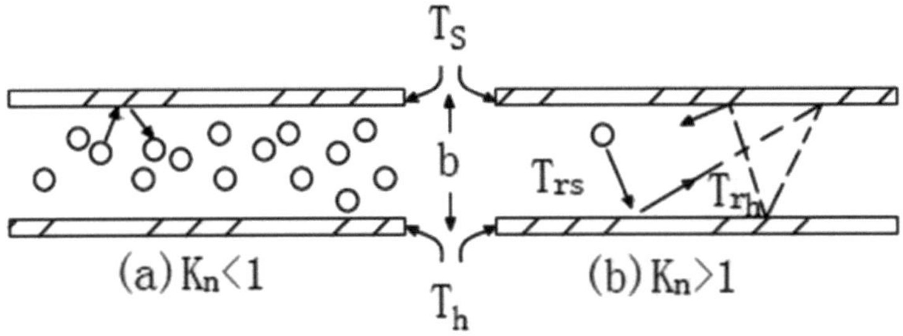

Figure 6.
Gas-conductive heat transfer between parallel plates at (a) low and (b) high Knudsen numbers, K.

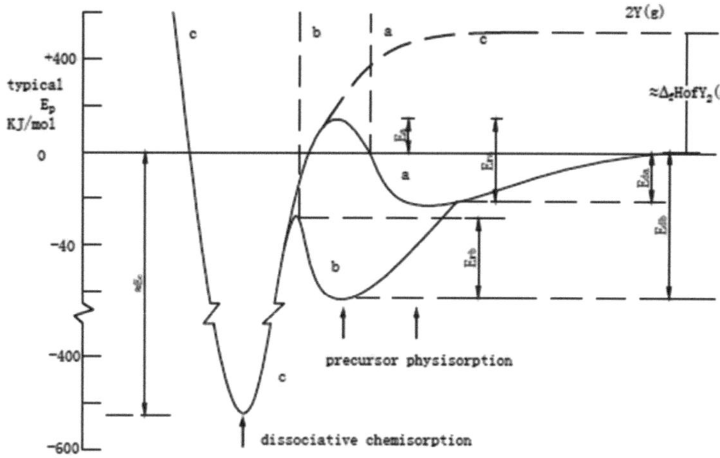

Figure 7.
Energetics of the precursor adsorption model. Energy scale is typical only.

chemisorbed state (c). By convention, the zero of E_p is set at the E_p of the element Y in its thermodynamic standard state, which we specify for this element to be the diatomic molecule in the gas phase. In fact, all gaseous elements except the inert gases have diatomic standard states. Note that lifting atomic Y out of its potential well along curve c results in a much higher E_p in the gas phase, which corresponds roughly to the heat of formation, $\Delta_f H$, of 2Y(g) from Y_2(g). [$\Delta_f H$ usually can be found in thermodynamic tables] [6–8]. The result of this high E_p for Y(g) is that curves a and c intersect at positive E_p, meaning that there is an activation energy, E_a, to be overcome for Y_2(g) to become dissociatively chemisorbed. For the deeper precursor well, b, chemisorption is not "activated," though there still is a barrier, as shown. The level of E_{ra} or E_{rb} and hence of E_s is determined by the degree to which the bonds within both the precursor and the surface must be strained from their relaxed condition before new bonds can be formed between the precursor and the surface.

There are two ways in which vapor can arrive at the surface having an $E_p > 0$. Gaseous molecules have their E_p raised by becoming dissociated. Solids and liquids have it raised by evaporating. If the E_p of the arriving vapor is high enough, curve c is followed, and direct chemisorption can occur without involving the precursor state. In the language of surface chemistry, direct reaction between an incoming species and a surface site or adsorbate is called the Eley-Rideal mechanism, whereas reaction among surface species is called the Langmuir-Hinshelwood mechanism.

A principal advantage of the energy-enhanced deposition processes is that they can provide enough energy so that the arriving molecules can surmount the E_a barrier and adsorb directly into the chemisorbed state. In other words, the arriving molecules immediately react with the surface to deposit the film. In sputter deposition, species arrive having kinetic energies of around 1000 kJ/mol as well as having $E_p > 0$ by having been vaporized. In plasma-enhanced deposition, vapor molecules become dissociated in the plasma and thus arrive along curve c, above the E_a barrier. Thus, an energy-enhanced process can supply E_a to the arriving species either as kinetic energy of accelerated molecules or as potential energy of dissociated ones.

Conversely, in thermally controlled deposition processes such as evaporation and CVD, the vapor often adsorbs first into the precursor state, that is, it falls to the bottom of the well on curve a or b. Thence, it may either chemisorb by overcoming the barrier $E_{r(a,b)}$ shown in **Figure 7** or it may desorb by overcoming the heat of physisorption, which is roughly $E_{d(a,b)}$. The competition between these two reactions results in a net rate of chemisorption whose behavior we would like to describe, since it is the basic film-forming reaction. We start with the conventional expression for the rate of a first-order chemical reaction, first-order meaning that rate is proportional to the concentration of one reactant; thus, $R_k = k_k n_s = k_k n_{s0}\Theta$, where R_k = rate of the kth surface reaction per unit surface area, mc/cm^2 s; k_k = rate constant, s^{-1}; n_s = surface concentration of reactant, mc/cm^2; n_{s0} = monolayer surface concentration, mc/cm^2; Θ = fractional surface coverage by reactant.

3.2 Diffusion of adsorbed atoms on substrate surface

Surface diffusion is one of the most important determinants of film structure because it allows the adsorbing species to find each other, find the most active sites, or find epitaxial sites. Various methods have been applied to measure surface diffusion rates of adsorbed molecules. The role of surface diffusion in thin films has mainly been inferred from observations of film structure. Scanning tunneling microscope (STM) gives us the extraordinary power to directly observe individual atoms on surfaces in relation to the entire array of available atomic surface sites. STM observation of the diffusion of these atoms should ultimately provide a wealth of data relevant to thin film deposition.

The expression of the surface diffusion rate will be derived using the absolute reaction rate theory [9]. Although this approach cannot provide a quantitative estimate of the diffusion rate, it will provide valuable insight into what factors determine this rate. **Figure** 7 showed that adsorbed atoms or molecules reside in potential wells on the surface, but it did not consider the variation in well depth with position, x, along the surface. **Figure 8a** shows that this depth is periodic, or corrugated, with a potential energy barrier of height E_s between surface sites. The top of the barrier is considered to be the “transition state” between surface sites, in

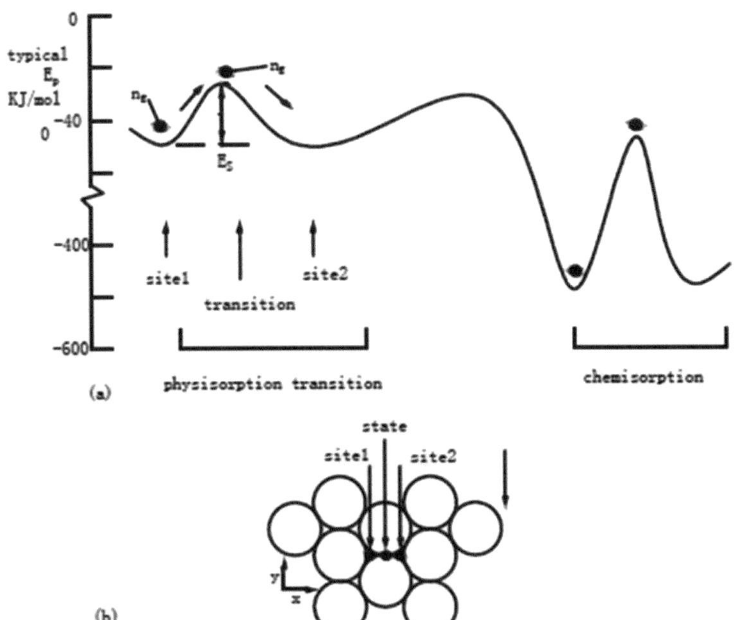

Figure 8.
Surface diffusion: (a) potential energy vs. position x along the surface and (b) typical adsorption sites on a surface lattice.

the language of reaction rate theory. **Figure 8b** illustrates a typical adsorbate situation leading to this corrugation. It is a hexagonally close-packed surface lattice on which the adsorption sites are the centers of the triangles of surface atoms and the transition state is the "saddle point" between them. Other bonding situations can lead to the adsorption sites being other points, such as the centers of the surface atoms. In the process of surface diffusion, the bond between adsorbate and surface should be partially destroyed, so that the adsorbate can migrate to the adjacent surface and form a new bond there. This process can be regarded as the basic form of chemical reaction, because any reaction involves the partial fracture of reactant bond and the formation of product bond when the atom moves through the transition state. Therefore, the principles discussed below apply to any chemical reaction, including those occurring in CVD.

There will be some flux, J_s (mc/cm s), of adsorbate across the E_s barrier between sites 1 and 2 in the x direction of **Figure 8b**. The flux here is in surface units, which are per linear cm of crosswise distance, y, instead of the previously encountered volume flux units, which are per cm^2 of cross-sectional area. If the distance between sites is a, then the rate of barrier crossing by transition state molecules, per unit area of surface, is:

$$R_s = J_s/a \left(\text{mc/cm}^2 \cdot \text{s}\right) \tag{4}$$

Considering the adsorbate to be a two-dimensional gas at thermal equilibrium, the Maxwell-Boltzmann distribution applies to these translating molecules. Thus, we may use $J_i = \frac{1}{4} n\bar{c}$ for the flux of molecules impinging on the barrier and $\bar{c} = \sqrt{\frac{8RT}{\pi M}}$ for the mean speed. Here, for simplicity, we ignore the small changes in the numerical proportionality factors that arise in going from a three-dimensional to a two-dimensional situation. (It turns out that these factors cancel each other, anyway.) Inserting these equations into Eq. $R_s = J_s/a$ (mc/cm^2 · s), we have:

$$R_s = \frac{1}{4} n_s^* \bar{c}/a = \frac{n_s^*}{a}\sqrt{\frac{RT}{2\pi M}} = \frac{n_s^*}{a}\sqrt{\frac{k_B T}{2\pi m}} \tag{5}$$

where n_s^* (mc/cm^2) denotes the surface concentration of adsorbate residing in the transition state. Now we must find the relation between n_s^* and n_s, the latter being the concentration of molecules in adsorption sites. At thermal equilibrium, statistical mechanics says that the concentration of molecules in a given state is proportional to the total number of ways of distributing the available thermal energy around a large system of molecules in that state. For each type of kinetic energy contributing to the thermal energy, the number of ways, Z, is equal to the sum over all of the quantized energy levels, ε_j, of the following products: the Boltzmann factor for each energy level times the number of ways of distributing energy at that level (the degeneracy of the level, g). Thus:

$$Z = \sum_j g_j e^{-\varepsilon_j/k_B T} \tag{6}$$

3.3 Nucleation

3.3.1 Surface energy

To understand nucleation, the concept of surface energy needs to be introduced. The familiar experiment of drawing a liquid membrane out of soapy water on a wire ring is illustrated in **Figure 9**. The force required to support the membrane per unit width of membrane surface is known as the surface tension, γ, expressed as N/m in SI units, or more commonly as dynes/cm in cgs units. For a wire of circumference b, the width of surface is $2b$, since the membrane has both an inner and an outer surface. Thus, the total force required to support the membrane is $F = 2b\gamma$. As the membrane is extended upward in the x direction, work $F(\Delta x)$ (N m or J) is done to create the new surface, and the surface area created is $A = 2b\Delta x$, assuming for simplicity a constant membrane circumference. The work is stored as surface energy (as in stretching a spring), so the surface energy per unit area of surface is:

$$F\Delta x/A\left(\mathrm{N}\cdot\mathrm{m}/\mathrm{m}^2\right) = (2b\gamma)\Delta x/2b\Delta x = \gamma\left(\mathrm{N/m\ or\ J/m^2}\right) \quad (7)$$

Thus, surface tension (N/m) and surface energy per unit area (J/m^2) are identical, at least for liquids. For solids at $T > 0$ K, surface Gibbs free energy is reduced by an entropy factor $[G = (U + pV) - TS = H - TS]$ which depends on the degree of surface disorder. For solids, there is also a quantity called surface stress, which differs from surface energy by a surface elastic strain term. Liquids cannot support such strain, because the atoms just rearrange to relax it.

The surface energy exists because the molecules in the condensed phase attract each other, which is the reason for condensation. The generation of a surface involves the removal of molecular contact (bond breaking) from above the surface, thus involving energy input. Therefore, the movement in the condensed phase can occur within a certain range, and this movement will continue to minimize the total surface energy, γA. In the liquid membrane case, where γ is fixed, this means minimizing A. Thus, when the wire is lifted far enough, the membrane snaps taut into the plane of the ring, and when a bubble is blown, it becomes spherical. In the case of solids, surface energy can be minimized by surface diffusion, which is the basis of the development of film structure. In the film growth, A and γ are varying. Area A depends on the surface morphology, and γ depends on many characteristics of the exposed surface, including chemical composition, crystal orientation, atomic reconstruction, and atomic-scale roughness.

For the deposition on foreign substrates, the substrate γ strongly affects the nucleation behavior. Here, the surface energy γ_s of the free surface of the substrate, the surface energy γ_i of the substrate film interface, and the surface energy γ_f of the free surface of the film should be considered. These three γ values generally depend

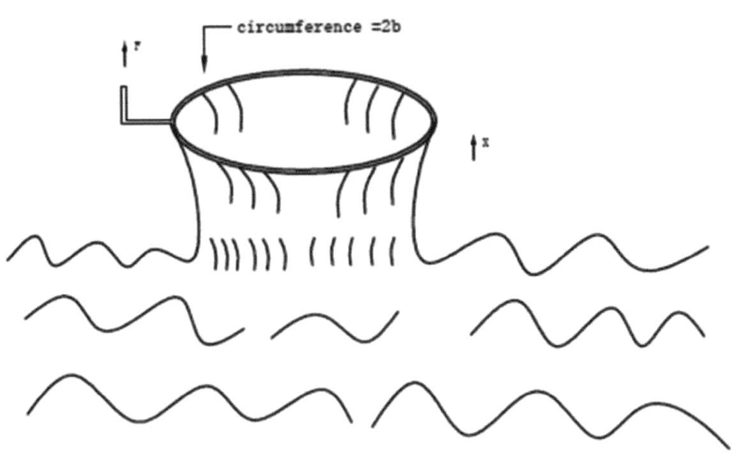

Figure 9.
Surface tension of a liquid membrane.

on crystal orientation, passivation, and other factors. It is assumed that there is sufficient surface diffusion to enable the deposited materials to rearrange to minimize γ, i.e., it is assumed that nucleation is not limited by dynamics and can approach equilibrium. For this reason, we must have $\Lambda \gg a$ (Λ is the diffusion length). On the contrary, if $\Lambda < a$, every atom will stick where it falls, and the growth behavior is "quenched." According to our hypothesis $\Lambda \gg a$, there are two kinds of nucleation on the bare substrate, as shown in the **Figure 10a** and **b**. In a, the film diffuses or "wets" the substrate on the substrate because:

$$\gamma_f + \gamma_i < \gamma_s \tag{8}$$

In other words, the total surface energy of the wetted substrate is lower than that of the bare substrate. This leads to the smooth growth of the atomic layer, which is the Frank-van der Merwe growth mode. To achieve this mode, there must be a strong enough bond between the film and the substrate to reduce the γ_i in Eq. (8). If there is no such bond, we will get $\gamma_i = \gamma_f + \gamma_s$, so spreading the film on the substrate will always increase the total surface energy $2\gamma_f$, as shown in the freestanding liquid film in **Figure 9**. Therefore, in the case of insufficient substrate bonding, Eq. (8) cannot be maintained, and the film does not wet the substrate, but forms a three-dimensional (3D) island, as shown in **Figure 10b**, which is called the Volmer-Weber growth mode. There is a third growth mode, Stranski-Krastanov growth mode, as shown in **Figure 10c**. In this mode, due to the change of the energy situation of the continuous single layer, the growth changes from one layer to an island after one or two layers. For liquids contacting solids, the degree of wetting is most easily observed by the rise or depression of a liquid column in a narrow tube (a capillary). Thus, film nucleation analysis in terms of degree of wetting is known as the "capillarity" model. Three-dimensional nucleation is usually undesirable, since it leads to rough, nonuniform films.

Different crystal shapes imply that underlying substrates critically influence the vapor phase growth mode. The substrate-dependent growth characteristics of various low-dimensional nanocrystals in both solution and vapor phase growth have been discussed for their growth mechanisms [10, 11].

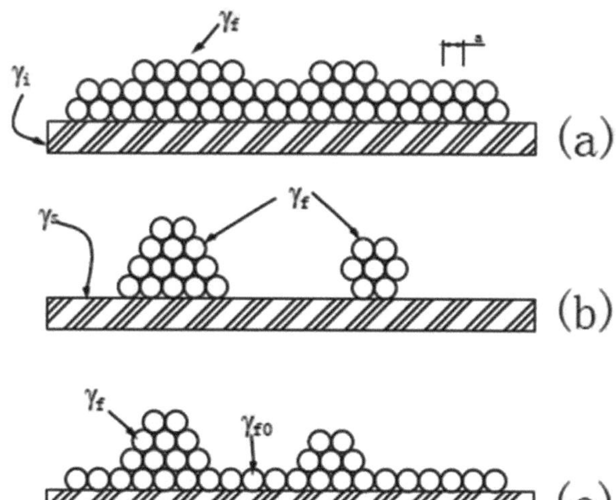

Figure 10.
Film growth modes: (a) Frank-Van der Merwe (layer), (b) Volmer-Weber (island), and (c) Stranski-Krastanov.

3.3.2 Kinetics vs. thermodynamics

In general, within the framework of the nucleation kinetics model [12], a gas phase growth reaction can be divided into two steps: (1) adsorption of vaporized precursors onto substrates and diffusion to the preferential growth sites and (2) incorporation of precursors into existing nuclei. The rate-limiting step in vapor phase crystal growth can be determined as either the diffusion-limited step or the reaction-limited step.

One way to achieve smooth growth is to reduce substrate temperature, T, to inhibit surface diffusion, thus "freezing" the nucleation and coalescence process. If the arriving species do not have enough heat energy to desorb or diffuse, they will stay where they land, leading to the aforementioned quenching growth. In this case, the nucleation process is kinetically inhibited by the surface diffusion activation energy barrier, E_s, in **Figure 8(a)**. This is also the case for ion-bombardment dissipation of 3D nuclei; the nuclei do not have time to reassemble themselves by surface diffusion before they are buried by depositing material.

The question of whether a process is approaching equilibrium or is instead limited by kinetics is an important one, and it arises often in thin film deposition. Process behavior and film properties are profoundly affected by the degree to which one or the other situation dominates. The answer is not always apparent in a given process, and this often leads to confusion and to misinterpretation of observed phenomena. Therefore, to elaborate briefly, the generalized mathematical representation of this dichotomy is embodied in Eq. (9):

$$R_s = R_{-s} \text{ or } n_s k_s = n_s^* k_{-s} \tag{9}$$

where $-s$ denotes the reverse reaction from the transition state back to adsorption site.

Eq. (9) describes the rate balance of a reversible reaction, and Eq. (10) defines its equilibrium constant:

$$K = \frac{k_s}{k_{-s}} = \frac{n_s^*}{n_s} e^{-\Delta_r G^0/RT} \tag{10}$$

Approach to equilibrium requires the forward and reverse rates to be fast enough so that they become balanced within the applicable time scale, which may be the time for deposition of one monolayer, for example. Then, the concentrations of reactant and product species are related by the difference in their free energies, $\Delta_r G^0$. If, on the other hand, the forward rate is so slow that the product concentration does not have time to build up to its equilibrium level within this time scale, then the product concentration is determined not by $\Delta_r G^0$ but, instead, by the forward rate. This rate is governed by Eqs. (11) and (12):

$$R_k = k_k n_s = k_k n_{s0} \Theta \tag{11}$$

where R_k = rate of the kth surface reaction per unit surface area, mc/cm^2 s; k_k = rate constant, s^{-1}; n_s = surface concentration of reactant, mc/cm^2; n_{s0} = monolayer surface concentration, mc/cm^2; Θ = fractional surface coverage by reactant.

$$k_k = \upsilon_{0k} e^{-E_k/RT} \tag{12}$$

where υ_{0k} = frequency factor or pre-exponential factor; E_k = reaction activation energy, kJ/mol in which E_k/T plays the dominant role. So it is that reactions can be frozen out and equilibration avoided if so desired, by lowering the T.

The difficulty of answering the question of kinetics versus thermodynamics arises from the fact that the applicable rate constants, k_k, are often unknown or not known accurately enough. The measurement of k_k is much more difficult than just measuring equilibrium concentration, both because it is a dynamic measurement and because it must be made in the absence of the reverse reaction.

3.3.3 Two-dimensional nucleation

When wetting is complete and Eq. (8) holds, the adsorbing atoms do not accumulate into 3D islands but, instead, spread out on the surface in a partial monolayer as shown in **Figure 10a**. Because total surface energy is reduced rather than increased by this process, there is no nucleation barrier in going from the vapor state to the adsorbed state, that is, the term in Eq. (13) is negative when the interfacial area is included:

$$\Delta G = -(\mu_v - \mu_c)\frac{V}{V_{mc}} + \gamma_f A_f = -\left(RTln\frac{p}{p_v}\right)\cdot\frac{(4/3)\pi r^3}{V_{mc}} + \gamma_f 4\pi r^2 \quad (13)$$

where p_v is saturation vapor pressure and V_{mc} is the molar volume of the condensate.

This means that deposition can proceed even in undersaturated conditions.

Assuming, as we did for 3D nucleation, that there is sufficient surface diffusion for equilibration, the partial monolayer of adsorbed atoms will behave as a 2D gas. By analogy to a 3D gas condensing into 3D nuclei, the 2D gas then condenses into 2D nuclei as illustrated in **Figure 11**. Here, only the top monolayer of atoms is drawn. The "atomic terrace" to the left represents a monolayer which is one atomic step (a) higher than the surface to the right. But unlike the 3D nucleation case, 2D nucleation from a 2D gas involves no change in any of the γ values, so one might expect there to be no nucleation barrier. However, the chemical potential, μ, of a 2D nucleus is higher than that of a continuous monolayer because of the exposed edge. This situation may be viewed in terms of an excess edge energy, β (J/m), which is analogous to the surface energy, γ, of the 3D case. The surface concentration of the 2D gas for which its μ is the same as that on the straight terrace edge of a continuous

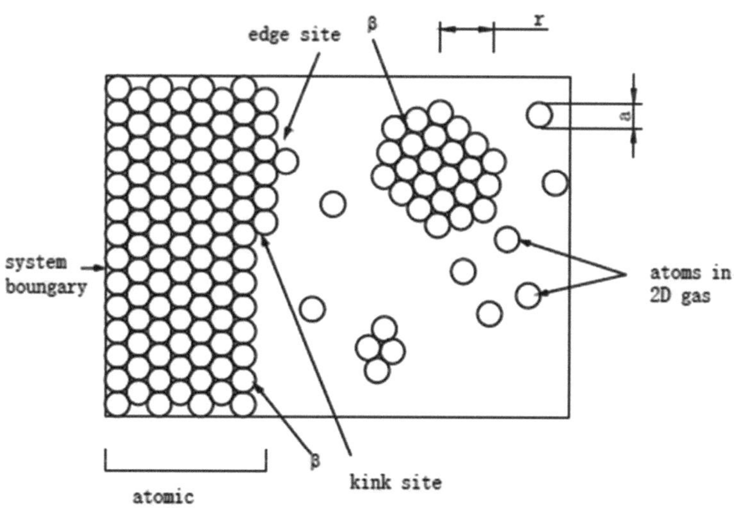

Figure 11.
Geometry of 3D nucleation, looking down at the surface.

monolayer (μ_c) may be thought of as the 2D saturation vapor concentration, n_v (mc/m^2). If n_s is the actual concentration of the 2D gas, then (n_s/n_v) becomes the 2D supersaturation ratio. By the same procedures as in the 3D case, we may then derive expressions for the critical nucleus as:

$$r^* = \frac{\beta}{a\left(\frac{RT}{V_{mc}}\right) ln\left(\frac{n_s}{n_v}\right)} \tag{14}$$

and

$$\Delta G^* = \frac{\pi\beta^2}{a\left(\frac{RT}{V_{mc}}\right) ln\left(\frac{n_s}{n_v}\right)} \tag{15}$$

Here, a is the monolayer thickness. Once supercritical nuclei form, the 2D gas continues to attach to their edges until coalescence occurs and the monolayer is complete. Meanwhile, the next monolayer is beginning to form, and the film continues to build up in this way, atomic layer by layer. In the special case of single-crystal film deposition (epitaxy), the surface may contain many atomic terraces with straight edges as shown in **Figure 11**. The "kink" sites shown in **Figure 11** are also important surface features. Attachment of a 2D gas atom to a random site on the straight edge involves an increase in total edge energy, because it increases the length of the edge. Conversely, attachment to the kink site makes no change in the length of the edge; this is therefore an energetically preferred site, and edge growth can most easily occur by attachment-driven motion of these kink sites along the edge.

It can be seen from the above that the surface energy depends not only on the facet direction discussed in Section 3.3.1 but also on the density of steps and kinks (Williams, 1994). The equilibrium densities of these two features increase with T because of their associated entropy (disorder), S. That is, when the T_s term for the Gibbs free energy, G, becomes larger; the internal energy term, U, also becomes larger to minimize G; and U here mostly consists of the potential energy of step and kink formation. This is the same T-driven tendency toward disorder that causes vapor pressure to rise with T.

During film deposition, if the surface diffusion rate is high enough and n_s is low enough so that the 2D gas atoms are more likely to attach to an edge than to form a critical nucleus within an atomic terrace, then edge attachment becomes the dominant growth mode, that is, we have $\Lambda > L$, where Λ is the surface diffusion length and L is the distance between terraces. This is called the "continuous" growth mode, as opposed to the nucleated mode. The continuous mode of 2D growth is analogous to the type of 3D nucleation in which nucleation is more likely to occur at active surface sites than by spontaneous nucleation elsewhere on the surface. Active sites and step edges, especially kinked edges, break the nucleation barrier by providing wetting at those sites.

Two-dimensional nucleation is usually preferred to 3D because it leads to smooth growth. In nonepitaxial growth, large grain size (coarse nucleation) may be desired in addition to smoothness. Unlike in the 3D nucleation case, here large grain size and smoothness are not incompatible. That is, if adatom mobility on the substrate is sufficient, large 2D nuclei will form before the first monolayer coalesces, and then subsequent monolayers will grow epitaxially on those nuclei. But there is another problem. High adatom mobility requires a low surface diffusion activation energy, E_s, in accordance with **Figure 8**, but E_s tends to increase with the strength of the adsorption, E_d or E_c, as suggested in **Figure 7b**. At the same time, good wetting requires low γ_i and therefore requires strong adsorption. As a result, it will not always be possible to achieve strong enough adsorption for wetting without immobilizing the adsorbate and preventing grain growth. Even so, small-diameter grains can become wider as the film grows thicker.

3.4 Texturing

The texturing described here refers to the crystal structure rather than the surface morphology, although they are often correlated. The degree of texturing is the degree to which the crystallites in a polycrystalline film are similarly oriented. In one limit, there is random orientation (no texturing), and in the other limit, there is the single crystal. A material in which the crystallites are nearly aligned in all three dimensions is called a "mosaic," and the limit of a perfect mosaic is a single crystal. The degree of texturing is best measured by X-ray techniques. Texturing can occur in one, two, or three dimensions. Epitaxy is the best way to achieve perfect three-dimensional texturing. Epitaxy occurs when the bonds of the film crystal align with the bonds of the substrate surface, making the interfacial energy, γ_i, very low, zero in the case of homoepitaxy; this is when the film material is the same as the substrate material. In other cases, when there is no such arrangement to operate, the most common form of film texture is two-dimensional texture, in which the crystal plane is arranged relative to the rotation of two axes on the substrate plane. This means that the film has a preferred growth plane parallel to the substrate, but has a random orientation relative to the rotation of the axis (i.e., azimuth) perpendicular to the substrate plane. It is often desirable to deposit the film on a substrate that cannot be crystallographically aligned, such as an amorphous substrate (such as glass) or a substrate with crystal symmetry or lattice size very different from the film. In this case, it is very ideal to realize two-dimensional texture when the required film properties are also crystal anisotropy.

3.5 Submonolayer and lattice mismatch

Because of the importance of atomically abrupt interfaces, we will focus next on physical and chemical vapor deposition processes which operate far from equilibrium in the sense that $J_v \ll R_r$. This is achieved by reducing the deposition T, T_s, until J_v becomes negligible. It does not mean, however, that all of the steps in the deposition process are operating far from equilibrium. Recall from Section 3.2 that deposition is a series of steps: adsorption, surface diffusion, reaction, nucleation, structure development, and interdiffusion. To obtain good deposition rate control, it is only important that either the adsorption or the reaction step be far from equilibrium. To prevent the broadening of interfaces after they are formed, it is important also that T_s be low enough so that interdiffusion is negligible during the total time of structure deposition. However, if T_s is too low, surface diffusion will become negligible, and structural equilibration will not occur. This is the "quenched growth" regime, and the crystallographic quality of epilayers is poorer in this regime than at higher T_s. Fortunately, there is often a T_s "window" within which good crystallography and sharp interfaces can both be obtained. Much of the development work in epitaxy has involved modifying processes to widen this window.

In addition to non-equilibrium growth, one must also have chemical compatibility and reasonably good lattice match between layers to obtain good heteroepitaxy. Now let us move on to chemical interactions. Epitaxy is particularly sensitive to degradation by impurities and defects. Moreover, complete disruption of epitaxy can occur if even a fraction of a monolayer of disordered contaminant exists on the substrate surface or accumulates on the film surface during deposition. This is because the depositing atoms need to sense the crystallographic order of the underlying material and chemical forces extend only one or two atomic distances. An island of surface contaminant becomes the nucleus for the growth of nonepitaxial material, and this region often spreads with further deposition, as shown in **Figure 12**, rather than being overgrown by the surrounding epilayer. Contamination can enter at any step in the thin film process. Removal of substrate contamination to improve adhesion is not discussed here. The additional substrate requirements that must be met to achieve epitaxy are of great concern. These include crystallographic order, submonolayer surface cleanliness, and chemical inertness toward the depositing species. Any crystallographic disorder at the substrate surface will be propagated into the depositing film. A few materials can be obtained as prepolished wafers with excellent surface crystallography. In other cases, careful preparation is necessary to remove the disorder introduced by wafer sawing and mechanical polishing. The crystallographic damage produced by polishing-grit abrasion extends into the crystal beneath the surface scratches, to a distance of many times the grit diameter, as shown by the dislocation line networks in **Figure 13a**. This damaged region must be removed by chemical etching. To promote uniform etching and prevent pitting, the "chemical polishing" technique is used. In this technique, the etchant is applied to a soft, porous, flat pad which is wiped across the wafer. If the depth of etching is insufficient, some damage will remain, as shown in **Figure 13b**, even though the surface may appear absolutely flat and smooth under careful scrutiny by Nomarski microscopy. However, these defects can be revealed by dipping the wafer in a "dislocation" etchant [13] that preferentially attacks them and thereby decorates the surface with identifying pits and lines. The crystallographic disorder at these defects, consisting of strained and broken bonds, raises the local free energy and thereby increases reactivity toward the etchant. After sufficient chemical polishing, the only remaining defects will be those grown into the bulk crystal, as shown at the etch pits in **Figure 13c**.

After crystallographic preparation of the substrate, surface contamination must be removed. In the final chemical cleaning step prior to wafer installation in the deposition chamber, one seeks to minimize residual surface contamination and also to select its composition so that it is more easily removed by the techniques available in the chamber.

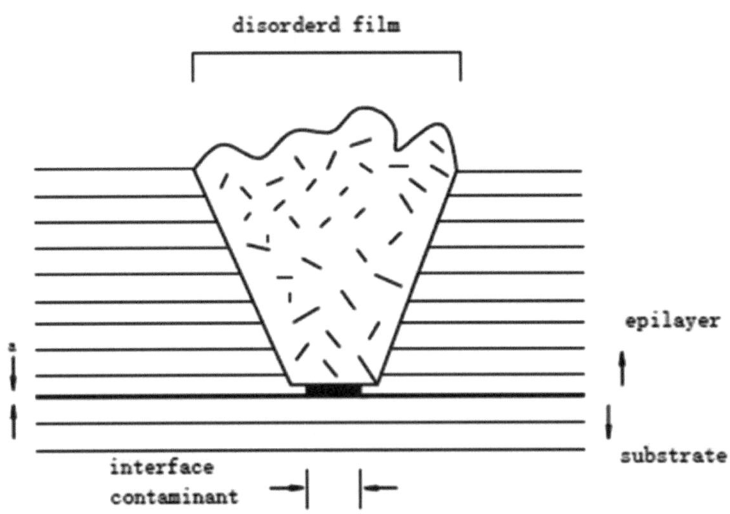

Figure 12.
Effect of submonolayer surface contamination on epitaxy.

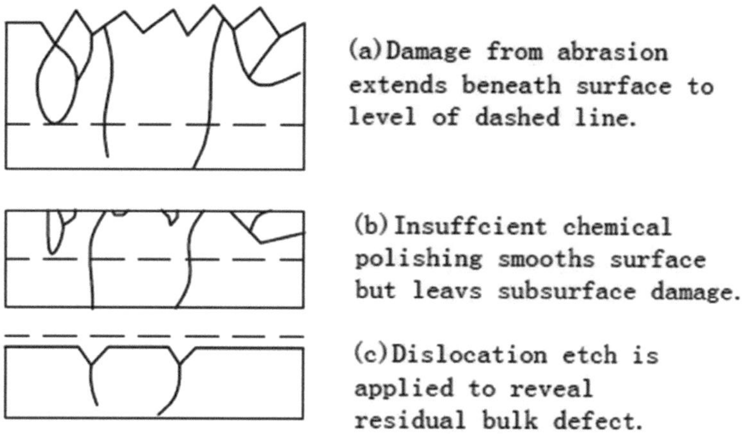

Figure 13.
Crystallographic damage due to wafer sawing and mechanical polishing.

Finally, the lattice mismatch is discussed. The expression of lattice mismatch factor is as follows:

$$f = \frac{(a_e - a_s)}{(a_e + a_s)/2} \approx (a_e - a_s)/a_s \tag{16}$$

Having now dealt with avoiding precipitates and controlling point defects, we can proceed to the problem of minimizing other crystallographic defects. It is useful to think of defects in terms of their dimensionality. Point defects are zero-dimensional (0D), while precipitates or disordered regions are 3D. Planar (2D) defects include grain boundaries, twin planes, stacking faults, and antiphase domain boundaries. Dislocations are line (1D) defects. We will see below how dislocations arise from the fractional lattice mismatch, f, at heteroepitaxial interfaces. For this purpose, we consider the simple square symmetry of cubic material growing in (001) orientation on a (001)-oriented substrate, although the same principles apply to other symmetries. **Figure 14** shows the various modes of

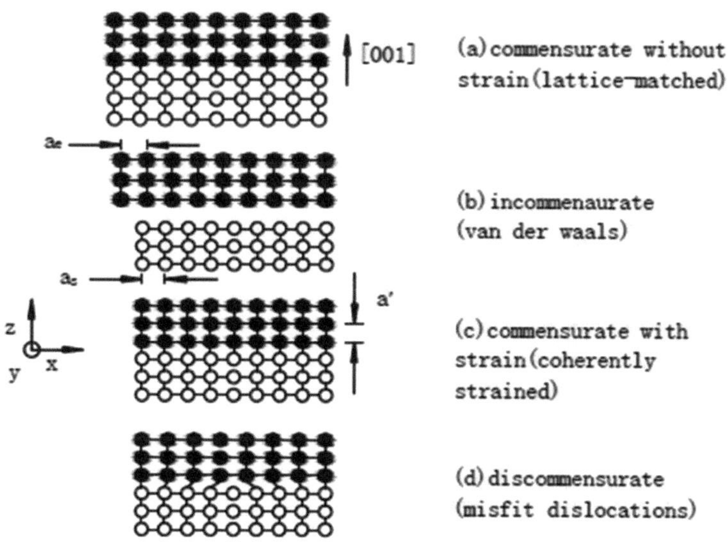

Figure 14.
Modes of accommodating epilayer lattice (solid circles) to substrate lattice (white circles).

mismatch accommodation. In the special case of perfect match (a), the lattices are naturally aligned, and the growth is therefore "commensurate" without requiring lattice strain. In (b–d), the atomic spacing of the epilayer, a_e, is larger than that of the substrate, a_s. In fact, f has been made quite large (0.14) here so that it may be readily observed, but it is much smaller in most heteroepitaxial systems of interest.

There are several ways in which lattice mismatch can be accommodated. In **Figure 14b**, bonding across the interface is weak, so that the epilayer "floats" on top of the substrate and is therefore "incommensurate" with it. This mode occurs, for example, with materials having a 2D, layered structure, such as graphite and MoS_2 [14]. In such compounds, there is no chemical bonding perpendicular to the hexagonally close-packed and tightly bonded basal plane, so that interaction of such a film with the substrate is only by Van der Waals forces. These weak forces are often strong enough to maintain rotational alignment with the substrate and to produce a small periodic compression and expansion in the epilayer lattice, but they are not strong enough to strain the epilayer so that it fits that of the substrate. There is a small periodic distortion in a_e as the lattices fall in and out of alignment periodically across the interface, and this produces a beautiful Moire pattern in STM images of the epilayer surface. Incommensurate growth can also occur when chemical bonding is weak because of a difference in bonding character between film and substrate. Chemical bonding can also be blocked by passivating the substrate surface.

In the more common situation, the epilayer is chemically bonded to the substrate, thus forming a unit called a "bicrystal." A thin epilayer with small f is likely to become strained to fit the substrate in *x* as shown in **Figure 14c** and similarly in *y*. This is sometimes referred to as "pseudomorphic" growth, but it really is not, because no change in crystal structure has occurred. It is properly termed "commensurate growth" or "coherent epitaxy." In **Figure 14c**, it is assumed that the substrate is much thicker than the epilayer, so that the substrate is rigid and all of the strain is in the epilayer. This "coherency" strain is then just $E_x = E_y = E_{x,y} = -f$, and the corresponding biaxial stress, σ_{xy}, is given by Eq. (18). The biaxial stress produces a strain in z, perpendicular to the growth plane, which is given by the three-dimensional form of Hooke's law:

$$\varepsilon_z = \frac{1}{Y}\left(\sigma_z - \upsilon\sigma_x - \upsilon\sigma_y\right) = \frac{-2\upsilon\sigma_{x,y}}{Y} = \frac{-2\upsilon\varepsilon_{x,y}}{1-\upsilon} \tag{17}$$

Here, the second equality was obtained by setting σ_z as it must be for the unconstrained direction, and the third was obtained using Eq. (18):

$$\varepsilon_x = \varepsilon_y = \varepsilon_{x,y} = \frac{(1-\upsilon)}{Y}\sigma_{x,y}\frac{\sigma_{x,y}}{Y'} \tag{18}$$

(where Y' is sometimes known as the biaxial elastic modulus. Poisson's ratio).

In **Figure 14c**, the epilayer is shown compressed in x and y and expanded in z in accordance with the above formula. This lattice is said to be "tetragonally" distorted, and the tetragonal strain is defined as:

$$\varepsilon_T = \varepsilon_z - \varepsilon_{x,y} = -\left(\frac{1+\upsilon}{1-\upsilon}\right)\varepsilon_{x,y} \tag{19}$$

X-ray diffraction measurement of the expanded atomic plane spacing a′ in z can be used with Eq. (17) to determine the fraction by which the epilayer lattice has compressed to fit the substrate in x and y. Electron diffraction can be used only when the change in a is larger than a few percent, because the peaks are much broader than in X-ray diffraction. The strain energy stored per unit area in the coherently strained epilayer U_ϵ is obtained by integrating force over distance as the film is compressed toward a fit to the substrate, starting from the relaxed state shown in **Figure 14b**. The force to maintain the compression is supplied from the rigid substrate by bonding across the interface. The integration can be done in one direction and then doubled to account for the orthogonal direction. The force, F, in the x (or y) direction, per unit width of film in y (or x), is given by Eq. (20):

$$F_f = F_s \text{ or } \sigma_f h_f = \sigma_s h_s \text{ or } \left(\frac{Y}{1-\upsilon}\right)_f \varepsilon_f h_f = \left(\frac{Y}{1-\upsilon}\right)_{fs} \varepsilon_s \qquad (20)$$

where $F_f, F_s; \sigma_f, \sigma_s$; and h_f, h_s are force, stress, and thickness of film and substrate, respectively, and the distance in x is the same as the strain E_x if we use a normalized film length of $L_x = 1$. Thus:

$$U_\epsilon\left(\frac{J}{m^2} \text{or} \frac{N}{m}\right) = 2U_x = 2\int_0^x F_x \frac{dx}{L_x} = 2\int_0^{\varepsilon_x} \left(\frac{Y}{1-\upsilon}\right) \varepsilon_x h d\varepsilon_x = \left(\frac{Y}{1-\upsilon}\right) \varepsilon_x^2 h \qquad (21)$$

The force of compression creates shear stresses in crystal planes that are not perpendicular to it, and along certain of these planes, the film will “slip” to relieve stress by breaking and then reforming bonds. After slippage, there will be extra rows of substrate atoms which are not bonded to the film, such as the one shown along y in **Figure 14d**. These features are known as misfit dislocations. Film stress is relieved by the development of a grid of such dislocations in the interface, the grid periodicity being determined by energy minimization, as we will see below. This growth mode is known as “discommensurate.” In addition to the bulk film strain energy, U_ε, there is an interface energy, γ_i, as discussed in Section 3.3.1, but since it does not depend on h, it will be neglected below.

Usually, defects of any dimensionality (0D through 3D) are undesirable within a film unless they are introduced for a specific purpose such as doping. Films in electronic applications are particularly sensitive to degradation by defects. They disturb the lattice periodicity and thus locally alter the band structure of a semiconductor crystal, often producing charge carrier traps or charge recombination centers within the band gap. Defects of 1D and 2D also provide paths for electrical leakage and impurity diffusion. Thus, in heteroepitaxial growth, it is important to know what conditions have to be met to avoid the generation of misfit dislocations. This situation needs to be analyzed based on the discussion of the properties of dislocations. It is not discussed here because of the space.

4. Conclusion

The above discussion has examined the factors determining epitaxy film structure, topography, interfacial properties, and stress. The kinetic mechanism of atom adsorption, diffusion, reaction, nucleation, and texture is given. The kinetic characteristics and related technological conditions of two-dimensional nucleation and layered ordered growth are described. A new optimized denotation index (a T_s “window” within which good crystallography and sharp interfaces can both be obtained) for epitaxy growth is proposed. Much of the development work in epitaxy has involved modifying processes to widen this window. Finally, two main factors in epitaxial growth are proposed. Two principal factors are the degree of interaction of the depositing vapor with the substrate and with itself and the amount of energy input to the deposition surface. When the energy input is thermal, care must be taken to achieve good substrate T control.

Author details

Hong Liu
Jiangsu Key Laboratory of Micro and Nano Heat Fluid Flow Technology and Energy Application, School of Mathematics and Physics, Suzhou University of Science and Technology, Suzhou, China

*Address all correspondence to: hliu03@sina.com

References

[1] Pham VP, Jang H-S, Whang D, Choi J-Y. Direct growth of graphene on rigid and flexible substrates: Progress, applications, and challenges. Chemical Society Reviews. 2017;**46**: 6276-6300

[2] Pham PV. Hexagon flower quantum dot-like Cu pattern formation during low-pressure chemical vapor deposited graphene growth on a liquid Cu/W substrate. ACS Omega. 2018;**3**: 8036-8041

[3] Oura K et al. Surface Science: An Introduction. Berlin Heidelberg: Springer; 2010. pp. 452

[4] Toth J. Adsorption: Theory, modeling, and analysis. Surfactant Science Series. 2001;**107**:509

[5] Needs RJ. Calculations of the surface stress tensor at aluminum (111) and (110) surfaces. Physical Review Letters. 1987;**58**:53

[6] Chase MW et al., editors. JANAF Thermochemical Tables. 3rd ed. Washington, DC: American Chemical Society; 1985

[7] Wagman DD et al., editors. NBS Tables of Chemical Thermodynamic Properties. Journal of Physical and Chemical Reference Data. 11(Suppl. no. 2). Washington, DC: American Chemical Society; 1982

[8] Haynes WM. CRC Handbook of Chemistry and Physics. 92nd ed. Boca Raton: CRC Press; 2007

[9] Cammack R, Atwood T, Campbell P, Parish H, Smith A, Vella F, et al. Oxford dictionary of biochemistry and molecular biology. 2nd ed. Oxford University Press; 2006

[10] Eaglesham DJ, Cerullo M. Dislocation-free stranski-krastanow growth of Ge on Si(100). Physical Review Letters. 1990;**64**:1943

[11] Guo M, Diao P, Cai S. Hydrothermal growth of well-aligned ZnO nanorod arrays: Dependence of morphology and alignment ordering upon preparing conditions. Journal of Solid State Chemistry. 2005;**178**:1864

[12] Hessinger U, Leskovar M, Olmstead MA. Olmstead, role of step and terrace nucleation in heteroepitaxial growth morphology: Growth kinetics of CaF_2/Si(111). Physical Review Letters. 1995;**75**:2380

[13] Walker P, Tarn WH. CRC Handbook of Metal Etchants. Boca Raton, FL: CRC Press; 1991 (also covers many compounds)

[14] Li M-Y, Shi Y, Cheng C-C, et al. Epitaxial growth of a monolayer WSe_2-MoS_2 lateral p-n junction with an atomically sharp interface. Science. 2015;**349**(6247):524-528. DOI: 10. 1126/ science.aab-4097

Chapter 2

Carbon Nanotubes: Synthesis, Properties and Applications

Aravind Kumar Jagadeesan, Krithiga Thangavelu and Venkatesan Dhananjeyan

Abstract

Recent discoveries of salient carbon nanoforms have paved tremendous interest among research and also toward their discrete applications in scientific fields. Various generation methods for carbon nanotubes (CNTs) involve chemical deposition of vapor, discharge using electric arc and laser ablation mechanism which were driven by functionalization, chemical addition, doping, and filing such that in-depth characterization and manipulation of CNTs were possible. The in-built elasticity, electromechanical, chemical, and optical properties of CNTs have a notable impact on its stability and reactivity. Perhaps, the flexibility along with its determined strength makes them to validate its potential application in diverse fields which enables that these CNTs will definitely procure a prominent role in nanotechnology.

Keywords: nanotechnology, nanotubes, synthesis, electromechanical, nanosensor

1. Introduction

Carbon materials can be grouped into three classifications based on their period of advancement: classical carbons, nano ones and new carbons. Cracking carbons incorporate engineered graphite squares principally utilized as anodes, carbon blacks, what are more, enacted carbons, for which creation systems were created before [1] the 1960s. During the 1960s, carbon materials not quite the same as these great carbons were designed: carbon filaments from different forerunners, including fume developed carbon filaments; pyrolytic carbons delivered by means of concoction fume testimony forms; glasslike carbons with high hardness and gas impermeability; high-thickness isotropic carbons created by isostatic squeezing; intercalation mixes with various functionalities, for example, high electrical conductivity; and precious stone like carbons as straightforward carbon sheets. These recently evolved carbon materials are grouped as new carbons [2]. Since the 1990s, different fullerenes with shut shell structure, carbon nanotubes with nanometer distances across, and graphene pieces of just a couple molecules' thickness has stood out from nanotechnology; these are ordered as nanocarbons.

On the off chance that these carbon materials are considered from the perspective of their surface, be that as it may, they might be ordered into two gatherings: nano-textured and nano-sized carbons [3]. Most carbon materials in the new carbon classification are delegated nano-textured carbon, in light of the fact that their nano texture is controlled by means of various procedures in their creation, notwithstanding the basic control. Then again, fullerenes, carbon nanotubes, and graphene can be delegated nano-sized carbon, the shell size of fullerenes, breadth of carbon nanotubes, and thickness of graphene, drops are on the nanometer scale [4]. Carbon blacks in great carbon are made out of nano-sized particles, yet they are not typically named nanocarbons since they have different applications as a mass, not as individual nano-sized particles [5].

Carbon, a basic chemical substance containing 6 as atomic number with 6 electrons tend to occupy s and p orbitals. It can able to undergo hybridization through three different forms such as sp/sp^2 or sp^3 means. Recent inventions of compact structured carbon materials such as fullerene [6], graphene [7], and carbon nanotubes [8] have envisaged prompt enquiries into this emerging field. Various physical properties of carbon nanotubes were mostly derived from base material (graphene). Such graphene involves the dense packing arrangement of carbon atoms in a regular sp^2 pattern bonded to honeycomb based atomic scale structure and especially this pattern is most suitable as a primary structure for other sp^2 materials [9]. Based on theoretical judgment, this CNT is explicitly distinct in the cylinder form fabricated of swirled up graphene thick sheet, which can delineate itself to single or multiple well. The single well nanotubes were known as single walled carbon tubes which were investigated during 1993 whereas multi-walled ones were found during 1991 itself [10].

CNTs have outstanding mechanical, thermal, electrical, and optical properties that are being used exclusively or in mix to deliver keen sensors or on the other hand multifunctional materials [11, 12]. They have high angle proportions that are perfect for long and persistent detecting. Their high surface region, for example, can be misused for storing materials to make half breed useful materials or functionalized to make cathodes for an assortment of uses [13]. CNTs are additionally known to display ballistic conductivity because of insignificant electron dispersing in their 1D structure with mean free-ways of the request of several microns [14, 15].

Mechanical strain may cause reproducible changes in the electrical properties of CNT filaments, making it conceivable to misuse them as electromechanical sensors [16, 17]. The partner changes incorporate inductance, capacitance, and obstruction which can be associated to the strain. Of incredible significance is that CNT filaments are receptive to elastic, compressive, flexural, and torsional strain [18].

The working standards of sensors produced using a CNT plainly visible get together incorporate difference in their electrical resistivity or obstruction because of mechanical strain known as piezoresistivity, change of their inductance and capacitance because of mechanical strain, change of their electrical resistivity because of variety in temperature known as thermoresistivity [19], change of their electrical obstruction because of variety in an attractive field known as magnetoresistance [20], and change in their electrical opposition with change of their mechanical thunderous recurrence because of variety of temperature, weight, mass, and strain [11]. The adjustment in conductance or obstruction is substantially more predominant than other variety in electrical properties. This is somewhat in light of the fact that charge transporters are handily isolated under simultaneous deformation prompting an expansion in obstruction. For extremely little strains, the total deformation has demonstrated to be flexible and the conductive system is completely recouped when the strain is evacuated, prompting an abatement in opposition [21]. Thus, the presented chapter highlights the synthesis details, associated properties and current applications of carbon nanotubes.

2. Discussion

2.1 Synthesis of carbon nanotubes

There are many methods to synthesize CNTs, but these three methods are most important and commonly used methods. They are as follows.

2.1.1 Chemical vapor deposition method

Chemical vapor deposition (CVD): CVD is a technique in which the vaporized reactants react chemically and forms a nanomaterial product that is deposited on the substrate **Figure 1**.

Sources for carbon: The precursor for carbon nanotubes are hydrocarbon gases such as acetylene, ethylene, methane, etc. [22].

Substrate used: Substrates are materials on which the CNTS are grown. The commonly used substrates in CVD method are zeolite, silica, silicon plate coated with iron particles, etc.

Catalyst used: To produce single-walled carbon nanotubes metal catalyst nanoparticles such as iron, cobalt, nickel, molybdenum, iron-molybdenum alloys, etc. are used.

Sources for CVD used: Based on the heating source, the CVD can be:

- Thermal activated CVD which is heated by IR radiation, RF heater, etc.
- Photo assisted CVD which is heated by Arc lamps, CO_2 laser, Argon ion laser, Nd:YAG laser, etc.
- Plasma assisted CVD which is heated by microwave radiation, etc.

Conditions maintained: The following conditions are maintained inside the furnace.

- Temperature: 500–900°C.
- Inert gas atmosphere: Argon gas.

2.1.2 Procedure for synthesis of CNTs by thermal CVD method

CNTs are synthesized by thermal CVD method by using hydrocarbon gas as carbon source. In this method, a quartz tube is placed inside a furnace maintained at high temperature (500–900°C) heated by RF heater. A crucible containing the substrate coated with catalyst nanoparticles is placed inside quartz tube filled with inert gas

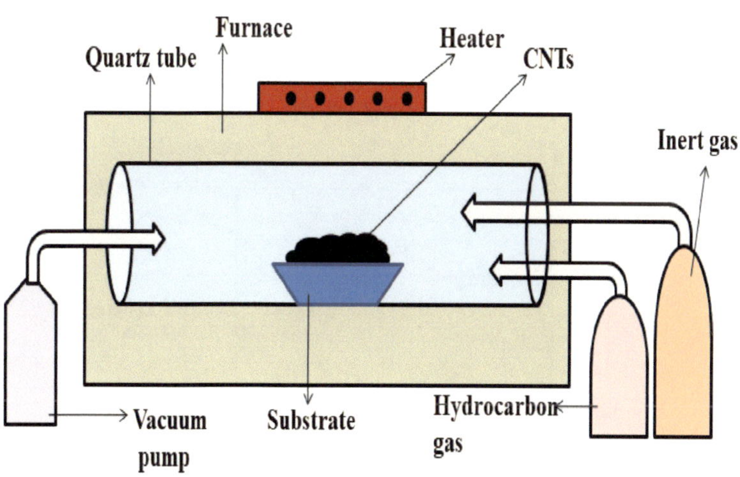

Figure 1.
CVD method.

such as argon gas. The hydrocarbon gas (carbon source) is pumped into the quartz tube which undergoes pyrolysis reaction and forms vapor carbon atoms. These carbon atoms bind to the substrate and join to eachother by Vanderwaal force of attraction and grow as multi-walled carbon nanotubes (MWCNTs) on the substrate [23]. To synthesize single-walled carbon nanotubes catalyst nanoparticles of Fe, Co, Ni are used. The obtained CNTs are further purified to get the pure form of CNTs.

2.1.3 Electric arc discharge method

Carbon nanotubes are synthesized by electric arc discharge method which is also called Plasma Arcing method.

2.1.3.1 Description

Electrodes: Pure graphite rods (both positive and negative electrode). The positive electrode is adjustable from outside to maintain the gap between the two electrodes.

Diameter of electrodes: 5–20 μm.

Gap between electrodes: 1 mm.

Current: 50–120 amperes.

Voltage: 20–25 V.

Inert gas pressure: 100–500 torr (No CNT formed below 100 torr). Inert gas is used for cooling and condensation of atoms to form the CNTs. Inert gas determines the structure of carbons to be present in CNTS. Commonly used inert gas is helium gas.

Temperature: 3000–3500°C.

Reactor: It contains a quartz chamber which is connected to vacuum pump and a diffusion pump to inert gas supply. Initially the chamber is made vacuum by the vacuum pump and then the chamber is filled with helium gas by the diffusion pump [24].

2.1.4 Procedure for synthesis of CNTs by Electric arc discharge method

In this method, a potential of 20–25 V is applied across the pure graphite electrodes separated by 1 mm distance and maintained at 500 torr pressure of flowing helium gas filled inside the quartz chamber **Figure 2**. When the electrodes are made to strike each other under these conditions it produces an electric arc. The energy produced in the arc is transferred to the anode which ionizes the carbon atoms of pure graphite anode and produces C^+ ions and forms plasma (Plasma is atoms or

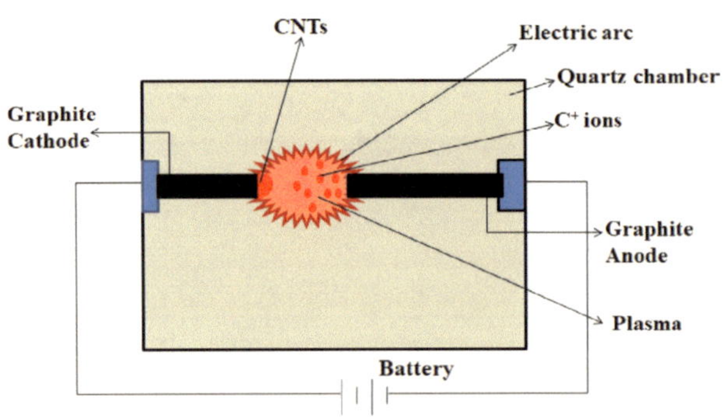

Figure 2.
Electric arc method.

molecules in vapor state at high temperature). These positively charged carbon ions moves towards cathode, gets reduced and deposited and grow as CNTs on the cathode. As the CNTs grow, the length of the anode decreases, but the electrodes are adjusted and always maintain a gap of 1 mm between the two electrodes. If proper cooling of electrodes are achieved uniform deposition of CNTs are formed on the cathode which is achieved by inert gas maintained at proper pressure [25]. By this method multi-walled carbon nanotubes are synthesized and to synthesize single-walled carbon nanotubes catalyst nanoparticles of Fe, Co, and Ni are incorporated in the central portion of the positive electrode. The obtained CNTs are further purified to get the pure form of CNTs.

2.1.5 Laser ablation method

Physical vapor deposition (PVD): PVD is a technique by which a material can be vaporized into gaseous form and then deposited on the surface of a substrate.

Target source: The most common carbon source target used is solid graphite which is irradiated by laser source and vaporized into vapor carbon atoms.

Laser source: Laser source used for vaporization of target material into target vapor atoms can be continuous laser source such as CO_2 laser or pulsed laser source such as Nd:YAG laser (Neodymium doped Yttrium Aluminum Garnet, $Nd{:}Y_3Al_5O_{12}$) .

Substrate used: The substrate used in this method is the water cooled copper collector on which the vaporized carbon atoms deposit and grow as CNTs.

Inert gas atmosphere: Argon gas is commonly used as inert gas which flows at a constant flow rate towards the water cooled copper collector.

2.1.6 Procedure for synthesis of CNTs by Laser Ablation method

Laser Ablation method is a Physical Vapor Deposition method in which graphite target is vaporized by laser source **Figure 3**. In this method the graphite target is placed at the center of quartz chamber filled with argon gas and maintained at 1200°C. The graphite target is vaporized by either continuous laser source or pulsed laser source. The vaporized target atoms (carbon) are sweeped toward cooled copper collector by the flow of argon gas. The carbon atoms are deposited and grown as CNTs on cooled copper collector. In case of continuous laser beam, the carbon atoms are continuously vaporized whereas in case of pulsed laser beam the amount of CNTs produced can be monitored as each shot of pulsed laser beam is directly proportional to the amount of carbon atoms vaporized [26]. By this method multi-walled carbon nanotubes are synthesized and to synthesize single-walled carbon

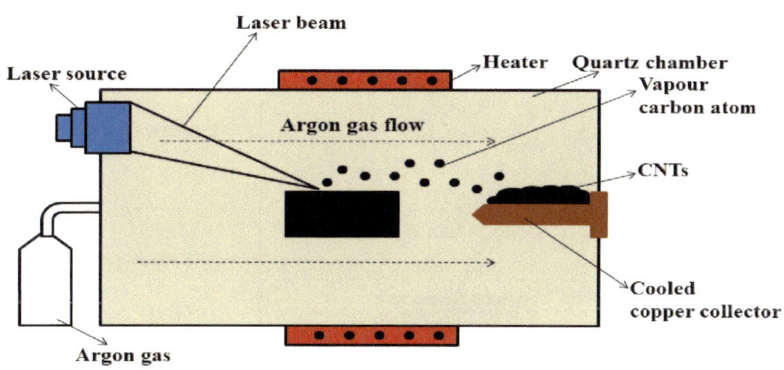

Figure 3.
Laser ablation method—schematic representation.

nanotubes catalyst nanoparticles of Fe, Co, Ni are used. The obtained CNTs are further purified to get the pure form of CNTs.

2.1.6.1 Procedure for pulsed laser deposition method

Pulsed Laser deposition is a thin film deposition technique in which the target material is vaporized by pulsed laser beam and vaporized target atoms are made to deposit on substrates **Figure 4**. The furnace contains a target at bottom and substrate mounted on the top. A pulsed laser beam from Nd:YAG laser source is made to strike the target to produce vaporized target atoms called the plume (plume is vaporized atoms at high temperature) [27]. The plume moves towards the substrate and it is deposited and grown as CNTs. Each shot of laser is directly related to the amount of material ablated, thus deposition rate can be controlled and calibrated.

2.1.6.2 Purification of CNTs

The synthesized CNTs can be separated from the amorphous carbon, carbon nanoparticles, residual catalyst and other impurities by various methods. The conventional methods of purification are not very successful but methods like gas phase, liquid phase and intercalation methods show good results.

Gas phase purification of CNTs: In this method the CNTs are subjected to a high temperature oxidation followed by repeated extractions with nitric acid and hydrochloric acid. This procedure makes the synthesized CNTs purer and high stability with fewer amounts of residual catalyst and other non CNTs forms.

Liquid phase purification of CNTs: A series of steps are followed in the liquid phase purification of synthesized CNTs. They are:

- Preliminary filtration to remove bulk graphite particles.
- Dissolution in both organic solvents and concentrated acids to remove the fullerenes and catalyst, respectively.
- Centrifugal separation of CNTs (Solid part) from the solution (containing impurities).

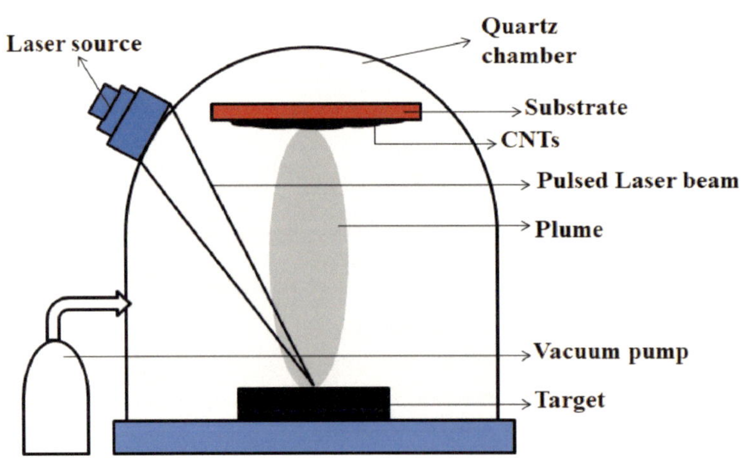

Figure 4.
Pulsed laser ablation method—Schematic representation.

- Microfiltration.

- Chromatography to isolate multi-walled carbon nanotubes, single-walled carbon nanotubes, etc.

Intercalation purification of CNTs: In this method the nanoparticle impurities present are oxidized by metallic copper which acts as oxidation catalyst formed from the reduction of copper chloride added during the process. This process introduces intercalate residues and damage CNTs during oxidation process.

2.2 Structure/properties of CNT

There are numerous mechanisms available to build up structures occluded with various characteristics. The sp^2 nature of carbon hybridization constructs a layered pattern of arrangement with weaker plane bonding of Vander Waals forces at the outside and strong forces at inner plane bounds. Few numbers of concentric cylinders were equipped with regular spacing of interlayers that are located around central hollow section and demonstrated as multi walled CNTs **Figure 5**. In general, the real time spacing of MWCNTs contain interlayer spacing in range of 0.35–0.40 nm. The inner diameter of multi-walled CNTs can even range from 0.40 nm to few nanometers [28]. Outer diameter can exist up to 25 nm. The tips on both sides were closed and protruding ends were capped using dome shaped half width fullerene molecules. Axial molecules can exist up to few centimeters. The primary function of half width fullerene molecules is to aid in shutting down the tubes at both the ends. Whereas SWCNT can exist up to 4 nm. Length is up to micrometer range. Such arrangement is organized in a hexagon shape so as to develop a crystal [29].

2.2.1 Structure of SWCNT and MWCNT

Based on wrapping mechanism, three different forms of SWCNTs include chiral, armchair, and zigzag pattern. The single walled structure is primarily characterized by a set of indices (n and m) which describes the vector mechanism of chiral and absolutely it impinges an impact on electrical tendency of both nanotubes **Figure 6**.

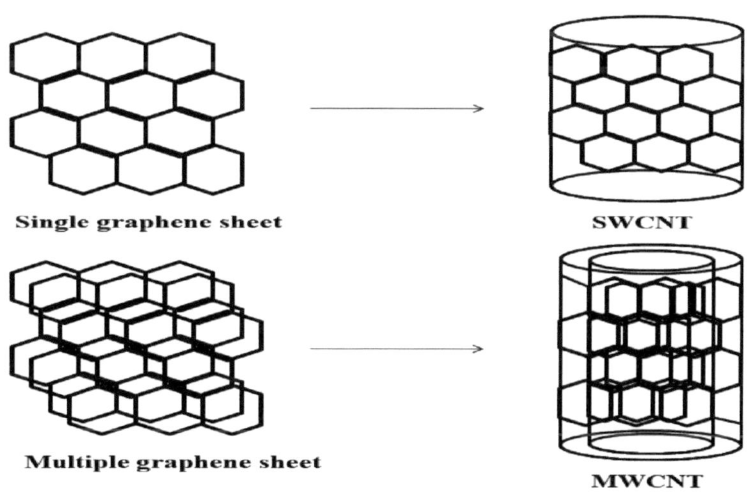

Figure 5.
Graphene to CNT.

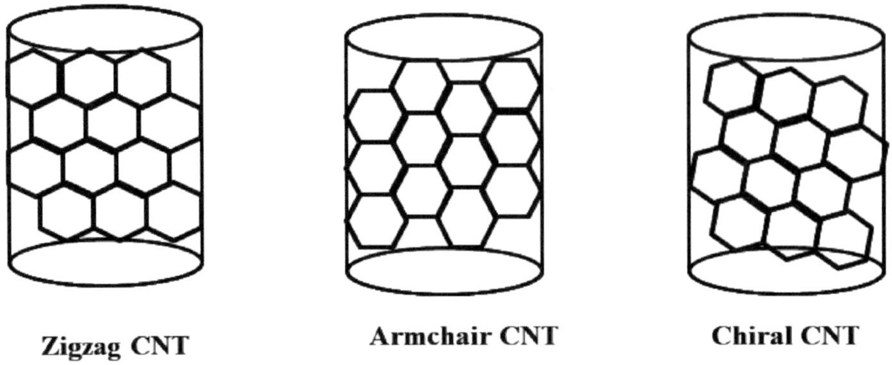

Figure 6.
Different forms of SWCNTs.

As a general predict, when n = m, these nanotubes are known as armchair ones and if m = 0, they are said to be zigzag and for other range as chiral pattern [30].

The vector value of chiral mechanism can be determined using $C = na_1 + ma_2$, where a_1 and a_2 represent the base vectors of graphite cell and also used to evaluate the tube radius and moreover this vector function also estimates the rolling direction of graphene sheet. Hence, the radius of carbon nanotube can be estimated using.

$$r = a\frac{m2 + mn + n2}{2\prod}$$

where a takes the lattice parameter in graphite sheet.

Whenever n – m = 3 times of any value, it indicates the carbon nanotube to be metallic or extremely conducting nature and if it is not so, it can be semi-metallic type or a semi-conductor. At most of the times, armchair type can be referred as metallic one whereas all other forms can be denoted as a semi-conductor. Various involved parameters and vector representations [31] can provide an impinging impact on structure of CNT as follows.

1. chiral vector = $na_1 + na_2 >> (n,m)$

2. Translational vector, $T = t_1a_1 + t_2a_2 >> (t_1, t_2)$

3. Chiral vector length, $L = a(n^2 + n.m + m^2)^{1/2}$ a is constant of lattice parameter.

4. angle of chiral vector = $(2n + m)/2^*(n^2 + n.m + m^2)^{1/2}$

5. radius = $L/2\prod$

6. Rotation angle, $\Psi = 2\prod/N$

7. vector of symmetry, $R = pa_1 + qa_2$

2.2.2 Formation of multi walled nanotubes

Such MWCNTs can be developed via two distinct models such as Russian doll and Parchment type model. If the diameter of outer CNT exceeds the inner tube, such a model is prescribed as Russian type model whereas, wrapping of a single graphite sheet to many a fold around itself constitutes the simple Parchment model. Both multi walled and single walled CNTs possess similar properties. Due to multi-layered arrangement of multi walled nanotubes, the outer portion not only cover the inner tubes from certain chemical reactions when contaminate with ambient substances but also exhibit greater tensile characteristics, which would be a drawback of single walled CNTs [32].

Owing to the presence of sp^2 bonds available betwixt indigenous carbon atoms, CNTs possess higher tensile property compared to steel as well as Kevlar. such a bond exhibits more strength rather than sp^3 bonding of diamond. Hence, SWCNTs possess maximum tensile property which may be nearly 100 times as that of steel [33].

2.2.3 Elasticity property

An amazing feature of CNTs is its elasticity. Under maximum force and high pressure by exposing it to greater compressive forces along axial direction, it can even bend, kink, twist and ultimately buckle without causing any damage to CNT. Thus nanocarbon tubes can retain its original geometric structure. But sometimes, elasticity tends to cope up with a limit and hence under the influence of stronger physical pressure forces, it can even undergo a temporary deformation to form the nanotube shape. Few defects may weaken its structure which includes the atomic defects or else rearrangement developed on the carbon bonds.

The elasticity property for both single walled and multi walled CNTs is examined by the term known as modulus of elasticity or elastic modulus. Such property of multi-walled CNTs can be analyzed using transmission electron microscopy (TEM). Using such an apparatus, the researchers examine and investigate the molecular vibrations owing to thermal forces created at both edges of tubes [34].

As the atomic bond strength is high, CNTs not only withstand elevated temperature levels but also act as excellent thermal conductors. Hence under vacuum atmospheric pressure ranges, they are able to withstand 2900°C and nearly 800°C at normal pressure conditions. But the prevailing tube temperature and ambient environment may have an impact on thermal conductivity of carbon nanotubes [35]. The prescribed physical properties were outlined in **Table 1**.

Various types of indigenous single walled CNTs obtained using chemical vapor deposition technique onto a supporting chemical agent are mostly of semi-conducting nature (I type). Such nanotube type depicts the impact of field transistor (FET) nature at atmospheric conditions and these have been recently attaining greater interest and also achieved extensive exploration towards their application as nano-electronic materials indulging logic circuit devices and electronic transistors. Such growing CNTs are seemed to be p-type containing doped holes with absolute hole depletion and reduced conductance values (100 kΩ to 1 MΩ) in specific to positive logic gate voltages. in the present context, it has been demonstrated that adsorption of molecular oxygen onto the CNTs is a contributing factor do drive the hole doping effect of SWCNTs. Oxygen removal can even lead to mere existence of semi-conducting nature. Instead, day by day investigations on CNTs reveals that the electrical properties of such carbon nanotubes are much sensitive to chemical doping impacts and charge transfer mechanism in spite of exhibiting extreme robustness [36].

The II type CNTs developed by CVD technique appears to be quasi metallic consisting smaller band gaps in the order of 10 meV. Such CNTs are not sensitive compared to semi-conducting type due to their electrostatic doping mechanism through gate potentials but exhibit a mere conductance dip occluded with that of smaller band gap. These CNTs origin towards a class of non-armchair single walled CNTs and band origin may be due to shift of sp^2 to sp^3 orbital hybridization which occurs prominently by the existence of non-flat hexagonal nature of tube walls. Quasi metallic types exhibit enhanced electrical conductivity at low temperature levels when subjected to temperature dependent experimental studies [37].

Physical properties	Parameter	Range	References
Structure during equilibrium	Mean diameter	1.3–1.5 nm	[31]
Density	Zig zag (16,0)	1.33 g/cm^3	[30]
	Armchair (10,9)	1.32 g/cm^3	[30]
	Chiral (12,5)	1.41 g/cm^3	[32]
Lattice parameter	Zig zag (16,0)	16.53 nm	[32]
	Chiral (12,5)	16.53 nm	[33]
	Arm chair (10,9)	16.55 nm	[32]
Interlayer distance	Zig zag	3.40 Å	[33]
	Chiral	3.38 Å	[33]
	Arm chair	3.37 Å	[34]
Elastic nature	Young's modulus	1.0–1.27 TPa	[35]
	Tensile strength	About 100 GPa	[35]
Thermal property	Mean free path	Around 100 nm	[36]
	Thermal conductivity	Around 2000 W/m-K	[37]
Electrical behavior	Current density	1015 A/m^2	[38]
	Conductance	13.0 (K.Ohms)$^{-1}$	[38]

Table 1.
Physical properties of CNTs.

Even quantum interfering impacts were also being observed: (1) phonon acts as the basic scattering mechanism existing in single walled CNTs at ambient conditions and (2) excellent levels of ohmic frequency contacts can be proliferated in the nanotubes with a probability of adequate transmission T = 1 and 3 electron transfer is explicitly phase coherent along with ballistic ability in CNTs at even low temperature levels. This also suggest a lengthy mean distance for ballistic electron transfer in super quality CVD developed SWCNTs [39].

2.2.4 Electromechanical properties

Schematic pattern of growth has been extensively used to obtain suspended CNTs in single wall across certain trenches along with normal nanotubes which may be electrically wired up with relative easiness. By manipulating a suspended CNT using an AFM probe while measuring its electrical conductivity, the impact of mechanical deformation on electrical characteristics of CNT can be judged. The wide scope of CNTs based on nanoelectro-mechanic (NEM) devices are invented to explore twisting pattern of single nanowires, pure stretching levels and also due to their high frequency characteristics of resonance measurements. Operated NEMs switches and accessible memory devices have also been envisioned in nearby future. Powerful control and deterministic mode of synthesis of CNT will further explore exciting opportunities and greater possibilities of finding novel nanomaterials and other devices [38].

2.2.5 Chemical properties and species interaction

SWCNTs are mostly inert in nature. The covalent attachment agglomerated the molecular species with fully bonded sp^2 hybridization onto sidewalls of CNT proves to be complex. The adsorbed molecules onto CNTs through the development of non-covalent forces has evidently turned to be facile and consequently lead to possible effects on their physical properties and also with their potential applications. Desorption of orientation molecules from single walled tubes can be achieved by heating the nanotubes to higher temperature levels [40].

Similarly, illumination of UV light at low photon intensity forces a drastic molecular desorption rate from SWCNTs at even ambient conditions whereas, wavelength governing measurements predict that photo-desorption process may occur due to sudden excitation of electrons occluded in the nanotubes and perhaps it is a non-thermal process. The excitation of electrons in specific by $\prod$ plasmons included in SWCNTs due to UV light results in electron/hole pair formation which occur through Landau damping. The studies portray that surface and photochemistry problems are much predominant to exhibit properties and to create molecular nano surface wires that possess ultrahigh surface distribution with each and every atom accommodating onto the surface. Therefore, surface science study can be evaluated at single wire level itself by incorporating both chemical and electrical properties of CNTs as thin probes [41].

2.2.6 Optical properties

Carbon nanotubes have helpful assimilation, photoluminescence (fluorescence), and Raman spectroscopy properties. Spectroscopic strategies offer the chance of speedy and non-dangerous portrayal of moderately a lot of carbon nanotubes. There is a solid interest for such portrayal from the mechanical perspective: various parameters of nanotube union can be changed, purposefully or accidentally, to modify the nanotube quality. As demonstrated as follows, optical assimilation, photoluminescence, and Raman spectroscopies permit brisk and solid portrayal of this "nanotube quality" as far as non-rounded carbon content, structure (chirality) of the delivered nanotubes, and auxiliary imperfections. These highlights decide about some other properties, for example, optical, mechanical, and electrical properties [42].

Carbon nanotubes are novel "one-dimensional frameworks" which can be imagined as moved single sheets of graphite (or all the more accurately graphene). This rolling should be possible at various points and ebbs and flows bringing about various nanotube properties. The width normally fluctuates in the range 0.4–40 nm (i.e., "just" ~100 times), yet the length can shift ~100,000,000,000 times, from 0.14 nm to 55.5 cm [43]. The nanotube perspective proportion, or the length-to-breadth proportion, can be as high as 132,000,000:1 [44] which is unmatched by some other material. Thusly, all the properties of the carbon nanotubes comparative with those of common semiconductors are incredibly anisotropic (directionally reliant) and tunable.

2.2.7 Outline information

While mechanical, electrical, and electrochemical (supercapacitor) properties of the carbon nanotubes are entrenched and have quick applications, the down to earth utilization of optical properties is yet muddled. The previously mentioned tunability of properties is conceivably helpful in optics and photonics. Specifically, light-discharging diodes (LEDs) and photograph detectors dependent on a solitary nanotube have been created in the lab. Their exceptional element is not the effectiveness, which is yet moderately low, however the limited selectivity in the frequency of discharge and recognition of light and the chance of its adjusting through the nanotube structure. What's more, bolometer and optoelectronic memory gadgets have been acknowledged on groups of single-walled carbon nanotubes [45].

Crystallographic absconds additionally influence the cylinder's electrical properties. A typical outcome is brought down conductivity through the flawed space of the cylinder. An imperfection in easy chair type tubes (which can lead power) can make the encompassing area become semiconducting, and single monatomic opening incite attractive properties.

3. Applications

3.1 Biomedical field

The characteristic properties ascertained to CNTs are really enthusiastic. in the last decades, many research studies have proposed potential uses of CNTs and also have remarkably portrayed promising applications when such newly developed materials are joined together with typical scientific products, for example, nanorods production using such CNTs as reactive template materials.

Applications of CNTs encompass major fields and various disciplines, which include nanotechnology, medicine, construction, manufacturing, electronics, peripheral hardware, software and so on **Figure** 7. The mentioned applications can be considered: actuators, composites with maximum strength, energy storage as well as energy conversion equipment, media for H_2 storage, nano-sensors and probes, electronic instruments and process catalysis. Anyway the forthcoming sections will highlight detailed applications of CNTs in biomedical field. There are three parameters which may act as barriers before using CNTs in the fields of biotechnology and biomedical based industry. These barriers have to be overcome: toxicity, pharmacology and functionalization perspectives of CNTs [46].

The most prominent barrier is toxic nature of CNTs. In general, the coexistence of maximum surface area rendered by CNT along with the intrinsic toxic nature of nano surface can become most important for the harmful impacts of aggregated nanoparticles. The toxic nature of CNTs can be influenced by the particle size of designed nanotubes. If the particles are less than a size of 100 nm, they are able to exhibit definite harmful effects such as enhanced potential hazard to the liver, lung, protein structure modification, escape from usual phagocytic powerful attacks, activation of immunological and inflammatory responses, and explicit redistribution strategy from their spot of tube deposition [47].

Another predominant barrier with CNT is the pharmacokinetics and biodistribution of aggregated nanoparticles which are in-turn influenced by distinct physicochemical attributes such as size, shape, aggregation capacity, chemical composition, solubility of surface and effective fictionalization. Previously made studies have demonstrated that CNTs of water soluble nature are much more biocompatible with the inbuilt human body fluids and also do not show any toxic ill-effects or abnormal mortality [48].

The most notable disadvantage of CNTs is the lack of aqueous solubility when exposed to any media and in order to eradicate such a problem, surface modification is introduced on the carbon nanotubes, that is, stable fictionalization of surface with suitable hydrophilic substituents and reaction chemistries which can improve both aqueous solubility as well as biocompatibility of CNT [49].

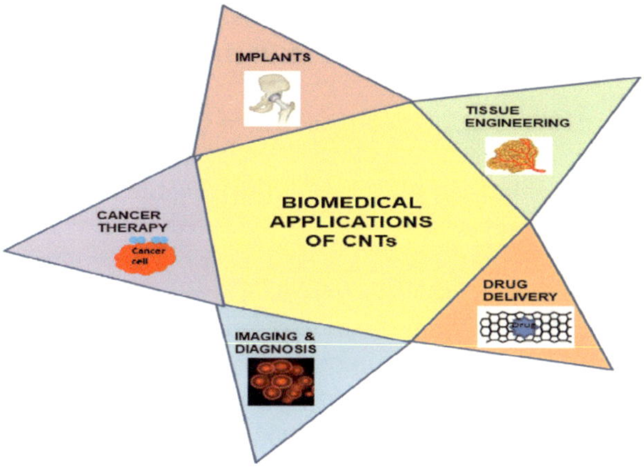

Figure 7.
CNT applications in biomedical field.

3.2 Artificial implant scopes

Nanomaterials portray their chosen probability and thrusted promise in the field of regenerative medicine due to their extraneous physical/chemical properties. In general, the rejected implant materials which may be the cause for post administration implant pain and in order to avoid such rejection, nanotubes were attached to amino acids and to proteins, thereby achieved a promising development. Both the single and multi-walled forms can be effectively utilized as implants which may be either artificial joints or else other implant materials without any kind of host rejection output response. Perhaps, due to its unique material properties such as maximum tensile strength, these can effectively act as implant materials for bone substitutes and if suitably filled with calcium, such implants can be shaped or arranged within the bone structure [50, 51].

It has also been invented that proliferation and cellular adhesion can increase with the availability of SW and MW carbon nanotube composites and hence, these can be essentially integrated into natural nanomaterials and synthetic type materials to fabricate suitable nanocomposites. The specific type of CNT accustomed to artificial implants was represented in **Table 2**.

3.3 Tissue engineering

The scope of tissue engineering lies in the substitution of damaged tissue using biological alternatives that can possibly replace/repair original and normal

Type of CNT	Background materials	Properties	References
SWCNT impregnated	Chitosan type	Enhancement in cell morphology and growth	[50]
SWCNT (pore type)	Membrane of polycarbonates	Increase cytoskeletal extensions	[51]
SWCNT	Sodium alginate	Elevate proliferation	[51]
MWCNT	Collagen type	Enhancement in cellular adhesion	[52]
MWCNT	Poly-acrylic acid type	Improves cellular discrimination	[53]

Table 2.
Nature of CNT in artificial implants application.

function. The recent advances in the emerging fields of material science have been promisingly supported in the growth of tissue engineering and regenerative medicine [54].

CNTs can be recommended for use in the field of tissue engineering under four different perspectives. They are cell tracking/labeling, sensing cellular nature and mechanism, augmenting the mechanism and enhancing the tissue matrices. Cell tracking/labeling is the specific capacity to identify implanted cell structures and to record the noteworthy improvement of tissue growth in vivo as well noninvasively [52]. Labeling of transferred cells by implants not only permits the evaluation of engineered tissue viability but also promotes a deep understanding of migration, bio-distribution, movement pathways and relocation of implanted cells. The non-invasive techniques nowadays become more familiar than traditional techniques such as cytometry owing to more time consumption and practical challenges associated with handling of usage. Hence, CNTs can be more feasible as contrast imaging agents for optical resolution, magnetic resonance behavior and also for radio tracer simulating models [54].

One of the prominent applications of CNTs in the study of tissue engineering is its ability to bio-distribution measurement and also it can be systematically varied using radiotracers applicable to gamma scintigraphy. The proper design of engineered tissue structures enhances and promotes monitoring of cellular physiology that includes protein/metabolite secretion rate, enzyme and other cofactor interactions, cellular growth/mechanism and molecules/ions transport. Novel nanosensors will be effectively utilized in such a way to determine continuous monitoring associated towards the working performance of engineered tissues.

Numerous and popular features involved in the structure of CNTs envisage them to become key elements for nanosensing devices owing to its maximum surface area and DNA or protein immobilizing capacity along with electrical properties [53].

Moreover, these carbon nanotubes possess distinctive electronic structures make the invention of redox active proteins and also amino acids thereby rendering the cell monitoring activity in engineered tissue patterns. In another research, MWNTs were combined with that of platinum micro nanoparticles and also able to identify thiols such as prescribed amino acids which include glutathione and L-cysteine observations in rat.

The cell matrix predominates its function in tissue engineering. Even though, PLA and PLGA have been utilized for tissue engineering, they tend to adopt the inbuilt mechanical strength and cannot be just functionalized pertaining to controversial version of synthetic polymer compounds. Hence, CNTs have potential applications as tissue scaffolds and able to ascertain the structural reinforcement. The only demerit of CNTs is such that they are not biodegradable. When CNTs are dissolved in certain quantity of polymeric substance, rapid enhancement in mechanical strength has been promptly noted. If these MWNTs combine with chitosan material it may lead to advancement in properties whereas SWNT on blending with natural collagen improves the cell growth of smooth muscles [43].

3.4 Cancer cells tracing

Nanodevices have been investigated that has the effective potential to generate new techniques in cancer treatment, diagnosis as well as detection. The geometric structures of these nanomaterials may be very small (< 100 nm) such that the body will evacuate it rapidly so as to become more efficient either through detection or by imaging and thereby enter the damaged cells and organelles inside the body to have interaction with DNA or protein molecules. The possible detection of cervical cancer causing cells among human beings can be improved by carrying out modification on the graphene electrode using peptide form of nanotube folic acid [55].

As large quantity of cancer types is truly asymptotic throughout their initial stage and also due to absence of specific morphologic modifications among most of the neoplastic disorders in preliminary stage, possibly traditional cancer imaging and clinical methods such as X-ray, CT scan and even MRI scan does not require any kind of spatial resolution for such disease detection in initial stages. Imaging analysis using single walled CNTs have been explored for the past few decades. Coupling of radioisotopes with single walled CNTs along with imaging techniques based on radio nucleotides can progress advancement in tissue sensitivity, penetration as well as nano medium spatial resolution.

Many sophisticated protein biomarkers are available which are often overexpressed in the interior of cancer cells and they offer an entry mark for preliminary diagnosis, maintaining surveillance, prognosis, curative surgery techniques, advancement in disease monitoring therapy and finally detecting therapeutic response. Special categories of tumor biomarkers have been tremendously applied and also conceivably utilized in diagnosis and treatment of hepatocellular carcinoma, pancreatic, colorectal cancer, prostate cancer, ovarian tumor on epithelial cells which includes CA19-9 (carbohydrate antigen), carcinoembryonic antigen (CEA), alpha fetoprotein (AFP), human chorionic gonadotropin (hCG), carcinoma antigen 125 (CA125) and specific antigen to prostate (PSA) [27].

3.5 Gene and drug delivery applications

There are various interim obstacles associated with conventional application and administration of chemo-therapic agents which includes system toxicity, lack of proper sensitivity/selectivity, minimum solubility, poor cellular distribution, lagging of certain clinical procedures, and inefficiency of specific drugs to overcome the cellular barriers for achieving the treatment of multidrug resistive cancer disease. Scientists and research experts have conceptualized various drug handling/delivery systems to combat these terrific issues using silica nanomaterials, quantum dots, polymeric materials, dendrimers, liposomes, emulsions, micelles and even molecular conjugates [56, 57].

As discussed above **Table 3**, CNTs possess specific features such as ultrapure/maximum surface distribution, which essentially provoke them to act as a promising tool for drug delivery, nucleic acids and peptides. The selective drug or desired gene can be effectively coupled with tips and walls of carbon nanotubes and can easily trace out specific cancer causing receptors that are available on the cell structure and thereby such CNTs can even over cross the cell membrane of mammals through the mechanism of endocytosis or other plausible procedures. Thus, it can recognize the therapeutic drugs/genes much more reliably and safely in the affected cells that are reluctantly inaccessible during previous procedures.

Recent discoveries of research experts paved the way to invent novel and efficient single walled CNT based drug delivery tool for targeting tumor which comprises of targeting ligands for tumor, drugs for anticancer treatment and also functionalized single walled CNTs. If such system has an interaction with cancer causing cells, it can provoke the receptor assisted endocytosis by identifying specific receptors to cancer cells onto the affected cell surface and perhaps specifically and efficiently release the active chemotherapeutic agents [58]. The release of drugs using CNTs were highlighted in **Table 3**.

3.6 Sensor-based biomedical applications

In cardiology field, the CNTs are used for artificial valves to heart blocks, pacemakers, etc. These CNTs may be involved either as individual or in combined form to synthesize smart sensors and various multi-functional materials. They possess maximum aspect ratios that make the CNT to be extremely ideal for operating longer duration and of course continuous sensing. Their greater surface distribution shall be definitely exploited for providing material deposition in order to generate hybrid variety of functional materials or better functionalized to formulate different electrodes for certain applications [59].

CNTs are much prone to promote its ballistic conductivity which may occur owing to minimum scattering of electrons in their one dimensional solid structure containing mean free paths in the measurable range of tens to microns. The exerted mechanical stress/strain can even aid few reproducible changes acquainted with the electrical properties involved in CNT fibers, thereby felicitating it extremely viable to act as electro mechanical sensor devices. The occluded changes are noted as capacitance, inductance and electric resistance which can be directly correlated to the impact of strain **Figure 8**. Moreover, such CNTs are hugely responsive to compressive, flexural, tensile /torsional strain [60, 61].

The working mechanism of sensors derived from a simple CNT macroscopic assembly inculcate the change of their electrical resistance or resistivity resulting from mechanical strain is said to be piezo-resistivity, change of its inductance/capacitance via mechanical strain, variation of its electrical resistivity with a

Nucleic acid/ drug-CNT type	Cell/tissue	Properties
Doxorubicin—SWCNT	Colon cancer	Efficient take up of cancer causing cells and then transfers to nucleus
Radionuclide-SWCNT	Burkitt lymphocytes	Targeted tumor both in vitro/in vivo
siRNA-SWCNT	Human neuro cells	Improves the working of siRNA assisted gastrin peptide receptor
Toxic siRNA—MWCNT	Human lung system	Inhibits discrete tumor growth
Cisplatin—SWCNT	Neuro Carcinoma	Tumor development—drastic recession
Paclitaxel—SWCNT	Breast cancer	Improved efficacy in treatment with minimal effects

Table 3.
Release of drugs from CNTs.

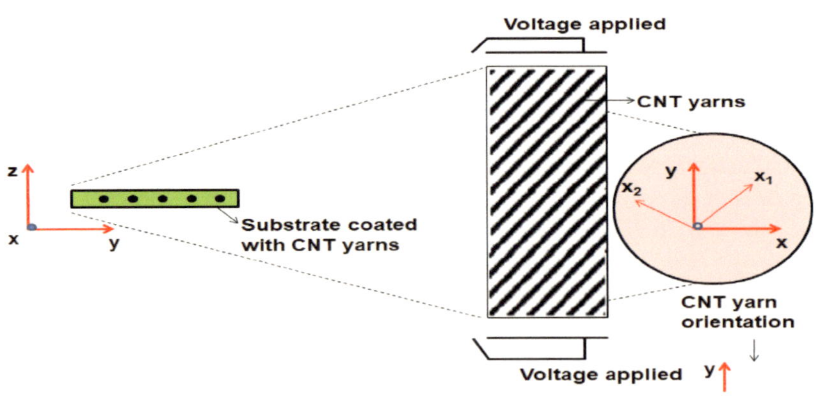

Figure 8.
Nanosensor schematic representation.

plausible variation with temperature is said to be thermo resistivity and a magnetoresistance may result due to variation in electrical resistance which may be due to varying magnetic field is known as magnetoresistance [62].

Even electrical resistance change can occur owing to variation in mechanical resonance frequency thereby resulting change in temperature, mass, pressure and strain. In contrast, the simultaneous change in electrical conductance/resistance is most predominant than any other mild variation accustomed in electrical properties. This happens partially due to selective and simple separation of charge carriers under the influence of temporary/permanent deformation which may lead to certain elevation in resistance. For every minute strain effects, the deformation is shown to be extremely elastic and the electrically conductive mechanism can be completely withdrawn due to its associated strain removal and thereby leads to a deduction in resistance [63]. However, plastic deformation has been proved to be much different. Though the resistance approaches zero due to strain removal and hence hysteresis curve has been fully observed. The various sensors involved with such CNT fiber yawns were strain type, pressure sensor devices, chemical sensors, and even mass sensors.

4. Conclusion

Nanomaterials provide an enriched knowledge on distinct probability and also definitely sound well in biomedical regenerative therapy for its uniqueness owing to its excellent physical as well as chemical properties. Thus, CNT both in modified and purified type have a definite potential of establishing promising applications in wide sectors of scientific fields. Perhaps, the prompt impregnation of other substituents in carbon nanoforms would confirm its strong perspective for their enhancing biomedical applications and in general medicine. Still then, there exist questions on unsolved issues whereas proximate homogeneity of the selected nanomaterial contains extensive distribution of nanotubes radius, unlike its geometric structures, classification of nanotubes, trace inclusion of residual elements, and a marked sensitivity to different species and other toxic gases.

Author details

Aravind Kumar Jagadeesan[1]*, Krithiga Thangavelu[2] and Venkatesan Dhananjeyan[1]

1 Department of Chemical Engineering, Sathyabama Institute of Science and Technology, Chennai, India

2 Department of Chemistry, Sathyabama Institute of Science and Technology, Chennai, India

*Address all correspondence to: arvindhjsbu@gmail.com

References

[1] Inagaki M. New Carbons: Control of Structure and Functions. Vol. 4. Japan: Elsevier Science; 2000. pp. 1003-1243. Available from: https://doi.org/10.1016/B978-0-08-043713-2.X5000-6

[2] Endo M. Jpn. Journal of Applied Physics. 2012;**51**:040001-040020

[3] HVance M, Kuiken T, Vejerano E, McGinnis S, Hochella M, Rejeski D, et al. Nanotechnology in the real world: Redeveloping the nanomaterial consumer products inventory. Beilstein Journal of Nanotechnology. 2015;**6**:1769-1780

[4] Bachilo SM, Strano MS, Kittrell C, Hauge RH, Smalley RE, Weisman RB. Structure-assigned optical spectra of single-walled carbon nanotubes. Science. 2002;**298**:2361-2366

[5] Stampoulis D, Sinha S, White J. Assay-dependent phytotoxicity of nanomaterials to plants. Environmental Science & Technology. 2009;**43**(24):9473-9479

[6] Kim H, Lee J, Kahng SJ, Son YW, Lee SB, Lee CK, et al. Direct observation of localized defect states in semiconductor nanotube junctions. Physical Review Letters. 2003;**90**(21):216107-216123

[7] Ouyang M, Huang JL, Cheung CL, Lieber C. Atomically resolved single-walled carbon nanotube intramolecular junctions. Science. 2001;**291**(5501):97-100

[8] Grobert N. Carbon nanotubes—Becoming clean. Materials Today. 2007;**10**(1):28-35

[9] Abbasi E, Sedigheh Fekri A, Abolfazl A, Morteza M, Hamid Tayefi N, Younes H, et al. Dendrimers: Synthesis, applications, and properties. Nanoscale Research Letters. 2014;**9**(1):247-255

[10] Shen Z, Wu A, Chen X. Iron oxide nanoparticle based contrast agents for magnetic resonance imaging. Molecular Pharmaceutics. 2016;**14**(5):1352-1364

[11] Alamusi A, Hu N, Fukunaga H, Atobe S, Liu Y, Li J. Piezoresistive strain sensors made from carbon nanotubes based polymer nanocomposites. Sensors. 2011;**11**(11):10691-10723

[12] Behabtu N, Young CC, Tsentalovich DE, Kleinerman O, Wang X, Ma AWK, et al. Strong, light, multifunctional fibers of carbon nanotubes with ultrahigh conductivity. Science. 2013;**339**:182-194

[13] Viet PP, Woo JY, Sik OJ, Min KS, Woo PJ, Hee KS, et al. Effect of plasma–nitric acid treatment on the electrical conductivity of flexible transparent conductive films. Japanese Journal of Applied Physics. 2013;**52**:075102

[14] Zhao H, Zhang Y, Bradford PD, Zhou Q, Jia Q, Yuan F, et al. Carbon nanotube yarn strain sensors. Nanotechnology. 2010;**21**:305502

[15] Foroughi J, Spinks GM, Aziz S, Mirabedini A, Jeiranikhameneh A, Wallace GG, et al. Knitted carbon-nanotube-sheath/spandex-core elastomeric yarns for artificial muscles and strain sensing. ACS Nano. 2016;**10**:9129-9135. DOI: 10.1021/acsnano.6b04125

[16] Ma X, Dong Y, Li R. Monitoring technology in composites using carbon nanotube yarns based on piezoresistivity. Materials Letters. 2017;**188**:45-47

[17] Abot JL, Anike JC, Bills JH, Onorato Z, Gonteski DL, Kvelashvili L, et al. Carbon nanotube yarn sensors for precise monitoring of damage evolution in laminated composite materials: latest experimental results and in-situ and

post-testing validation. In: Proceedings of the American Society for Composites: Thirty-Second Technical Conference, Purdue University, West Lafayette, Indiana. Lancaster, PA, USA: DEStech Publications, Inc., Electronic product. 23-25 October 2017. p. 6

[18] Abot JL, Wynter K, Mortin SP, Borges de Quadros H, Le HH, Renner DC, et al. Localized detection of damage in laminated composite materials using carbon nanotube yarn sensor. Journal of Multifunctional Composites. 2014;**2**:217-226

[19] Anike JC, Bajar A, Abot JL. Time-dependent effects on the coupled mechanical-electrical response of carbon nanotube yarns under tensile loading. Journal of Carbon Research. 2016;**2**:3-15

[20] Zhao H, Zhang Y, Bradford PD, Zhou Q, Jia Q, Yuan F, et al. Carbon nanotube yarn strain sensors. Nanotechnology. 2010;**21**:305502-305513

[21] Bonfanti P, Moschini E, Saibene M, et al. Do nanoparticle physico-chemical properties and developmental exposure window influence nano Zn embryotoxicity in *Xenopus laevis*? International Journal of Environmental Research and Public Health. 2015;**12**(8):8828-8848

[22] Ebbesen TW. Carbon Nanotubes—Preparation and Properties. Technology and Engineering – Book chapter. United Kingdom: CRC Press; 1996. pp. 123-175

[23] Pham VP, Jang S-H, Whang D, Choi J-Y. Direct growth of graphene on rigid and flexible substrates: Progress, applications, and challenges. Chemical Society Reviews. 2017;**46**:6276-6300

[24] Pham VP. Hexagon flower quantum dot-like Cu pattern formation during low-pressure chemical vapor. Deposited Graphene growth on a liquid Cu/W substrate. ACS Omega. 2018;**3**:8036-8041

[25] Torres T. Carbon nanotubes and related structures: Synthesis, characterization. In: Functionalization and Applications. Vol. 135. Weinheim: Wliey-VCH; 2010. pp. 215-228. Available from: https://doi.org/10.1002/anie.201006930

[26] Haris PJF, Hirsch A, Backes C. Carbon Nanotubes Science: Synthesis, Properties and Applications. Vol. 102. Germany: Cambridge University Press; 2009. pp. 210-230

[27] Mirakabad FST, Akbarzadeh A, Zarghami N, Zeighamian V, Rahimzadeh A, Alimohammadi S. PLGA-cased nanoparticles as cancer drug delivery systems. APJCP Asian Pacific Journal of Cancer Prevention. 2014;**15**(1):517-535

[28] Hirlekar R, Yamagar M, Garse H, Vij M, Kadam V. Carbon nanotubes and its applications: A review. Asian Journal of Pharmaceutical and Clinical Research. 2009;**2**(4):17-27

[29] Avouris P. Molecular electronics with carbon nanotubes. Accounts of Chemical Research. 2002;**35**:1026-1034

[30] Eatemadi A, Daraee H, Karimkhanloo H, Kouhi M, Zarghami N, Akbarzadeh M, et al. Carbon nanotubes: Properties, synthesis, purification, and medical applications. Nanoscale Research Letters. 2014;**9**:393-405. DOI: 10.1186/1556-276X-9-393

[31] Zhang M, Li J. Carbon nanotube in different shapes. Materials today. 2009;**12**(6):12-18

[32] He ZB, Maurice JL, Lee CS, Cojocaru CS, Pribat D. Nickel catalyst faceting in plasma-enhanced direct current chemical vapor deposition of carbon nanofibers. The Arabian Journal for Science and Engineering. 2010;**35**(1C):11-19

[33] Varshney D, Weiner BR, Morell G. Growth and field emission study

of a monolithic carbon nanotube/ diamond composite. Carbon. 2010;**48**(12):3353-3358

[34] The Transmission Electron Microscope (TEM) Images of a MWNT [Online]. 2014

[35] Obitayo W, Liu T. A review: Carbon nanotube-based piezoresistive strain sensors. Journal of Sensors. 2012;**41**:652438-652450. DOI: 10.1155/2012/652438

[36] Derycke V, Martel R, Appenzeller J, Avouris P. Carbon nanotube inter- and intramolecular logic gates. Nano Letters. 2001;**1**:453-456

[37] Liu X, Lee C, Zhou C, Han J. Carbon nanotube field-effect inverters. Applied Physics Letters. 2001;**79**:3329-3331

[38] Chu LL, Que L, Gianchandani YB. Measurements of material properties using differential capacitive strain sensors. Journal of Microelectromechanical Systems. 2002;**11**:489-498

[39] Zhou C, Kong J, Dai H. Intrinsic electrical properties of single walled carbon nanotubes with small band gaps. Physical Review Letters. 2000;**84**:5604-5607

[40] Shim M, Javey A, Kam NWS, Dai H. Polymer functionalization for air-stable n-type carbon nanotube field-effect transistors. Journal of the American Chemical Society. 2001;**123**:11512-11513

[41] Seidel RV, Graham AP, Kretz J, Rajasekharan B, Duesberg GS, Liebau M, et al. Sub-20 nm short channel carbon nanotube transistors. Nano Letters. 2005;**5**(1):147-150

[42] Chen R, Zhang Y, Wang D, Dai H. Non-covalent sidewall functionalization of single-walled carbon nanotubes for protein immobiloization. Journal of the American Chemical Society. 2001;**123**:3838-3839

[43] Zhang X, Meng L, Lu Q, Fei Z, Dyson PJ. Targeted delivery and controlled release of doxorubicin to cancer cells using modified single wall carbon nanotubes. Biomaterials. 2009;**30**(30):6041-6047

[44] Lin J, He C, Zhang L, Zhang S. Sensitive amperometric immunosensor for α-fetoprotein based on carbon nanotube/gold nanoparticle doped chitosan film. Analytical Biochemistry. 2009;**384**(1):130-135

[45] Rezaei-Sadabady R, Zarghami N, Barzegar A, Eidi A, Akbarzadeh A, Rezaei-Tavirani M. Studies of the relationship between structure and antioxidant activity in interesting systems, including tyrosol, hydroxytyrosol derivatives indicated by quantum chemical calculations. Soft. 2013;**2**:13-18

[46] Rao CNR, Cheetham AK. The Chemistry of Nanomaterials, Synthesis, Properties and Applications. 1st ed. John Wiley & Sons: Oxford University; 2006

[47] Tibbetts GG, Meisner GP, Olk CH. Hydrogen storage capacity of carbon nanotubes, filaments, and vapor-grown fibers. Carbon. 2001;**39**(15):2291-2301

[48] Eatemadi A, Daraee H, Zarghami N, Hassan Melat Y, Abolfazl A. Nanofiber: Synthesis and biomedical applications. Artificial cells, nano-medicine, and biotechnology. 2014;**43**(7):1-11

[49] Castillo JJ, Svendsen WE, Rozlosnik N, Escobar P. Detection of cancer cells using a peptide nanotube-folic acid modified graphene electrode. The Analyst. 2013;**138**(4):1026-1031

[50] Bian Z, Wang RJ, Wang WH, Zhang T, Inoue A. Carbon-nanotube-reinforced Zr-based bulk metallic glass composites and their properties. Advanced Functional Materials. 2004;**14**(1):55-63

[51] Marquis FD. Fully integrated hybrid polymeric carbon nanotube composites.

In: 2nd International Conference on Advanced Materials Processing, Materials Science Forum. Vol. 437-504; 2003

[52] Abarrategi A, Gutierrez MC, Moreno-Vicente C, Ramos V, Lopez-Lacomba JL, Ferrer ML, et al. Multiwall carbon nanotube scaffolds for tissue engineering purposes. Biomaterials. 2008;**29**(1):94-102

[53] Panini NV, Messina GA, Salinas E, Raba J. Integrated microfluidic systems with an immunosensor modified with carbon nanotubes for detection of prostate specific antigen (PSA) in human serum samples. Biosensors & Bioelectronics. 2008;**23**(7):1145-1151

[54] Zhang L, Webster TJ. Nanotechnology and nanomaterials: Promises for improved tissue regeneration. Nano Today. 2009;**4**(1):66-80

[55] Kim H, Lee J, Kahng SJ, Son YW, Lee SB, Lee CK, et al. Direct observation of localized defect states in semiconductor nanotube junctions. Physical Review Letters. 2003;**90**(21):216107-216114

[56] Ebrahimnezhad Z, Zarghami N, Keyhani M, Amirsaadat S, Akbarzadeh A, Rahmati M, et al. Inhibition of hTERT gene expression by silibinin-loaded PLGA-PEG-Fe3O4 in T47D breast cancer cell line. BioImpacts: BI. 2013;**3**:67-74

[57] Dhar S, Liu Z, Thomale J, Dai H, Lippard SJ. Targeted single-wall carbon nanotube-mediated Pt (IV) prodrug delivery using folate as a homing device. Journal of the American Chemical Society. 2008;**130**(34):11467-11476

[58] Abbasi E, Milani M, Sedigheh Fekri A, Mohammad K, Abolfazl A, Hamid Tayefi N, et al. Silver nanoparticles: Synthesis methods, bio-applications and properties. Critical Reviews in Microbiology. 2014;**46**(6):1-8

[59] Chen J, Chen S, Zhao X, Kuznetsova LV, Wong SS, Ojima I. Functionalized single-walled carbon nanotubes as rationally designed vehicles for tumor-targeted drug delivery. Journal of the American Chemical Society. 2008;**130**(49):16778-16785

[60] Zhu Z, Song W, Burugapalli K, Moussy F, Li YL, Zhong XH. Nano-yarn carbon nanotube fiber based enzymatic glucose biosensor. Nanotechnology. 2010;**21**:165501-165512

[61] Zhu Z, Garcia-Gancedo L, Flewitt AJ, Moussy F, Li YL, Milne WI. Design of carbon nanotube fiber microelectrode for glucose biosensing. Journal of Chemical Technology and Biotechnology. 2012;**87**:256-262

[62] Kahng SK, Gates TS, Jefferson GD. Strain and temperature sensing properties of multiwalled carbon nanotube yarn composites. In: NASA Technical Report. SAMPE ''08 Fall Technical Conference, Memphis, TN, United States; 2008

[63] Kim HH, Haines CS, Li N, Kim KJ, Mun TJ, Choi C. Harvesting electrical energy from carbon nanotube yarn twist. Science. 2017;**357**:773-778

Chapter 3

Technological Background and Properties of Thin Film Semiconductors

Orkut Sancakoglu

Abstract

Especially with the development of nanotechnology and polymer science, interest in research and production of both efficient and lower-cost semiconductor thin film materials is increasing day by day. The use of nano-structured thin films for efficient use of solar cells in production of n-type semiconductor materials is one of the most important sources of energy and new-generation energy. Considering the indicated trends and energy requirements, it has been important to transfer this technology in detail regarding the surface technologies related to the semiconductor materials produced with thin film technologies instead of bulk materials. With this aim, this book chapter "Technological Background and Properties of Thin Film Semiconductors" includes a brief story about semiconductors, band gap theory, thin film applications, and besides traditional thin film processing methods finally a new technology called aerosol deposition technique which allows room temperature processing of several materials for semiconductor applications, respectively. It is thought that it will make important contributions to the relevant field and bring a new perspective and direct scientific research in "process-structure–property-performance" relation.

Keywords: thin film semiconductors, transparent conducting oxide (TCO), coating, room temperature aerosol deposition, transparency, resistivity

1. Introduction

Semiconductor materials are defined as materials in solid form whose conductivity has a value between conductors and insulators. Semiconductor materials play an important role in the advances in the modern electronics industry in the twenty-first century and in the industrial applications of many electronic devices. These materials include many materials such as silicon, gallium arsenide, germanium, cadmium sulfide, and cadmium telluride, which are widely used today. From the first silicon integrated circuits produced in semiconductor technology, high-tech microprocessors, solar cells, and many other electronic devices have developed rapidly to the present day [1].

Especially with the development of nanotechnology and polymer science, interest in the research and production of both efficient and lower-cost semiconductor thin film materials is increasing day by day. The use of nano-structured thin films for the efficient use of solar cells in the production of n-type semiconductor

IntechOpen

materials is one of the most important sources of energy and new-generation energy. In short, semiconductor devices for nanotechnology and polymer science have taken the advancement of research in semiconductors to a new step, aiming to improve the chemical and physical properties of these materials. To understand the nature of these crucial engineering materials, the difference and theory between conductors, insulators, and semiconductors must be fully understood. In addition, basic concepts such as band theory, doping processes, and p-n connection theory of solids are theoretical bases that will give a general idea of understanding semiconductors.

2. Technical background

It is important to assimilate the band theory that explains the nature of semiconductors in terms of energy levels between the valence and the conduction bands in order to gain knowledge about semiconductor materials. The main difference between metallic materials and semiconductors is that the current is provided by electron flow in metallic conductors, whereas in semiconductors this flow occurs not only by the electron flow but also by the flow of positively charged holes.

Electrical conductivity is directly related to the band structure of a material. If we look at the basis of the theory, atomic energy levels of each atom are equal when two different atoms are sufficiently distant from each other. However, as these atoms approach each other, differences in the original energy levels of the atoms are observed, and as a result of these differences, an interaction occurs that creates molecular bands between the atoms. Therefore, it is for this reason that materials with different band structures show different conductivity properties. In conductors, the energy difference between partially filled energy levels and empty levels is very low. Therefore, when a potential is applied to metals, electron mobility between the filled and empty levels, which takes place by using very low energy, is easily realized, and the flow of electrons is provided. Therefore, it would not be the right approach to talk about an obstacle between the levels defined as valence band and conduction band in metals. As known, an electron must have an empty energy level to move, otherwise electrons cannot move in solid material. Based on this approach, it can be clearly understood why electrical current is not observed in insulating structures. In insulators, the valence band is the highest band fully filled by electrons, and the conduction band is the lowest empty band, with a forbidden band gap of about 5–10 eV between these two bands [2]. This broadband between the valence band and the conduction band prevents the transmission of electrons to the conduction band, and no electrical current is generated in the insulating materials. Similar to the band structure of the insulators, semiconductors have a valence band occupied by electrons and a conduction band ready to be filled with electrons. In semiconductors, just like insulating materials, there is a band gap between these two bands. The main difference between the band gap in these two groups excluding conductors is that the band gap value is much smaller in semiconductors (1.1 eV for silicon) than in the insulator [3]. Since the thermal energy in semiconductors creates the driving force for the movement of electrons, the conductivity of these materials is directly related to the temperature. The conductivity of a semiconductor material as a result of a decrease in resistivity can be associated with increased kinetic energy with temperature. **Figure 1** schematically shows the energy band gaps in conductors, semiconductors, and insulators.

In semiconductors, the valence band that is below the forbidden band gap is almost completely full. On the other hand, there is a nearly empty conduction band over the prohibited band gap. When a semiconductor material is excited, if the

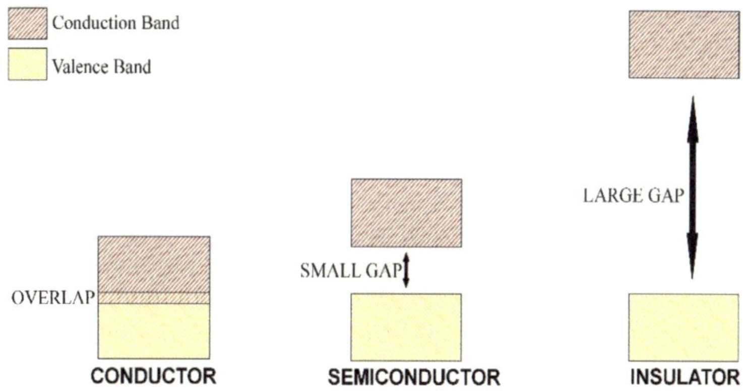

Figure 1.
Energy band gaps in materials.

energy of the photon is greater than the band gap, the electrons will easily pass to the conduction band, and electrical current will occur. According to the theory, as a result of the transition of an electron to the conduction band, the excited electron leaves a hole in place that can flow through the material and act as a positively charged particle. Here, carrier production and recombination can be defined as two basic factors in the creation of charge-carrying electrons and positively charged holes [4].

3. Types of semiconductors

In general, depending on the level of doping, semiconductors can be classified into two main groups such as *intrinsic semiconductors* and *extrinsic semiconductors*. The intrinsic semiconductors are pure semiconductors and no addition is made. In this type of semiconductor, conductivity is provided by the thermal stimulation of electrons. At the same time, the number of excited electrons and positively charged holes is equal. The behavior here appears as a result of the carrier production and recombination steps [5].

On the other hand, extrinsic semiconductors have low conductivity values, and an important process called *doping* is applied to overcome the problems encountered in applications and to increase the conductivity [6]. This process can be explained simply by adding small amounts of impurities in the concentrations of charge-carrying electrons and positively charged holes, thereby increasing the conductivity level. The aim is to change the electronic structure by impurity addition into the structure without changing the crystal structure. For example, arsenic with five valence electrons to an atom and germanium with four valence electrons will cause the arsenic atom to covalently bond with the germanium atom. The extra fifth electron of the arsenic atom will have the electrical conductivity as it will have the freedom to move from one atom to another [4]. Such semiconductors, which the dopant element donates an electron, are called n-type semiconductors. In addition to producing free electrons in n-type doping, an equal number of positive charges are also produced in pairs with free electrons. As a result, the doped semiconductor material remains electrically neutral. However, these positive charges should not be understood as positively charged holes. These charges occur in the absence of free electrons, but do not contribute to a current flow. Another contribution of free electrons to pure semiconductors is that the donor electron is much closer to the conduction band than an electron in the valence band of the original atom.

In another saying, the energy level of the donor electron is at another level that is much narrower than the energy level for valence electrons and facilitates the flow of current in the n-type semiconductor.

In another type of extrinsic semiconductor, if a pure semiconductor with trivalent electrons (e.g., germanium) is replaced by gallium, this type of semiconductor is called p-type semiconductors. Three of the four covalent bonds are occupied in the structure; the bond remains empty and acts as a relatively moving hole in the opposite direction of a moving electron, which is not a real move. The main phenomenon is defined as a relative movement caused by the movement of electrons from one bond to another and leaving a hole after. Similar to the energy band structure of n-type semiconductors, which differs from pure semiconductors, p-type semiconductors also have a higher acceptor energy level than the valence band.

4. Application fields and thin film technology

Materials are produced with developing semiconductor technology; it has been used for more than 50 years in the production of a wide variety of devices such as diodes, transistors, sensors, microprocessors, and solar cells. On the other hand, thin film nano- or microstructures are frequently used in applications developed in semiconductor technology. The most common area that benefits semiconductor thin film technology refers to photovoltaic applications. Simply, this application is the technology field for using solar cells to produce solar energy. The importance of photovoltaics comes from the fact that photovoltaics are not only important in terrestrial applications to gain renewable and clean energy but also enable the benefits of solar energy which can be used as a major source of electrical power for space systems [7]. Besides, all the concerns based on environment, renewability, and sustainability have increased attention for using alternative, new energy sources, where photovoltaics are seen as the new trend.

In general, semiconductor thin films are produced in one or more thin layers. Common applications of such structures include many electronic materials such as transistors, sensors, and photovoltaic devices. The structural, chemical, and physical properties of semiconductor thin films are directly related to the production technique, and their thickness ranges from a few nanometers to hundreds of micrometers [8].

Especially in recent years, semiconductor thin films can be obtained compared to the use of bulk materials, making it a promising ideal candidate in the electronic material industry due to its wide variety of properties. First of all, a wide range of chemical, electrochemical, and physical deposition techniques (see Section 5) enables the low-cost production of semiconductor material on large areas of the desired geometry and structure. In addition, single- or multi-crystalline structures with complex geometry and even microstructures of nanocrystalline thin films can be easily obtained by changing the method, temperature, substrate, and other production parameters depending on the method. Industrial applications of electronic materials come to life with different junction types between different semiconductor materials to improve the electrical properties of thin films.

Thin film solar cells, which we have already mentioned and attracted great interest from researchers, consist of materials of different structure and properties in the form of several thin layers deposited on various substrates. To draw a general schematic, the common structure of a solar cell consists of a hard or flexible substrate, a transparent conductive oxide (TCO) layer, a window layer made of n-type semiconductor, p-type semiconductor layer, and a metal contact and absorbent layer. In order to make the structure more understandable, the schematic of a solar panel is examined in **Figure 2** [9].

Additionally, the flowchart of solar energy diagram using the panels is given below sequentially:

i. Solar panels convert sunlight to DC current.

ii. Inverter converts DC electricity to AC.

iii. Electricity then gets used by home applications.

iv. Extra electricity gets credited into the grid.

In a solar panel, each of the components given in **Figure 2** has different physical and chemical properties. The performance of the solar cell depends on the properties of the previously mentioned layer components and the compatibility between them. The layers described should be mechanically and chemically stable as well as compatible with each other based on their properties. The high processing temperature, which is a requirement of most processes, results in the need of a hard substrate such as glass, because flexible substrates (can be polymeric, etc.) do not have high temperature stability. On the other hand, it is reported that the choice of the substrate has a direct effect on the efficiency of the cells and the grain structure of the films grown on this substrate [10–12].

When it comes to the transparent conductive oxides, which are among the trend research subjects of recent years (see Section 6), it provides an integration between the substrate and the window layer with good electrical conductivity and high optical permeability in the visible region. In order to reduce the total resistance of the system, high transparency is required for good conductivity feature and higher degree of light passing through the absorbent layer. The window layer, on the other hand, has the function of directing the maximum amount of incoming light to the junction point and the absorbent layer in a hetero combination with an absorbent

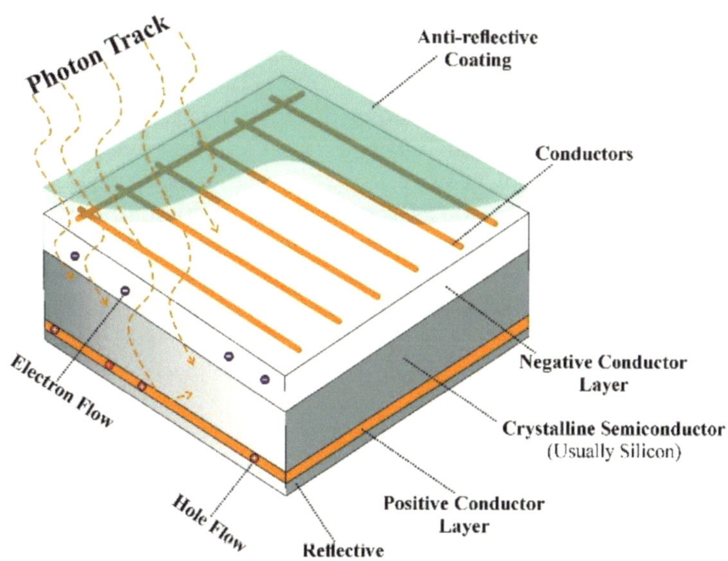

Figure 2.
Schematic of a laminated solar panel.

layer. Moreover, the compatibility of the window layer with the absorbent layer and the lattice structure is of great importance for device performance.

5. Thin film processing

The coating process is the basis for the production of devices based on thin film technology. In improving device performance, high-quality thin films produced by using rapid, economical, and effective techniques are required. The semiconductor industry, which is almost entirely based on solid thin films, has demonstrated the importance of coating technology and has contributed to its rapid development, especially in recent years. In this context, there are various methods for creating thin films on nano-micro scale. Basically, deposition techniques can be divided into two groups as chemical and physical methods [13].

Subgroups of chemical and physical deposition methods can be defined as (i) chemical vapor deposition (CVD), including chemical bath deposition, electroplating, molecular beam epitaxy, and thermal oxidation, and (ii) physical vapor deposition (PVD), including evaporation and spraying. It is possible to mention that CVD and PVD have an important place in new-generation thin film technologies.

Of these techniques, CVD is used to create high-purity and effective solid thin films [14, 15]. In this process, the substrate is placed inside a reactor where it is exposed to volatile gasses, and a solid layer is formed on the substrate surface by chemical reactions between the gas used and the substrate. CVD method can produce high-purity single- or polycrystalline or even amorphous thin films. Another feature of CVD is that both pure and complex materials can be synthesized at the desired purity at low temperatures. The chemical and physical properties of the forms to be created can be easily adjusted by properly adjusting the reaction and precipitation parameters such as temperature, pressure, gas flow rate, and gas concentration.

As another technique, PVD can be defined as thin film production by condensing the evaporated materials released from a source (target material) on the substrate surface. Sub-methods of this technique are spraying and evaporation, respectively.

Spraying method is based on the principle of removing surface atoms from a target and depositing them on the substrate. Among the advantages of the spraying method, thin films with the desired composition closest to the composition of the target material can be easily obtained. Besides, the method provides better adhesion to the surface, and evaporation of materials with a high melting point is possible.

Evaporation is another common PVD method in thin film production. In this technique, the evaporated particles are condensed on the substrate to form a solid thin film, while the source material (mostly metallic) is evaporated in a vacuum tube. The great advantage of evaporation compared to sputtering is that the process is relatively faster.

6. Future technology based on TCO semiconductors

Due to the wide range of applications mentioned in Section 4 and the unique optical and electrical properties that can be combined, TCO materials are used in transparent and flexible device applications such as flat panel displays, plasma screen panels, organic light-emitting diodes, solar cells, and gas sensors. As the most common TCOs, indium, tin, cadmium, and group III and VII doped zinc oxide can be mentioned. It consists of semiconductor metal oxides with a wide band gap.

These materials show high conductivity, transparency, and reproducibility, and the possibility of producing very flat films and tin-doped indium oxide (ITO) is the most employed TCO. Due to the cost and difficulty of finding indium in TCO, cost-effective and alternative materials with the same properties are needed. In addition to high conductivity TCO, the development of materials other than ITO has increased in few years. ITO's cost and potential supply limit are among the reasons for research in this field. Current candidate materials for high conductivity TCOs include CdO, SnO, In_2O_3, and ZnO with these binary combinations along with alternative additives. ZnO as the current candidate is considered to be the primary candidate of binary oxide TCO materials, with its high conductivity and low cost. However, it is very sensitive to oxygen, and process control is more difficult; the development of ZnO with alternative impurity continues.

ZnO:Al (AZO) produced with pulsed laser deposition (PLD) shows at least equivalent resistivity with ITO. However, PLD is not considered to be a suitable deposition method for large area applications. Metallic and ceramic targets are used to develop robust process control and to scale wide area bases in the production magnetron sputtered (MS) ZnO:Al (AZO) films. AZO and ZnO:Ga (GZO) are both commercially applicable in thin film solar cells and can be used as translucent electrodes [16].

In ZnO film growth, several techniques with their own advantages and disadvantages are used such as evaporation, radio-frequency magnetron sputtering (RF-MS), direct current magnetron sputtering (DC-MS), ion beam sputter, spray pyrolysis, sol–gel process, PLD, and CVD. Transparent n-type ZnO semiconductors can easily be fabricated by B, Al, In, and F doping. According to the results of several research in which RF-MS, DC-MS, and MOCVD are used, ZnO:Al with relatively low resistivity values is found to be 1.4×10^{-4}, 3.4×10^{-4}, and 3.0×10^{-4} Ωcm, respectively. These resistivity values are quiet close to the electrical properties of ITO. The optoelectronic properties of these thin film TCOs may change after thermal annealing in an inert gas or reactive gas atmosphere [17].

Lastly, if we consider PVD as one of the trending methods in the past few years and even a more innovative sub-method, the magnetron sputtering technique, the applicability problems brought by the researcher who has practical experience in this field and the applicability of the method to large surfaces as an assessment of the literature-scale results are known. In such semiconductor materials, homogeneity is another problem to be solved and mostly at high-temperature processes after coating (additional heat treatments). This makes the aforementioned method disadvantageous. In addition, while many traditional methods used in semiconductor technologies are generally suitable for coating applications on metal or ceramic substrate, the phase transformation and/or defect removal with additional heat treatments makes these techniques unsuitable for developing polymer-based semiconductor applications for high-technology semiconductor field.

From this point of view, the search for alternative methods that will meet the advancing technology and meet technological requirements not only for metal and ceramic substrates but also polymer-based ones encouraged researchers to work on new techniques and methods. "Ceramic powder aerosol deposition method," which is pointed, is not only requiring additional high-temperature heat treatment or sintering steps but also homogeneous and continuous applicability to large areas. It has become a trend that draws the attention of researchers on semiconductor material technologies annually. As a result of all these advantages, the aerosol deposition method has been deemed worthy of examination both in terms of its technical details and applicability, and because it is a virgin technology that is still open to development in the field, it may meet the need in new-generation semiconductor technologies.

Aerosol deposition technique, in terms of homogeneous, continuous, and applicability to large area TCO, is a convenient method for industrial production of materials that has not been studied much before. As it is known, new-generation ceramic integration technologies require device performance improvement for various applications such as size reduction and/or micro-mechanical systems, image technologies, fuel cells, and optical devices. Common industrial applications of these devices are high-speed coatings and productions that bring low processing temperatures and thin lithographic processes in ceramic coatings. "Ceramic powder aerosol deposition method" has many advantages to meet many of the requirements mentioned above when compared to other conventional film/coating applications.

While ceramic material productions generally require sintering over 1000°C, this makes it impossible to integrate metal, glass, and even polymers with relatively low melting point with functional ceramics. On the other hand, high-temperature film applications trigger high-temperature crack formations and substrate-coating incompatibilities and discontinuities, as well as long processing times. Especially in film-type applications, the substrate-coating incompatibilities and surface discontinuities will adversely affect the optical, electrical, and mechanical properties of the device for the application and also decrease its performance. Similar results were observed in our previous research studies in the form of surface cracks and unexpected decrease in electrical properties as a result of ZnO-based coatings obtained by magnetron scattering method, which requires additional heat treatment up to 700°C.

The main output of the aerosol deposition method in the development of a new technique and industrial product with high-speed coating efficiency, low processing temperature, and even at room temperature reduces costs and production time and adversely affects properties.

The basic principle of the technique is based on the conversion of some of the kinetic energy of the previously prepared and suitable form of fine ceramic particles with the substrate at high speeds, into both interparticle and interparticle-binding energy. In this context, while the aerosol deposition method offers many advantages compared to traditional thin/thick film methods and thermal spray coating techniques, it has not been fully understood as a new method, revealing the open points for the research. A schematic of this method is given in **Figure 3**.

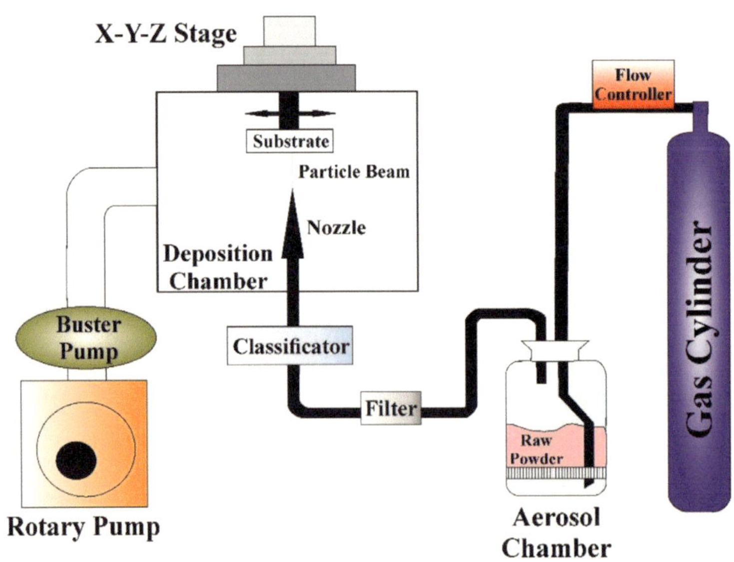

Figure 3.
Room temperature aerosol deposition system.

Simply, aerosol deposition method is based on the principle of ultra-pressure ceramic particles coming together with the surface of the substrate and solidification with effective pressure and shock loading. First, ceramic particles are mixed with a gas in the system chamber to form an aerosol. This aerosol is then sprayed through a nozzle onto various targets to create a film layer with the aid of pressure. During the process, some of the kinetic energy of the particles is converted into thermal energy, which partially increases the temperature in the interaction zone. In this way, the bonding between the substrate and the particles and also the particles with each other is provided and results in a high-density continuous coating layer.

In general terms, aerosol deposition technique is as follows:

- Appearing as a suitable method for all types of substrates.
- As a horizon for the electronics industry.
- Not only for energy conversion applications but also for various semiconductor technology applications.
- Does not require additional heat treatment.
- Most importantly, as it is a low-temperature process, in terms of being a feasible method without changing or destroying the substrate form.

It is thought that this new technology will make important contributions to the relevant field and bring a new perspective and direct scientific research in process-structure–property-performance relation.

7. Conclusions

Due to the cost and difficulty of finding indium in TCO, cost-effective and alternative materials with the same properties are needed. In addition to high conductivity TCO, the development of materials other than ITO has increased in few years. ITO's cost and potential supply limit are among the reasons for research in this field.

Besides, not only for the metallic and/or ceramic substrates but also for the substrates with low melting point or polymers need to be focused on new semiconductor fabrication technologies. At this point, room temperature aerosol deposition method seems to be an alternative technique and may bring a new perspective in high-technology semiconductor applications.

Author details

Orkut Sancakoglu
Metallurgical and Materials Engineering Department, Dokuz Eylul University, Izmir, Turkey

*Address all correspondence to: orkut.sancakoglu@deu.edu.tr

References

[1] Moss SJ, Ledwith A. Chemistry of the Semiconductor Industry. 1st ed. Netherlands: Springer; 1989. p. 426. ISBN: 9780216920057

[2] Giancoli DC. Physics for Scientists & Engineers with Modern Physics. 6th ed. London, England: Pearson Education; 2004. p. 1040. ISBN: 9780130911833

[3] Roulston D, An J. Introduction to the Physics of Semiconductor Devices. 5th ed. Oxford University Press: New York; 1999. p. 320. ISBN: 0195114779

[4] Fishbane PM, Gasiorowicz S, Thornton ST. Physics for Scientists & Engineers with Modern Physics. 3rd ed. New Jersey, USA: Pearson Prentice Hall; 2005. p. 1232. ISBN: 978-0131420946

[5] Kakani SL, Kakani S. Modern Physics. 2nd ed. Anshan, China: MV Learning Anshan; 2015. p. 1536. ISBN: 9788130932156

[6] Pham VP, Yeom GY. Recent advances in doping of molybdenum disulfide: Industrial applications and future prospects. Advanced Materials. 2016;**28**:9024-9059. DOI: 10.1002/adma.201506402

[7] Romeo A, Bätzner DL, Zogg H, Tiwari AN. Potential of CdTe thin film solar cells for space applications. In: Proceedings of the 17th European Photovoltaic Conference and Exhibition. 22-26 October 2001; Munich, Germany: WIP-Renewable Energies; 2005 pp. 2183-2186

[8] Chopra KL. Thin-Film Phenomena. 1st ed. New York: McGraw-Hill; 1969. p. 736. ISBN: 9780070107991

[9] Coventry Climate Action Network Solar Energy Diagram [Internet]. 2020. Available from: https://covcan.uk/how-do-solar-panels-work/ [Accessed: 02 February 2020]

[10] Nakada T, Kume T, Mise T, Kunioka A. Superstrate-type Cu(In, Ga)Se2 thin film solar cells with ZnO buffer layers. Japanese Journal of Applied Physics. 1998;**37**:499-501

[11] Ramanathan K, Contreras MA, Perkins CL, Asher S, Hasoon FS, Keane J, et al. Progress in photovoltaics research and applications. Thin-Film Solar Cells. 2003;**11**:225-230. DOI: 10.1002/pip.494

[12] Gee JM, Buck ME, Schubert WK, Basore PA. Progress on the emitter wrap-through silicon solar cell. In: Proceedings of the 12th European Photovoltaic Solar Energy Conference. 11-15 April 1994; Amsterdam, Netherlands: Commission of the European Communities; 1994. pp 743-1746

[13] Schuegraf KK, editor. Handbook of Thin-Film Deposition Processes and Techniques. 2nd ed. Norwich, New York: Noyes; 1988. p. 629. ISBN: 0815514425

[14] Adams AC. Dielectric and polysilicon film deposition. In: Sze SM, editor. VLSI Technology. 1st ed. McGraw-Hill: New York; 1983. pp. 93-129. ISBN: 9780070626867

[15] Geittner P. Preparation of optical fibres by means of the PCVD process. In: Proceedings of the Ninth International Conference on Chemical Vapor Deposition. 4 November 1988; New Jersey: The Electrochemical Society; 1984. p. 479-496

[16] Bright CI. Review of transparent conductive oxides (TCO). In: Mattox DM, Mattox VH, editors. 50 Years of Vacuum Coating Technology and the Growth of the Society of Vacuum Coaters. Tuscon, Arizona: 3M Company; 2007. pp. 38-45

[17] Stadler A. Transparent conducting oxides - An up-to-date overview. Materials. 2012;**5**:661-683. DOI: 10.3390/ma5040661

Section 2

Etching and Lithography of Thin Film

Chapter 4

The New Etching Technologies of Graphene Surfaces

Phuong V. Pham

Abstract

Recently, graphene nanomaterial has drawn great interest due to its excellent electrical and optoelectrical properties. The etching of graphene based on plasma engineering to achieve atomically thin layer and extremely clean surface is a hot issue, which is highly desirable for industrial applications. The resided contaminants with high intrinsic roughness create the degradation of performance. The impurities are removed via surface cleaning method and layer-by-layer plasma etching via top-down lithography. Recently, new plasma technology-based etching causes no damage and secures its π-binding, which plays a key role in conductivity and other characteristics. Thus, this chapter presents the recent advances in new etching technologies for nanomaterials (e.g., graphene) as well as emerging applications based on these technologies.

Keywords: graphene etching, plasma, ion beam, neutral beam, inductively coupled plasma (ICP), atomic layer etching (ALE), reactive ion etching (RIE), chemical vapor deposition (CVD)

1. Introduction

An atomical graphene layer, which was invented in 2004 [1, 2], is considered as one of the best candidates for a broad application range with novel electronic and optoelectronic behaviors [1–21]. Unfortunately, the conductive graphene with no bandgap prevented its outstanding physical and chemical potentials. Thus, its bandgap tuning via various approaches is highly desired for extreme performance devices (**Figure 1**).

Recently, etching technologies are emerging as one of the best efficient tools to tune a device's performance, thereby extending to many different fields in broadband [20–30]. The new approaches include the following: (i) inductively coupled plasma (ICP), neutral beam-based atomic layer etching (ALE), ion beam and reactive ion etching (RIE) [22–29], (ii) chemical vapor deposition (CVD) [30, 31] and (iii) thermally activated nanoparticles [32]. Plasma has used Si-integrated circuits for etching [22]. Among breakthroughs, plasma etching represents an important role in Si and non-Si (metal)-assisted devices. This chapter will present recent advances in new graphene etch technologies and their related applications.

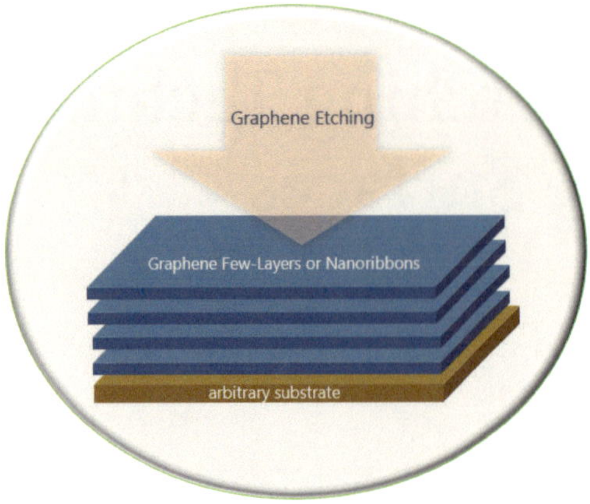

Figure 1.
Etch processing of graphene few-layer or graphene nanoribbon on various substrates through plasma, physic, chemistry to tune its electronics and optoelectronics.

2. Emerging etch technologies of graphene surfaces

Graphene layers have a number of independent bandgaps, e.g., single layer has no bandgap, but a bilayer has bandgap, and could be utilized to make transistor a superior performance. The layer-by-layer graphene etching would form (i) a cleaner surface with removed residues and (ii) thinner graphene film leading to smaller bandgap value until there is no bandgap at a single layer. Depending on the types of defects such as disorder [33], doping [34], external field [35] and mechanic strain [36–40], the etching can make the host material (e.g., graphene) very useful (high conductivity, high mobility, high work function, etc.) [41]. As a result, bandgap can be higher (or lower) depending on the types of vacancy defects and etching rates [41].

2.1 O_2 plasma etching

Plasma etch technology presents many advantages such as easy scale-up, manipulation and mass production. Under O_2 plasma exposure, graphene multi-layers were well-etched on SiO_2 [23, 25] or SiC [24]. In 2014, the etching of host bilayer graphene was carried out by O_2 using ICP and RIE apparatuses on the vertical and horizontal etch directions (**Figure 2a, b**) [23]. However, this approach formed defects during the use of RIE, but the defects were very few in the ICP case because of the high damage energy of RIE. Raman data provided the proof through disorder characteristics based on I_D/I_G ratio (0.94 and 1.18) when utilizing RIE and ICP, respectively [23]. Treating another substrate, SiC, the contact angle changed from 92.7° (multilayer), 91.9° (bilayer) and 92.5° (single layer) down to 70° when one layer epitaxial graphene etched away at 10 W and 2 min (**Figure 2c, d**) [24]. In 2011, through nanosphere lithography with low-power O_2 plasma, Liu et al. found out the etched ordering of graphene nanoribbons (GNRs) on SiO_2, which performed well in various shapes such as branches, chains, connected rings and circular rings (**Figure 2e**) [25].

2.2 N_2 plasma etching and postannealing

Yang et al. utilized N_2 plasma and postannealing (Ar/O_2, 900°C), another technology in integration of layer-by-layer thinned plasma and post-annealing.

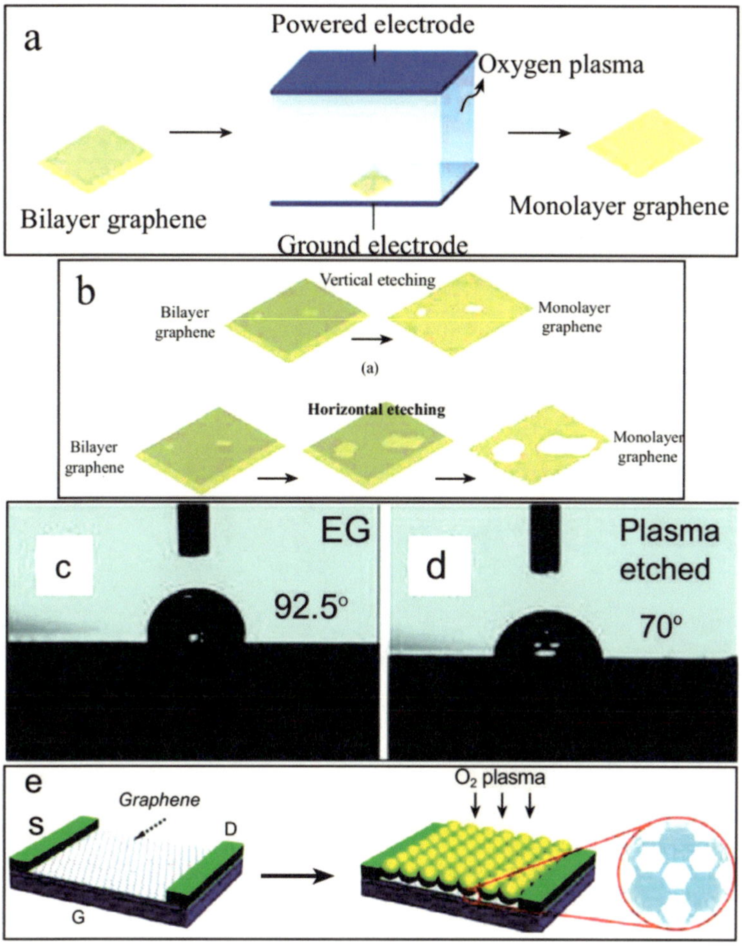

Figure 2.
(a) Sequence of graphene etching via O_2 plasma. (b) Schematic of two etching mechanisms of O_2 plasma: vertical etching and horizontal etching. (c, d) The contact angle of graphene/SiC without/with O_2 plasma, respectively. (e) On-chip device assisted by O_2-etched nanosphere graphene ((a) and (b) are reproduced with permission from [23], Copyright 2014, Springer; (c) and (d) are reproduced with permission from [24], Copyright 2010, American Chemical Society; (e) is reproduced with permission from [25], Copyright 2011, Wiley-VCH Verlag GmbH & Co. KGaA, Weinheim).

As a result, this dry-etching thinned regarding layer-by-layer easily from intrinsic multilayer graphene (**Figure 3a–c**) [26]. In another innovative etch technology by Lim et al. [27] and Kim et al. [28], Lim et al. utilized a neutral beam ALE via two-step process of O_2 radical absorption and Ar neutral beam desorption, and multilayer graphene was well-etched for each layer (**Figure 3d**). Although this etching was much more effective than the previous study [24–27], defects formed slightly on graphene lattice as high Raman D-peak (**Figure 3c**) [27].

2.3 Cyclic etching (O_2 adsorption and Ar desorption by ion beam)

In 2017, Kim et al. newly innovated by adding two mesh grids between the plasma source and the substrate holder in the ICP chamber (**Figure 3e–j**) [28]. Consequently, the damage on graphene surface disappeared after the two-step plasma etching process of chemical absorption of O_2 radical and physical desorption of Ar ion beam at optimized plasma energy (11.2 eV).

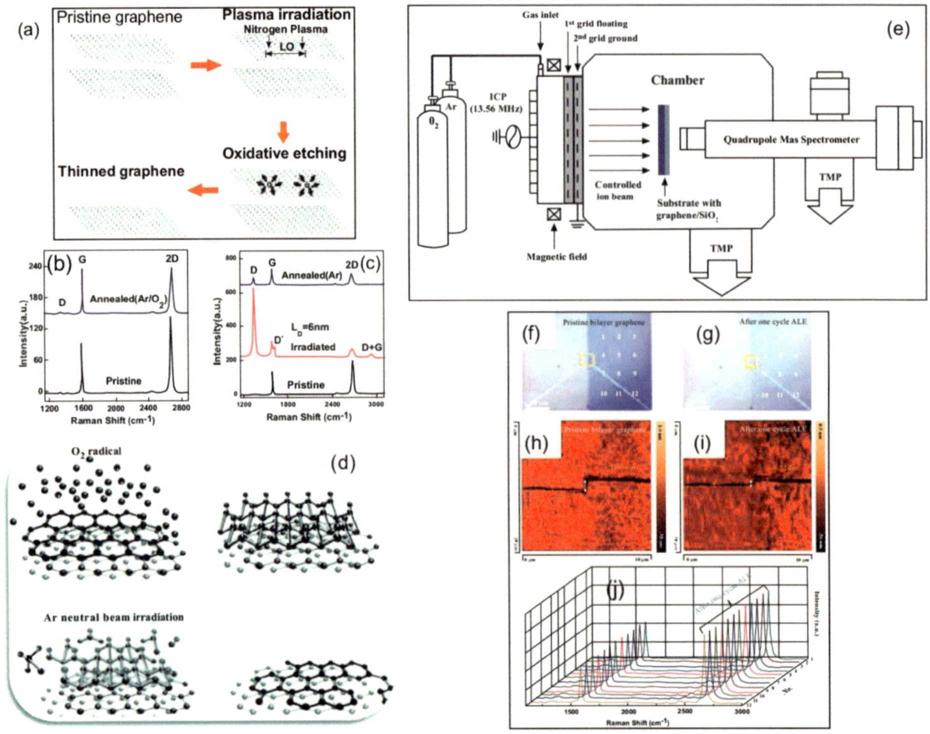

Figure 3.

(a) Sequence of layer-by-layer etching via N_2 irradiative and oxidative etch. (b, c) Raman data of pristine, irradiated and annealed multilayer graphene. (d) Schematic of ALE process of graphene via O_2 radical absorption and Ar neutral beam desorption. (e) A double-grid ICP ion beam apparatus for graphene etching via chemical O_2 absorption and physical low-energy Ar ion beam desorption. (f, g) OM and (h, i) AFM data of pristine bilayer graphene, and after an ALE cycle. (j) Raman data of pristine bilayer graphene after an ALE cycle for white dots of (f, g) ((a–c) are reproduced with permission from [26], Copyright 2011, IOP Publishing; (d) is reproduced with permission from [27], Copyright 2012, Elsevier; (e–j) are reproduced with permission from [28], Copyright 2017, Nature Publishing Group).

2.4 Others (RIE, H_2, CH_4/H_2 and Fe NPs)

In addition, there are still strategies for graphene surface etching such as Ar/H_2 mixture in reactive ion etching (RIE) (**Figure 4a**) [29], H_2 etching during CVD graphene growth (**Figure 4b–e**) [30], CH_4/H_2 etching during CVD graphene growth (**Figure 4f**) [31] or thermally activated Fe nanoparticles (NPs) (**Figure 4g, h**) [32]. However, the demonstrated results showed high defects through very high D-peak intensity in Raman spectra [29] or the random and nonuniform nanoribbon-etched graphene [28] and nanotrench-etched graphene based on Fe NPs [32]. Compared with the developed etch technologies above, the etching method by Kim et al. [28] revealed to be the best to date because of perfectly no damage and layer-by-layer etching from an innovative ion beam ICP.

3. Applications based on etched graphene

In **Table 1**, applications associated with the above etched-graphene investigations are briefly summarized. A chip utilized nanosphere-etched GNRs by O_2 plasma at low power [25] and revealed the high-performance electronic device

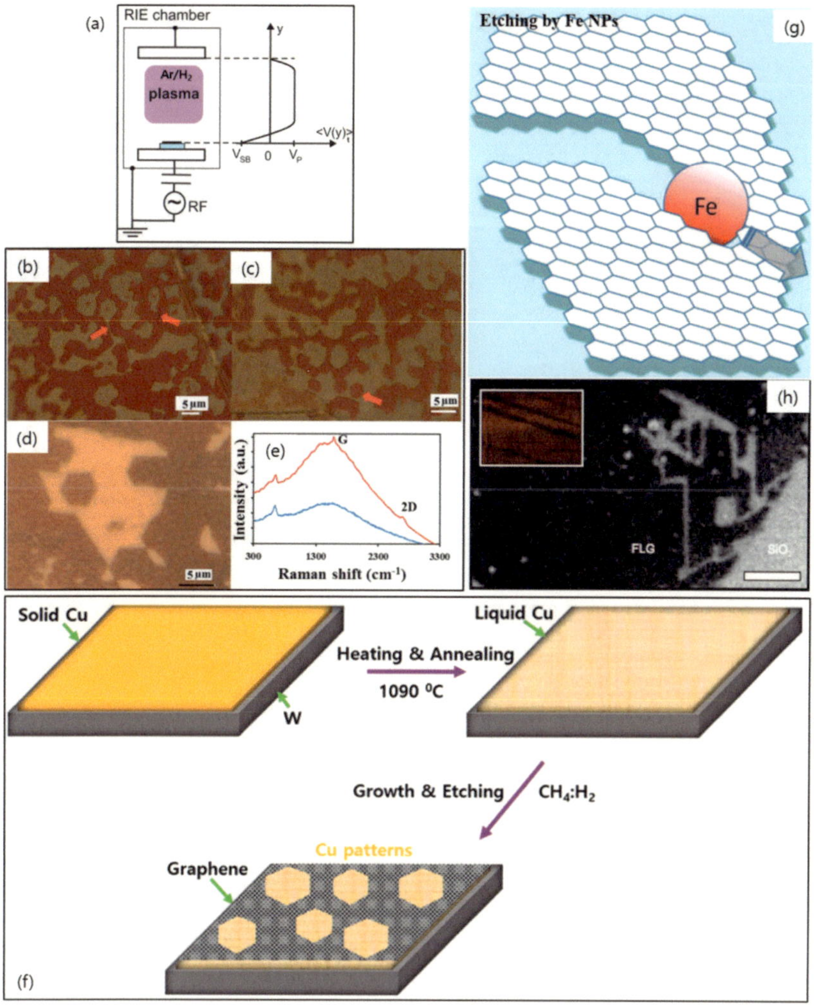

Figure 4.
(a) An RIE apparatus using Ar/H_2 to etch graphene on SiO_2. OM data of H_2-etched graphene during CVD growth and then annealing (Ar/H_2, 1000°C) under ambient atmosphere (b–d). (e) Raman data of etched graphene and partially oxidized Cu surface. (f) Schematic of few-layer graphene etching by thermally activated iron nanoparticles. (g) SEM data of etched few-layer graphene as nanotrench; tiny dots are iron NPs; scale bar is 0.8 μm. Inset is AFM data of few-layer graphene after being etched ((a) is reproduced with permission from [29], Copyright 2011, AIP Publishing; (b–e) are reproduced with permission from [30], Copyright 2015, the Royal Society of Chemistry; (f) is reproduced with permission from [31], Copyright 2018, American Chemical Society; (g) and (h) are reproduced with permission from [32], Copyright 2008, American Chemical Society).

with the exotic GNR architectures (chain, branch and circle ring). In another application, a metal oxide semiconductor (MOS)-like transistor was made; although the etched effect was formed, it simultaneously produced a high-energy plasma damage that induced poor electrical characteristics [27]. One more application related to the monolayer deep patterning was fabricated by etching (N_2) and annealing (Ar/O_2), and this pattern presented good quality for slight defects [26]. In 2015, Papon et al. fabricated the Y- and Z-shaped GNRs during CVD graphene growth, and the etching effect vehemently happened at high H_2 concentrations [30]. But these shapes were random and not well-controlled mechanically.

Etching methods	Substrate	Applications of Etched- Graphene	Results	Ref.
Nanosphere lithography using low-power O_2 plasma etching	SiO_2	Chip device based on Nanosphere-shaped graphen e nanoribbons (GNRs)	Superior electronic quality, and achieved GNRs architectures included chains, bra nches, circular rings, and connected ring s at low cost	[25]
O_2 plasma etching	SiC	NA	NA	[24]
O_2 plasma etching by ICP-RIE	SiO_2	NA	NA	[23]
N_2 plasma + post annealing (Ar/O_2)	SiO_2	Monolayer-deep patterns	Thinned graphene with good quality wit h few defects	[26]
O_2 absorption + Ar etching by neutral beam	SiO_2	Metal-oxide semiconductor (MOS) devices	Poor electrical characteristic due to high energy damage	[27]
O_2 absorption + Ar etching by ion beam	SiO_2	NA	The best atomically layer-by-layer etching candidate to date	[28]
Reactive ion etching (RIE) system using Ar/H_2 mixture	SiO_2	NA	NA	[29]
H_2 etching during CVD graphene growth	Solid Cu foil	Y- and Z-shaped GNRs	Obtained Y- and Z-shapes with controlle d-H_2 etch process	[30]
CH_4/H_2 etching during CVD graphene growth	Liquid Cu foil	NA	Obtained hexagon flower shape with integrated growth/etch process	[31]
Fe nanoparticles	SiO_2	NA	NA	[32]

Table 1.
Graphene etching methods and their applications. Source: "NA" is "not applicable".

4. Conclusions

Generally, there are many unexploited huge potentials from the etched-graphene products, but the perspectives are bright. If these etching technologies are extended to other nanomaterials such as transition metal dichalcogenides (TMDs) or transition metal carbides, nitrides, and carbonitrides (MXenes), and black phosphorous [42], it will definitely achieve high-quality electronics and optoelectronics. Bandgap tuning for nanomaterials will significantly improve the on/off current ratio, photoresponsivity, quantum efficiency, conductivity and others. Layer-by-layer etching for multilayer materials by low-energy plasma technology (double mesh grids inserted for plasma apparatus such as chlorine-radical ICP, neutral-beam ICP and ion beam ICP) with no physical damage would be the next research direction and can be applied to the other low-dimensional materials [28, 43] to achieve ultrahigh performance of electronic and optoelectronic devices [28, 43]. For instance, in the latest report in 2019, Kim et al. utilized a chlorine ICP innovative plasma apparatus that has no physical damage effect by inserting double mesh grids for cyclic ALE process on intrinsic multilayer MoS_2 and successfully fabricated heterostructured photodetector with ultra-responsivity (~10^6 A/W) in the visible range [43].

Conflict of interest

There are no conflicts of interest to declare.

Author details

Phuong V. Pham[1,2,3]

1 SKKU Advanced Institute of Nano Technology (SAINT), Sungkyunkwan University (SKKU), Suwon, South Korea

2 Center for Multidimensional Carbon Materials, Institute for Basic Science, Ulsan, South Korea

3 School of Information Science and Electronic Engineering, College of Microelectronics, Zhejiang University, Hangzhou, China

*Address all correspondence to: pvphuong@skku.edu

References

[1] Novoselov KS, Geim AK, Morozov SV, Jiang D, Zhang Y, Dubonos SV, et al. Electric field effect in atomically thin carbon films. Science. 2004;**306**:666-669. DOI: 10.1126/science.1102896

[2] Pham VP, Jang HS, Whang D, Choi JY. Direct growth of graphene on rigid and flexible substrates: Progress, applications and challenges. Chemical Society Reviews. 2017;**46**:6276-6300. DOI: 10.1039/c7cs00224f

[3] Pham VP, Nguyen MT, Park JW, Kwak SS, Nguyen DHT, Mun MK, et al. Chlorine-trapped CVD bilayer graphene for resistive pressure sensor with high detection limit and high sensitivity. 2D Materials. 2017;**4**:025049. DOI: 10.1088/2053-1583/aa6390

[4] Pham VP, Kim KN, Jeon MH, Kim KS, Yeom GY. Cyclic chlorine trap-doping for transparent, conductive, thermally stable and damage-free graphene. Nanoscale. 2014;**6**:15301-15308. DOI: 10.1039/c4nr04387a

[5] Pham VP, Kim KH, Jeon MH, Lee SH, Kim KN, Yeom GY. Low damage pre-doping on CVD graphene/Cu using a chlorine inductively coupled plasma. Carbon. 2015;**95**:664-671. DOI: 10.1016/j.carbon.2015.08.070

[6] Pham VP, Mishra A, Yeom GY. The enhancement of Hall mobility and conductivity of CVD graphene through radical doping and vacuum annealing. RSC Advances. 2017;**7**:16104-16108. DOI: 10.1039/c7ra01330b

[7] Pham VP, Kim DS, Kim KS, Park JW, Yang KC, Lee SH, et al. Low energy BCl_3 plasma doping of few-layer graphene. Science of Advanced Materials. 2016;**8**:884-890. DOI: 10.1166/sam.2016.2549

[8] Kim KN, Pham VP, Yeom GY. Chlorine radical doping of a few layer graphene with low damage. ECS Journal of Solid State Science and Technology. 2015;**4**:N5095-N5097. DOI: 10.1149/2.0141506jss

[9] Pham VP. Chemical vapor deposited graphene synthesis with same-oriented hexagonal domains. Eng Press. 2018;**1**:39-42. DOI: 10.28964/EngPress-1-107

[10] Pham VP. How can the nanomaterial surfaces be highly cleaned? Edelweiss Applied Science and Technology. 2018;**2**:184-186. DOI: 10.33805/2576-8484.136

[11] Pham VP. Layer-by-layer thinning of 2D materials. Edelweiss Applied Science and Technology. 2018;**2**:36-37. DOI: 10.33805/2576.8484.111

[12] Pham VP. Plasma-related graphene etching: A mini review. Journal of Science Engineering and Advance Technology. 2018;**17**:91-106. DOI: 10.18642/jmseat_7100121943

[13] Pham VP. Cleaning of graphene surface by low pressure air plasma. Royal Society Open Science. 2018;**5**:172395. DOI: 10.1098/rsos.172395

[14] Pham VP. A library of doped-graphene images via transmission electron microscopy. C-Journal of Carbon Research. 2018;**4**:34. DOI: 10.3390/c4020034

[15] Pham PV. Direct Growth of Graphene on Flexible Substrates towards Flexible Electronics: A Promising Perspective. London, UK: IntechOpen; 2018. DOI: 10.5772/intechopen.73171

[16] Pham PV. Atmospheric Pressure Chemical Vapor Deposition of Graphene. London, UK: IntechOpen; 2019. DOI: 10.5772/intechopen.81293

[17] Tomas H, Jan P, Richard K, Pavel S, Petr D, Martin W, et al. Atmospheric dry hydrogen plasma reduction of inkjet-printed flexible graphene oxide electrodes. ChemSusChem. 2018;**11**:1-8. DOI: 10.1002/cssc.201702139

[18] Ferrari AC, Bonaccorso F, Fal'ko V, et al. Science and technology roadmap for graphene, related two-dimensional crystals, and hybrid systems. Nanoscale. 2015;**7**:4587-5062. DOI: 10.1039/c4nr01600a

[19] Butler SZ, Hollen SM, Cao L, Gupta JA, Guitierrez HR, Heinz TF, et al. Progress, challenges, and opportunities in two-dimensional materials beyond graphene. ACS Nano. 2013;**7**:2898-2926. DOI: 10.1021/nn400280c

[20] Geim AK, Novoselov KS. The rise of graphene. Nature Materials. 2007;**6**: 183-191. DOI: 10.1038/nmat1849

[21] Zhang H, Yang P, Prato M. Grand challenges for nanoscience and nanotechnology. ACS Nano. 2015;**9**:6637-6640. DOI: 10.1021/acsnano.5b04386

[22] Donnelly VM, Kornblit A. Plasma etching: Yesterday, today, and tomorrow. Journal of Vacuum Science and Technology A. 2013;**31**:050825. DOI: 10.1116/1.4819316

[23] Al-Mumen H, Rao F, Li W, Dong L. Singular sheet etching of graphene with oxygen plasma. Nano-Micro Letters. 2014;**6**:116-124. DOI: 10.5101/nml.v6i2.p116-124

[24] Shin YJ, Wang Y, Huang H, Kalon G, Wee ATS, Shen Z, et al. Surface-energy engineering of graphene. Langmuir. 2010;**26**:3798-3802. DOI: 10.1021/la100231u

[25] Liu L, Zhang Y, Wang W, Gu C, Bai X, Wang E. Nanosphere lithography for the fabrication of ultranarrow graphene nanoribbon and on-chip bandgap tuning of graphene. Advanced Materials. 2011;**23**:1246-1251. DOI: 10.1002/adma.201003847

[26] Yang X, Tang S, Ding G, Xie X, Jiang M, Huang F. Layer-by-layer thinning of graphene by plasma irradiation and post-annealing. Nanotechnology. 2011;**23**:025704. DOI: 10.1088/0957-4484/23/2/025704

[27] Lim WS, Kim YY, Kim H, Jang S, Kwon N, Park BJ, et al. Atomic layer etching of graphene for full graphene device fabrication. Carbon. 2011;**50**:429-435. DOI: 10.1016/j.carbon.2011.08.058

[28] Kim KS, Ji YJ, Nam Y, Kim KH, Singh E, Lee JY, et al. Atomic layer etching of graphene through controlled ion beam for graphene-based electronics. Scientific Reports. 2017;**7**:2462. DOI: 10.1038/s41598-017-02430-8

[29] Wojtasszek M, Tombros N, Caretta A, Loosdrecht PHMV, Wees BJV. A road hydrogenating graphene by a reactive ion etching plasma. Journal of Applied Physics. 2011;**110**:063715. DOI: 10.1063/1.3638696

[30] Papon R, Sharma S, Shinde SM, Thangaraja A, Kalita G, Tanemura M. Formation of graphene nanoribbon and Y-junctions by hydrogen induced anisotropic etching. RSC Advances. 2015;**5**:35297-35301. DOI: 10.1039/c5ra03268g

[31] Pham PV. Hexagon flower quantum dot-like pattern formation during low-pressure chemical vapor deposited graphene growth on a liquid Cu/W substrate. ACS Omega. 2018;**3**:8036-8041. DOI: 10.1021/acsomega.8b00985

[32] Datta SS, Strachan DR, Khamis SM, Johnson ATC. Crystallographic etching of few-layer graphene. Nano Letters.

2008;**8**:1912-1915. DOI: 10.1021/nl080583r

[33] Biel B, Blase X, Triozon F, Roche S. Anomalous doping effects on charge transport in graphene nanoribbons. Physical Review Letters. 2009;**102**:096803. DOI: 10.1103/PhysRevLett.102.096803

[34] Huang B, Yan Q, Zhou G, Wu J, Gu BL, Duan W, et al. Making a field effect transistor on a single graphene nanoribbon by selective doping. Applied Physics Letters. 2007;**91**:253122. DOI: 10.1063/1.2826547

[35] Son YW, Cohen ML, Louie SG. Half-metallic graphene nanoribbons. Nature. 2006;**444**:347-349. DOI: 10.1038/nature05180

[36] Ferralis N, Maboudian R, Carraro C. Evidence of structural strain in epitaxial graphene layers on 6H-SiC(0001). Physical Review Letters. 2008;**101**:156801. DOI: 10.1103/PhysRevLett.101.156801

[37] Teague M, Lai A, Velasco J, Hughes C, Beyer A, Bockrath M, et al. Evidence for strain-induced local conductance modulations in single-layer graphene on SiO_2. Nano Letters. 2009;**9**:2542-2546. DOI: 10.1021/nl9005657

[38] Topsakal M, Cahangirov S, Ciraci S. The response of mechanical and electronic properties of graphane to the elastic strain. Applied Physics Letters. 2010;**96**:091912. DOI: 10.1063/1.3353968

[39] Gao Y, Hao P. Mechanical properties of monolayer graphene under tensile and compressive loading. Physica E: Low-dimensional Systems and Nanostructures. 2009;**41**:1561-1566. DOI: 10.1016/j.physe.2009.04.033

[40] Moslemi MR, Sheikhi MH, Saghafi K, Moravej-Farshi MK. Electronic properties of a dual-gated GNR-FET under uniaxial tensile strain. Microelectronics and Reliability. 2012;**52**:2579-2584. DOI: 10.1016/j.microrel.2012.05.009

[41] Jokar Z, Moslemi MR. Effects of position and shape of atomic defects on the band gap of graphene nano ribbon superlattices. International Journal of Electronics and Communication Engineering. 2015;**9**:162-166. DOI: 10.1999/1307-6892/10000474

[42] Pham VP, Yeom GY. Recent advances in doping of molybdenum disulphide: Industrial applications and future prospects. Advanced Materials. 2016;**28**:9024-9059. DOI: 10.1002/adma.201506402

[43] Kim KS, Ji YJ, Kim KH, Choi S, Kang DH, Heo K, et al. Ultrasensitive MoS_2 photodetector by serial nanobridge multi-heterojunction. Nature Communications. 2019;**10**:4701. DOI: 10.1038/s41467-019-12592-w

Chapter 5

Surface Science of Graphene-Based Monoliths and Their Electrical, Mechanical, and Energy Applications

Mujtaba Ikram, Sana Arbab, Bilal Tariq, Rayha Khan, Husnain Ahmad, Abdullah Khan Durran, Muhammad Ikram, Muhammad Aamir Iqbal and Asghari Maqsood

Abstract

Ceramic monoliths are applied in many insulating and high resistive engineering applications, but the energy application of ceramics monoliths is still vacant due to less conductivity of monolithic ceramics (for example, in silica- and alumina-based hybrids). This book chapter is a significant contribution in the graphene industry as it explains some novel and modified fabrication techniques for ceramics-graphene hybrids. The improved physical properties may be used to set ceramics-graphene hybrids as a standard for electrical, mechanical, thermal, and energy applications. Further, silica-rGO hybrids may be used as dielectric materials for high-temperature applications due to improved dielectric properties. The fabricated nano-assembly is important for a technological point of view, which may be further applied as electrolytes, catalysts, and conductive, electrochemically active, and dielectric materials for the high-temperature applications. In the end, this chapter discussed porous carbon as a massive source of electrochemical energy for supercapacitors and lithium-ion batteries. Carbon materials which are future of energy storage devices because of their ability to store energy in great capacity, so sustainability through smart materials got a huge potential, so hereby keeping in view all the technological aspects, this chapters sums up important contribution of graphene and porous carbon for applied applications.

Keywords: graphene, ceramics, superior energy storage, chemical activation, supercapacitors, lithium-ion batteries

1. Introduction

Decades ago, scientists believed that carbon comes in two basic forms, that is, graphite and diamond, but in the last three decades, scientists have discovered new forms of carbon known as advanced carbon materials including fullerene, carbon nanotubes (CNTS), and graphene, respectively [1–6]. In recent years, graphene is considered as an outstanding candidate for enhancing the structural, electrical, mechanical, and thermal properties of materials (for example, metals, ceramics, and polymers) [7, 8]. In hybrid nanostructures, the physical property enhancement may be possible due to excellent physical properties of the graphene. Excellent physical properties included higher thermal conductivity (5000 $Wm^{-1} K^{-1}$), electrical conductivity (10^6 $S m^{-1}$), and Young's modulus (1 TPa), which are a driving force for enhancement in the physical properties of hybrids. Among the various types of graphene materials, graphite oxide-derived graphene plays an important role in increasing the physical properties of hybrids because of its surface functionalization and its ability of large-scale production at any level. Even a tiny amount of graphene in hybrids (either polymers or ceramics or metals) may alter its physical properties to a great extent. In case of graphene, the compatible structural properties and how it makes bond with various types of nanostructures are reasons for improved properties in the end product (hybrids or composites). For example, reduced graphite oxide (rGO)-polystyrene composites with a low threshold content of 0.1 (volume %) rGO have shown greatly improved electrical conductivity (approx. 0.1 $S m^{-1}$); this has been possible due to good dispersion of rGO in the polymer composite matrix. Similarly, in inorganic hybrids, rGO has been used for the deposition of Co_3O_4 particles for increased catalytic effects, which may have been used for the decomposition of ammonium perchlorate because of the complex properties of GO and Co_3O_4. In another research, rGO was used to improve the mechanical properties of the bulk silicon nitride (i.e. toughness is enhanced by up to 235%), which may be used for high-performance mechanical and structural applications [8, 9]. In short, graphene being the toughest, strongest, lightweight material may act as a wonder material for future scientific revolution in every aspect of life. Even if it is combined with polymers, metals, and ceramics, it may play a significant role in improving physical properties due to its versatile surface, morphology, chemistry, and physical properties. In this chapter, we will discuss graphene combination with various ceramics and how it has been used to improve their physical properties, and porous carbon for energy storage, respectively. This book chapter will be a significant contribution to advance studies on physical properties and technological applications.

2. Highly conductive graphene-alumina-based hybrid monoliths for electrical, thermal, and mechanical applications

Low strength and brittle attributes are the main properties of ceramics. The most widely used structural ceramic is alumina, due to its good thermal conductivity and the shape stability [7]. Alumina has a wide range of applications, some of the fields include dental implants, high speed cutting tools, chemical insulators, electrical insulators, and wear resistant coatings. Scientists have observed that mechanical properties of alumina may be improved using carbon nanotubes, for example, fracture toughness (by 94%), flexural strength (6.4%), and hardness (by 13%), respectively. On the addition of graphene platelets, about 40% enhancement in the fracture toughness of the ball milled alumina/zircon/graphene have been noticed. In another research study, the alumina-rGO core shell nanocomposites were fabricated using the method known as the sol-gel method, and it was found through this study that the BET surface area of the rGO is essential to enhance the surface charge properties of the hybrids. In another study, alumina graphene composite films were reported with a low optical gap of about 1.53 eV. Alumina-rGO nano-composites obtained via deposition during the process showed a unique morphology of aluminum nanoparticles with low prosperity and BET surface area of 242.4 $m^2 g^{-1}$.

Moreover, scientists have found that in a microwave preparation of alumina-rGO composites, the grain size of the alumina matrix was reduced from 475 to 180 nm, which was obtained from the conventional sintering process, leading to an increase in the Young's modulus from 148 to 180 GPa. Scientists have found that using solvothermal-hot press processing route, highly conductive alumina-rGO hybrids may be obtained, which consist of Al_2O_3 nanorods and rGO, respectively [7, 8, 10]. The same solvothermal method was used to form hybrids from cross-linked Al_2O_3 nanorods and reduced graphite oxide (rGO) platelets. Then after hot pressing, the hybrid monoliths were obtained, which were utilized for the systematical study of improved physical properties of hybrids. Using the same method, it is noticed that with the 3 h-calcinated hybrid, the Al_2O_3-rGO monoliths show enhanced electrical conductivity (changes from 5.1×10^{-10} to 6.7×10^{1} S m^{-1}), mechanical tensile strength (90% increase), thermal conductivity (80% increase), and a much higher dielectric constant (12 times) than the bare Al_2O_3. The highest values of electrical conductivity (8.2×10^{1} S m^{-1}), thermal conductivity (2.53 Wm^{-1} K^{-1}), dielectric constant (10^{4}), and Young's modulus (3.7 GPa) are determined for the alumina-rGO hybrid which is calcinated for about 1 h. It was noticed that the functional groups that contain oxygen on GO were useful for the adsorption of aluminum isopropoxide, leading to the dispersion of rGO and the Al_2O_3, which were obtained during the solvothermal process by the hydrolysis of the aluminum isoropoxide [7]. The improvement in the mechanical properties was caused due to the elongated Al_2O_3 nanorods, which was indicated by the study of aspect ratio of the nanorods. Graphene platelets, functional groups present, and their surface properties are driving forces for enhancement in the physical properties of alumina-rGO hybrids.

2.1 Preparations of highly conductive graphene-alumina-based hybrid monoliths

In the past, alumina rGO hybrids have been prepared using sol gel, molecular level mixing, and powder coating methods, but scientists have tried some conventional preparation methods followed by high temperature treatments [7, 11]. Such methods have shown great enhancement in physical properties of hybrids. Here, we discuss one of such advanced methods, that is, the preparation of Al_2O_3-rGO hybrids using solvothermal-hot press processing route. Al_2O_3-rGO was prepared by the mixing of GO and with cyclohexane and the aluminum isopropoxide ($C_9H_{21}AlO_3$), which was followed by the solvothermal process. The procedure involves 0.1 g of GO being first dispersed in 35 mL of cyclohexane, then 3.5 mL of aluminum isoproperoxide ($C_9H_{21}AlO_3$) being added dropwise. The mixture is then stirred continuously at room temperature at the rate of 1000 rpm for several days until the GO powder is dispersed homogeneously but the color of the suspension changes with time. Then the products are separated by centrifugation, and the products were then washed several times with cyclohexane. The solid sample obtained are denoted as $Al[O]_x$/GO. Then it was sent for the hydrothermal treatment. For this, it was dispersed in 50 mL cyclohexane and then transferred to the 100 mL Teflon-lined stainless steel autoclave, after which the reaction was carried out for about 6 h at about 453 K, and the resultant sample was again centrifuged and dried at about 303 K, which is denoted as $Al[O]_x$/rGO. It was then calcined at 723 K for about 1–3 h to form Al_2O_3/rGO hybrids. The condition of calcination is limited supply of air. The calcination process was controlled by using the quartz tubular furnace with open ends that will allow the calcination to occur in the limited supply of air; also, the furnace was heated to the desired temperature of about 723 K for the calcination times for 1–3 h. In the initial stage, the temperature was increased by using heating rate of about 15°C min^{-1}. As a result, the free standing

Al_2O_3 nanorods were formed as a result of calcination treatment and also the GO was reduced to rGO. The physical properties were studied by obtaining the Al_2O_3/rGO hybrid powder samples consisting of 16.707, 12.830, and 7.705 wt% using the same solvothermal process. The same process was then used for the preparation of the pure Al_2O_3 without the addition of GO. The calcination temperature was altered and was set at different temperatures for the processing time of about 1 h. For the analysis of crystallinity, it was set as 500, 600, 650, 700, 750, and 800 K and for the analysis of the effect of calcination temperature and time on the nanorods structure, the calcination time was set as 1, 2, 3, 4, and 5 h for the temperature of 723, 823, and 923 K. Hot pressing of powder samples was carried out in a vacuum furnace. The furnace was fitted with a hydraulic press which compresses the samples in a graphite pressing die. The heating temperature was made such that to increase from the room temperature at the heating rate of 10°C min^{-1} up to 900°C, which was then maintained constant for about 60 min. When the hybrids reached the set temperature, the pressure of about 25–30 MPa was then applied to the hybrids.

2.2 Improved physical properties of the highly conductive graphene-alumina based hybrid monoliths

In the case of hybrids, higher the rGO, higher will be the enhancement in the physical properties such as electrical, thermal, dielectric, and mechanical properties. In the hybrids, the surface area has been increased, and as a result, greater will be the interfacial interaction of the rGO [7, 12]. The higher rGO platelets will improve the physical properties because it provides a large surface area for interfacial interactions at nano-level. Due to higher surface area of graphene, BET surface area has been improved in the hybrids as represented in the **Table 1**, in comparison with various fabrication methods [7]. Scientists believe that the higher mechanical strength is caused due to the elongated dimensions of nanorods in alumina-rGO hybrids. From the literature, it is found that 90% increase in tensile strength and 75% increase in compressive strength occur when the content of rGO is increased up to 7.707% in the hybrid. The addition of rGO affects the dielectric constant, and it increases by four orders of magnitude through a second percolation threshold [7, 8].

Further, the hot press processing sustains the quality of rGO in the hybrids. An increase in calcination temperature resulted in enhanced crystallinity in the Al_2O_3 nanorods and rGO hybrids as also shown in XRD of hybrid (**Figure 1a**). From the surface science point of view, this may cause enhancement in the diameters and lengths of the nanorods in the hybrid as shown in the **Figure 1b**. TEM images showing variations in diameters of nano-rod structures with various calcination temperatures are presented in **Figure 1c–h**. As a result, after calcination and hot-press processing, Al_2O_3-rGO monoliths were obtained with enhanced physical properties. Researchers have found that with very little rGO in the alumina hybrid, higher electrical conductivity (8.2×10^1 S m^{-1}), higher dielectric constant by four orders of magnitude, and improved thermal conductivity (1.4 $Wm^{-1} K^{-1}$) have been achieved [7]. Hot pressing at 900°C ensured the complete reduction of GO and the higher crystallinity of Al_2O_3, resulting in enhanced physical properties. The elongated and fine Al_2O_3 nanorod morphology, atomic-level layered structure, and excess surface free electrons of rGO resulted in the best reported BET surface area (408 $m^2 g^{-1}$ in the 2 h-calcinated alumina–rGO), best thermal conductivity (2.53 $Wm^{-1} K^{-1}$ in the 1 h-calcinated alumina—rGO), and relatively small density (0.92 g cm^3 in the 1 h-calcinated alumina–rGO) and high strength (3.7 GPa in the 1 h-calcinated alumina–rGO), respectively [7].

DOI: http://dx.doi.org/10.5772/intechopen.93318

Sample type	BET surface area ($m^2 g^{-1}$)				Bulk density (g/cm^3)			
Solvothermal-hot press processing method	γ-Al_2O_3 (1 h calcination time)	γ-Al_2O_3-rGO (3 h calcination time)	γ-Al_2O_3-rGO (2 h calcination time)	γ-Al_2O_3-rGO (1 h calcination time)	γ-Al_2O_3 (1 h calcination time)	γ-Al_2O_3-rGO (3 h calcination time)	γ-Al_2O_3-rGO (2 h calcination time)	γ-Al_2O_3-rGO (1 h calcination time)
	280	361	408	379	2.75	1.61	1.37	0.92
Meso-porous Al_2O_3-rGO	Al_2O_3		Al_2O_3-rGO		Al_2O_3		Al_2O_3-rGO	
	243		327		2.40		1.65	
Core-shell flakes Al_2O_3-rGO	Al_2O_3		Al_2O_3-rGO		Al_2O_3		Al_2O_3-rGO	
	286.62		119.71		2.816		0.003	
In situ deposition Al_2O_3-rGO	Al_2O_3		Al_2O_3-rGO		Al_2O_3		Al_2O_3-rGO	
	N/A		242.4		N/A		N/A	

Table 1.
BET surface area and density comparison for γ-Al_2O_3-rGO (1, 2 and 3 h calcination time) and pure γ-Al_2O_3 (1 h calcination time), compared with various fabrication methods [7].

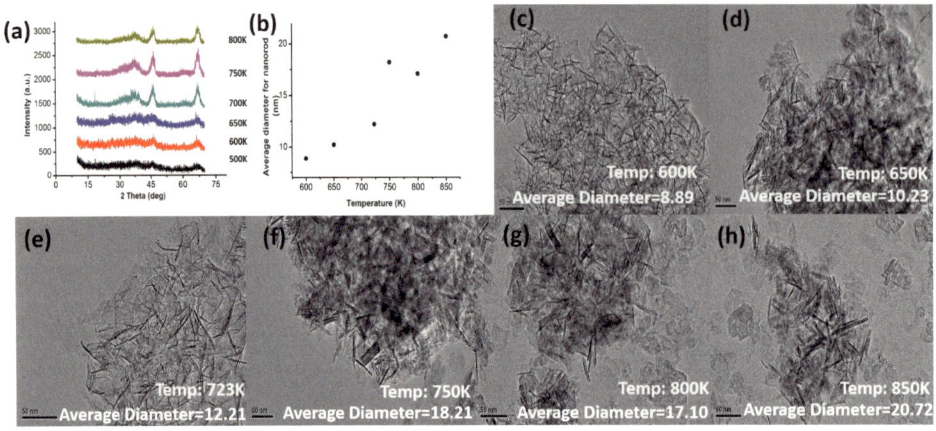

Figure 1.
(a) XRD of γ-Al_2O_3-rGO hybrids taken from 500 to 800 K, (b) average diameter of nano rods (nm) as function of calcination temperature (K), and (c–h) TEM images showing variations in diameters of nano rod structures with various calcination temperatures. Units for diameters are in (nm).

Figure 2.
SEM images of hot pressed samples: (a) γ-Al_2O_3-rGO (1 h calcination time), (b) γ-Al_2O_3-rGO (2 h calcination time), (c) γ-Al_2O_3-rGO (3 h calcination time), and (d) pure γ-Al_2O_3 (1 h calcination time).

Hot press processing may have an impact on the physical properties of hybrids; SEM images of hot pressed samples have shown particle-like morphology, as represented in **Figure 2**.

Moreover, well-aligned, elongated, and fine nanorod morphology of alumina is the reason for improvement in the mechanical strength [7, 13, 14]. Aspect ratio

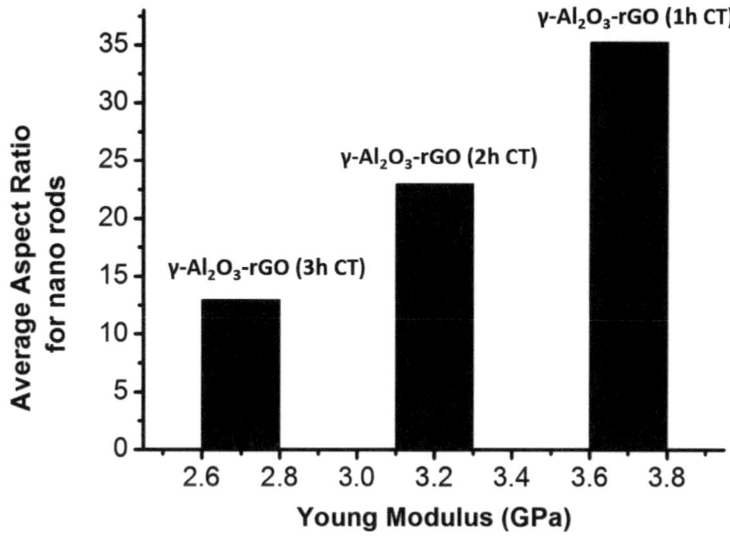

Figure 3.
Young's modulus as a function of average aspect ratio of nano-rods in hot pressed samples γ-Al_2O_3-rGO hybrids with 1, 2, and 3 h calcination time.

studies have confirmed that alumina-rGO hybrids (1 h calcinated) have more strength compared to hybrids that are calcined at more time (2 and 3 h), as shown in **Figure 3**.

Thus, nano-hybrids of alumina monoliths and rGO can be further applied as electrolytes, catalysts, and electrochemically active materials because of nanometer dimensions and improved physical properties [7, 15].

3. Highly conductive graphene-silica-based hybrid monoliths for dielectric applications

Improved physical properties may be achieved for SiO_2 -rGO monoliths using calcination followed by hot press processing. If adsorption of ethyl silicate($C_{10}H_{20}O_4Si$) is required as if the group on graphene oxide contains oxygen, then it is beneficial for adsorption because it helps in uniform dispersion of rGO within the SiO_2 matrix, which can be obtained during the hydrothermal reaction by hydrolysis of ethyl silicate [8, 16]. If SiO_2 spheres in hybrids become more crystalline, then good physical properties in the hybrid can be obtained. Crystallinity in the SiO_2 spheres can be enhanced by increasing calcination temperature and further hot press processing at 750°C. Graphene is a material having good physical properties. As experimentally proved by scientists worldwide, the thermal, electrical, and mechanical properties of polymers, metals, and ceramics may be improved using graphene. Graphite oxide-derived graphene has tunable surface functionalization and the potential for large scale production, so it can be used to enhance the physical properties of hybrids. For the decomposition of ammonium perchlorate, Co_3O_4 can be used as a catalyst [8, 17]. To increase the catalytic effect of Co_3O_4 , rGO can be used. Basically, rGO helps in uniform deposition of Co_3O_4 in inorganic hybrids. Silica has good functionalized ability and is very stable, so it can be used as an additive in numerous applications. In biomedical, polymer, and ceramics engineering, silica is used for various purposes. rGO can be used to enhance the physical properties of silica. Epoxy- SiO_2 -rGO hybrids have enhanced thermal conductivity (0.452 W m^{-1} K^{-1}), storage modulus (3.56), and dielectric constant (77.23). In another study, rGO has been used to improve the gas sensing performance, which was obtained by an electrostatic self- assembly approach. If dispersion, corrosion resistance, and barrier properties of hybrids are required to be enhanced, then the presence of rGO is essential, and this was found when nanocomposites of silica-graphene oxide were fabricated using in situ gel method. SiO_2 -graphene hybrids are better gas sensors as compared to rGO-based sensors [18, 19]. Toward 50 ppm NH_3 for 850 s, the gas sensing response of rGO-based sensors is 1.5% and that of SiO_2 -graphene-based sensors is 31.5%, respectively. It has been shown that SiO_2 -rGO composites having enhanced BET surface area (676 m^2 g^{-1}) can be obtained by one-step hydrothermal method. On addition of ultrathin graphene, SiO_2 -polyvinyldiene fluoride having high dielectric constant (72.94) and low dielectric loss (0.059) is obtained. Composites of epoxy-silica-graphene oxide have enhanced tensile strength, which leads to an increase in fracture toughness and Young's modulus [8]. The physical properties of SiO_2 -rGO such as dielectric, electrical, mechanical, and thermal properties need to be improved further. So, in this chapter, we are discussing various physical properties of silica-rGO hybrids for dielectric applications.

Figure 4.
SEM images of (a) SiO_2-rGO-6.75% (sample b) and (b) SiO_2-rGO-10.80% (sample c) fabricated at calcination temperature of 800 K for 1-h. (c) TEM images of the same SiO_2-rGO-10.80% (sample c) at lower magnification and (d) at higher magnifications.

3.1 Preparation of highly conductive graphene-silica based hybrid monoliths for dielectric applications

In brief preparation using solvothermal-hot press processing route [8], GO is mixed with cylcohexane and ethylsilicate ($C_{10}H_{20}O_4Si$) , followed by hydrothermal reaction to form SiO_2 -rGO hybrids. For the preparation, 0.1 g of GO is dispersed in 50 mL cyclohexane, followed by addition of 5 mL of ethylsilicate dropwise. Then GO powder is homogeneously dispersed by stirring the mixture at the speed of 1500 rpm for several days at room temperature. The products are then separated by using centrifugation. The products are then washed with cyclohexane for several times. The obtained solid samples are represented as $Si(O)_x$ /GO. From suspension of 0.1, 0.2 and 0.3 g GO, $Si(O)_x$ /GO powders were obtained, respectively. The solid samples obtained are then dispersed in 75 mL cyclohexane, and then for hydrothermal reaction, the mixture is transferred to a 100 mL Teflon-lined stainless-steel autoclave. The reaction is then carried out at 420 K for 4 h, and the sample obtained is then centrifuged and dried at 303 K. The sample obtained is denoted as $Si(O)_x$ -rGO. Then calcination of $Si(O)_x$ -rGO is carried out at 800 K for 1 h. To study the physical properties, using the same method, hybrids of SiO_2 -rGO consisting of 1.55, 6.75, and 10.82 wt% rGO are obtained. 1.55, 6.75, and 10.82% rGO is referred to as sample-a, sample-b, and sample-c, respectively. Without adding GO by using the same procedure, pure SiO_2 (referred to as sample-d) is obtained. The calcination temperature is then set as 500, 600, 700, and 800 K for a processing time of 1 h to study the effect of calcination temperature on crystallinity.

3.2 Surface and physical properties of highly conductive graphene-silica-based hybrid monoliths for dielectric applications

Researchers have developed a hydrothermal-hot press processing technique, a simple and efficient method that can improve the thermal, electrical, dielectric, and mechanical properties of the hybrid [8, 20]. By a hydrothermal reaction, GO is dispersed in cyclohexane and ethylsilicate to produce hybrids composed of rGO and silica monoliths [20, 21]. The SEM morphology of hybrids has shown sphere-particle-like morphology with thin layers of rGO, which act as a support for elongated matrix, as shown in **Figure 4**.

SEM images of SiO_2-rGO-1.55% (sample a) at various temperatures are shown in **Figure 5**. At all temperatures, hybrids have shown sphere-like morphology, but sphere size changes at various temperatures [9, 22].

The solvothermal-hot press processing method shows the best reported electrical conductivity (0.143 $S\,m^{-1}$), thermal conductivity (1.612 $W\,m^{-1}\,K^{-1}$), and higher dielectric constants for SiO_2 -rGO monoliths. Thus, due to enhanced physical properties of the nano hybrids, it can be applied as electrolytes, catalysts, conductive and electrochemically active materials, and dielectrics for high-temperature applications [8, 12, 23].

Table 2 have shown BET surface area and mesoporous volume % analysis for the hybrids. From the table, it is confirmed that BET surface area has been increased with more rGO in the hybrids, while mesoporous volume % increased with more silica [8, 24].

The dielectric properties of the SiO_2-rGO hybrids and bare SiO_2 were measured using an LCR meter as shown in **Figure 6**. The dielectric properties of the hybrids were measured at a frequency of 1 kHz. For SiO_2, its dielectric constant is found

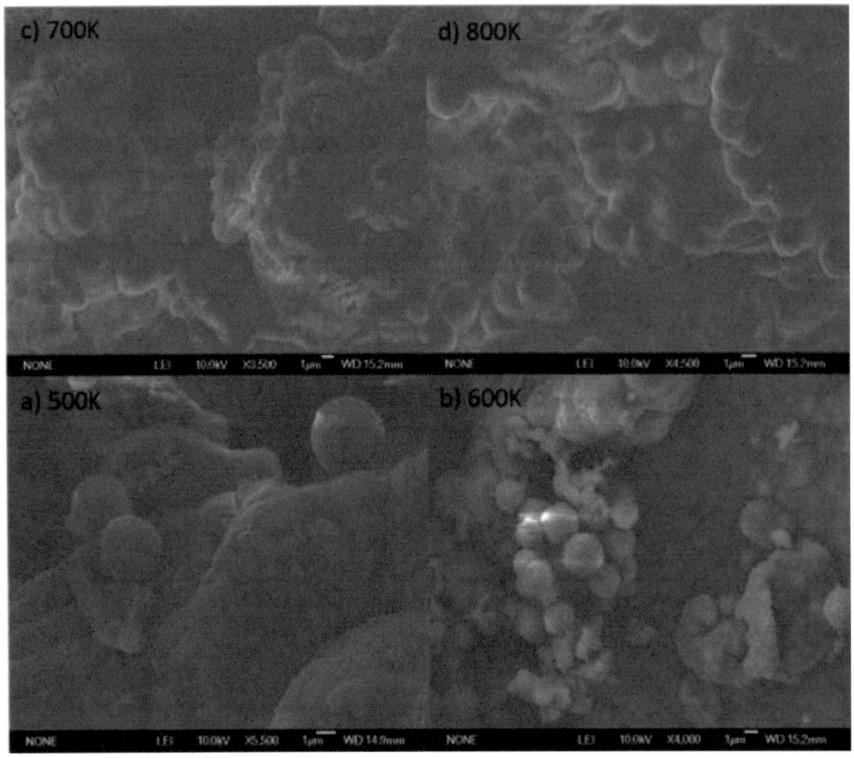

Figure 5.
SEM images of SiO_2-rGO-1.55% (sample a) at a calcination temperature of (a) 500 K, (b) 600 K, (c) 700 K, and (d) 800 K, respectively.

Sample type	BET surface area ($m^2 g^{-1}$)	Total volume ($cm^3 g^{-1}$)	Mesoporous volume ($cm^3 g^{-1}$)	Microporous volume ($cm^3 g^{-1}$)	Mesoporous volume (%)
Pure SiO_2	333.07	0.3821	0.3459	0.0362	90.52
SiO_2-rGO-1.55	611.21	0.4580	0.3694	0.0886	80.65
SiO_2-rGO-6.75	677.53	0.5521	0.3571	0.1950	64.68
SiO_2-rGO-10.8%	712.01	0.6812	0.3891	0.2921	57.11

Table 2.
BET surface area, mesoporous volume % of SiO_2 (sample d), SiO_2-rGO-1.55% (sample a), SiO_2-rGO-6.75% (sample b), and SiO_2-rGO-10.82% (sample c).

to be around 3.79, which is closer to that of pure silica. For sample a, the dielectric constant significantly increased by a value of 497, which indicates the presence and proximity of a first percolation threshold.

The enhanced dielectric constant (up to order of 105 and 107) was determined for samples b and c, which is much higher compared to that for sample d. Formation of conductive pathways is one of the main reasons for an increase in the overall dielectric constants. In sample c, significant leakage current leads to higher dielectric loss (300). By further increasing the rGO, the dielectric constant increased by seven orders of magnitude, indicating the presence of a second percolation

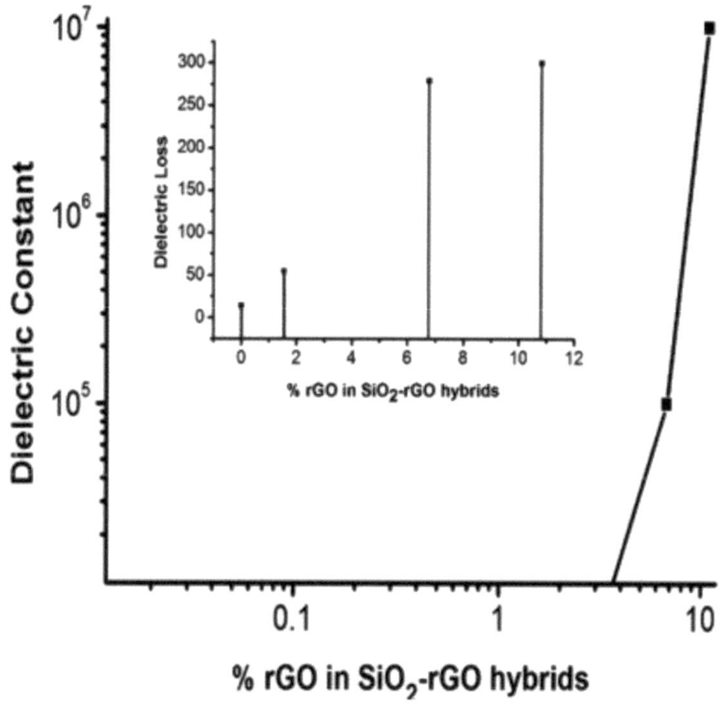

Figure 6.
Dielectric constant as a function of % rGO in the hybrid; inset is dielectric loss as a function of % rGO in the hybrid.

threshold, which is achieved through the higher value of dielectric constant. Similarly, the dielectric loss indicates very similar behavior in the real part of the dielectric constant as shown in the inset of **Figure 6**. Scientists have experimentally proved that a small amount of rGO in hybrids can enhance dielectric properties to a great extent. The existence of a double percolation threshold in SiO_2 and the rGO hybrids can be significant for applied applications because it can be used to enhance the dielectric permittivity (up to 107) with the addition of a small percentage of rGO in the hybrids. Silica-rGO hybrids may be used as dielectric materials for high-temperature applications due to better dielectric properties [7, 8, 23, 25].

4. Porous carbon for superior energy storage

A process called one-step carbonization-activation which is used to transform frozen tofu, mainly a source of carbon (C) and nitrogen (N), into a co-doped porous carbon having N (0.6–6.7 wt%) and O (3.6–9.5 wt %) and bearing a specific area of about 3134 $m^2\ g^{-1}$. Mesopores and micropores constitute a high volume of this hierarchy carbon, i.e. 1.11 $cm^3\ g^{-1}$ consists of mesopores and 0.71 $cm^3\ g^{-1}$ of micropores with a regular pore size appropriation somewhere in the range of 0.8–4 nm [9, 26]. When used as electrodes in supercapacitors, this porous carbon shows a specific capacitance of 243 F g^{-1} with sulfuric acid used as electrolyte and retains 93% of its initial capacitance after 10,000 cycles. In 1-butyl-3-methyl-imidazolium tetrafluoroborate ($BMIMBF_4$), a specific resistance of 170 F g^{-1} and a reliable rate capability can be observed using above prepared carbon which also provides an energy density of 72 W h kg^{-1} (calculated at an average power density of 889 W kg^{-1}). A total of 25 light emitting diodes (LEDs) which are connected in parallel fashion may be empowered immediately for more than 2 min in the wake of being charged for 25 s, using supercapacitors comprising of porous carbon, at a current density of 10 A g^{-1}. What's more, the porous carbon displays a high reversible charge capacity of 2120 mA h g^{-1} in the first cycle (estimated at 0.1 A g^{-1}) or 1035 mA h g^{-1} after 300 cycles (estimated at 1 A g^{-1}), when used as an anode for Li-ion batteries [9, 26, 27].

4.1 Preparation of porous carbon for superior energy storage

Devices having energy due to chemical reactions are getting more fame than other energy storage devices due to their considerable potential applications [28]. An instantaneous charging and discharging capability, which leads toward an efficient power density of about 10 kW kg^{-1}, can be observed in supercapacitors. The fabrication flowchart for porous carbon is shown in **Figure** 7.

4.2 Surface and physical properties of porous carbon for superior energy storage

On the other hand, in spite of having longer charging time, a high energy density of about 100–200 W h kg^{-1} can be referred in lithium-ion batteries (LIBs). There are two fundamental processes by which the energy can be stored in supercapacitors, which are as follows: (i) pseudocapacitive electrodes store ions based on quick faradaic reactions at the electrode-electrolyte interface, and (ii) electrical double-layer capacitive electrodes store energy by the adsorption and desorption of ions on the large surface area of the porous material [29, 30]. The working of lithium-ion batteries depends upon the transfer of lithium ions in between the cathode and the anode. The mechanism by which the lithium ions are stored or released, in lithium-ion batteries, depends upon the nature of the material of which the electrode is made [31]. High electrical conductivity, tailored porosity, and chemical stability are the main features of carbon materials that make their extensive use in many devices such as commercial supercapacitors and lithium-ion batteries (LIBs) [32]. Scientists reveal that in supercapacitors, mesopores and micropores are the main constituents of porous carbon as they provide ion buffering reservoirs, movement of ions and then storage site for ions, respectively. In lithium-ion batteries, the reversible Li-ion storage capacity is retained to an approximation of 372 mA h g^{-1}, for graphite, using graphite anode in lithium-ion batteries which interacts with Li-ions to produce a compound, LiC_6, that retains the reversible storage capacity to its mention value. In addition, because of the permeable structure, the use of carbon materials as a

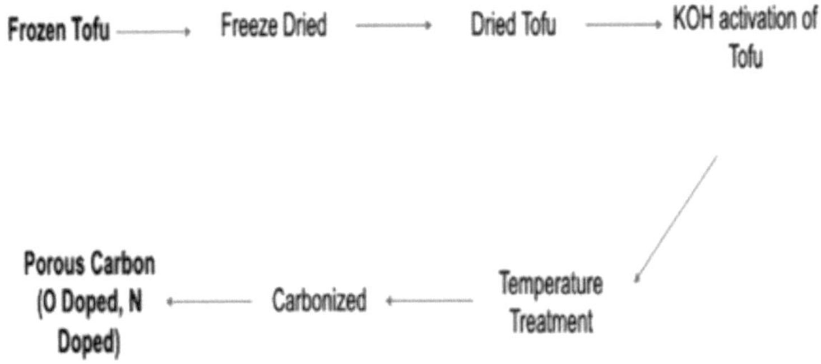

Figure 7.
The fabrication flow chart for porous carbon through one-step carbonization [9].

framework of electrodes, that is, in lithium-ion batteries and other energy storage devices, is increasing nowadays [33]. It is believed that the supercapacitors cannot fulfill the energy requirements of future electrical devices because of their low energy density (less than 6 W h kg^{-1}). Also, the capacity and rate capability of electrodes in LIBs are below to standards. To approach the above-mentioned requirements, porous carbon having good electrical conductivity and a modified 3-dimensional structure is required [34].

In the recent decades, a number of techniques named activation, self-assembly, and templating have been used for the production of porous carbon materials. But activation exceeds other techniques owing to the fact that it tends to produce a carbon of a large specific surface area of about 200 m^2 g^{-1} and other useful properties. Activation can also play an important role in the production of novel carbon by doing a proper processing of nanostructured carbon precursors [35]. For instance, graphene platelets can be rebuilt thoroughly to a 3D porous carbon having a specific surface area of approximately 3100 m^2 g^{-1} and pore size appropriation somewhere in between 0.6 and 5 nm, during the activation of microwave-exfoliated graphite oxide in the presence of KOH. Moreover, in graphite grids, the n-type can be brought up using the atoms like nitrogen which has the ability to donate electrons. The carbon doped with nitrogen finds its applications as anode in lithium-ion batteries because the hybridization between the lone pair electrons of nitrogen with π electrons of carbon can assist lithium lodging [36, 37]. Porous carbon materials derived from biomass are more sustainable than derived from other materials like coal, pitch, polymers, etc. Scientists have indicated that porous carbons for energy stockpiling applications can be acquired from different biomass sources, for example, rice husks, rice straw, algae, what's more, water bamboo. For instance, lithium and other confrere elements experience a one-step pyrolysis-activation synthesis to transform willow catkin into a cross-linked layered porous carbon which is co-doped with two metals, that is, nitrogen (N) and sulfur (S). The carbon thus produced exhibits some outstanding features related to chemical performance like it shows a specific capacitance of 298 F g^{-1} at 0.5 A g^{-1} in 1 molar solution of Na_2SO_4 with the great cycling stability along with the capacitance loss of only 2% when checked after 10,000 cycles at 5 A g^{-1}.

Tofu, rich in moisture, proteins, sugars, and follow sums of minerals, is a bounteous asset and has been viewed as a characteristic source of carbon and nitrogen [9, 36, 38]. It is obvious from the above discussion that tofu is a favorable predecessor material in the manufacture of carbon materials used for energy storage devices, but further developments are required for better performance like enhanced capacitance in symmetric supercapacitors and rate capability/cyclic stability in lithium-ion batteries. The features like large surface area, hierarchical (permeable) porous structure, and heteroatomic doping make the use of porous carbon samples (obtained from tofu) suitable for the material used as an anode in Li-ion batteries [9, 35, 38].

5. Conclusions

This study presented some novel and modified fabrication techniques for ceramics-graphene hybrids. The improved physical properties may be used to set ceramics-graphene hybrids as a standard for electrical, mechanical, thermal, and energy applications. Further, silica-rGO hybrids may be used as dielectric materials for high temperature applications due to improved dielectric properties. The fabricated nano-assembly is important for a technological point of view, which may be further applied as electrolytes, catalysts, and conductive, electrochemically active, and dielectric materials for the high-temperature applications. In addition, the porous carbon as a massive source of electrochemical energy for supercapacitors and lithium-ion batteries is also addressed.

Conflict of interest

The authors have declared no 'conflict of interest'.

Author details

Mujtaba Ikram[1]*, Sana Arbab[1], Bilal Tariq[1], Rayha Khan[1], Husnain Ahmad[1], Abdullah Khan Durran[1], Muhammad Ikram[2], Muhammad Aamir Iqbal[3] and Asghari Maqsood[4]

1 Applied Physics Lab, Institute of Chemical Engineering and Technology (ICET), University of the Punjab (PU), Lahore, Pakistan

2 Department of Physics, Government College University (GCU), Lahore, Pakistan

3 Center for Solid State Physics, University of the Punjab, Lahore, Pakistan

4 Nanoscale Laboratory, Department of Physics, Air University, Islamabad, Pakistan

*Address all correspondence to: mujtaba.icet@pu.edu.pk

References

[1] Novoselov KS, Fal'Ko VI, Colombo L, Gellert PR, Schwab MG, Kim K. A road map for graphene. Nature. 2012;**490**:192-200

[2] Novoselov KS, Geim AK, Morozov SV, Jiang D, Zhang Y, Dubonos SV, et al. Electric field effect in atomically thin carbon films. Science. 2004;**306**:666-669

[3] Pham VP, Jang H-S, Whang D, Choi J-Y. Direct growth of graphene on rigid and flexible substrates: Progress, applications, and challenges. Chemical Society Reviews. 2017;**46**:6276-6300

[4] Pham PV. The new etching technologies of graphene surfaces. ACS Omega. 2018;**3**:8036-8041

[5] Pham VP, Nguyen MT, Park JW, Kwak SS, Nguyen DHT, Mun MK, et al. Chlorine-trapped CVD bilayer graphene for resistive pressure sensor with high detection limit and high sensitivity. 2D Materials. 2017;**2**:4

[6] Pham VP, Jo YW, Oh JS, Kim SM, Jhon MS, Yeom GY. Effect of plasma–nitric acid treatment on the electrical conductivity of flexible transparent conductive films. Japanese Journal of Applied Physics. 2013;**52**:7R

[7] Ikram M, Tao Z, Ye J, Qayyum HA, Sun X, Xu J. Enhanced physical properties of γ-Al_2O_3–rGO hybrids prepared by solvothermal and hot-press processing. RSC Advances. 2018;**8**: 8329-8337

[8] Ikram M, Qayyum HA, Ali S, Tan Z, Ahmad M, Xu J. Hydrothermal-hot press processed SiO_2-rGO hybrid with enhanced physical properties. Journal of Solid State Chemistry. 2018;**265C**:364-371

[9] Sun X, Ye J, Pan F, Xu J, Cheng T, Wang X, et al. Hierarchical porous carbon obtained from frozen tofu for efficient energy storage. New Journal of Chemistry. 2018;**42**:12421-12428

[10] Balazs AC, Emrick T, Russell TP. Nanoparticle polymer composites: Where two small worlds meet. Science. 2006;**314**:1107-1110

[11] Yang X, Zhang X, Ma Y, Huang Y, Wang Y, Chen Y. Superparamagnetic graphene oxide–Fe_3O_4 nanoparticles hybrid for controlled targeted drug carriers. Journal of Materials Chemistry. 2009;**19**:2710-2714

[12] Stankovich S, Dikin DA, Dommett GHB, Kohlhaas KM, Zimney EJ, Stach EA, et al. Graphene-based composite materials. Nature. 2006;**442**:282-286

[13] Zhang Z, Pinnavaia TJ. Mesostructured γ-Al_2O_3 with a Lathlike framework morphology. Journal of the American Chemical Society. 2002;**124**:12294-12301

[14] Ahmad I, Cao H, Chen H, Zhao H, Kennedy A, Zhu YQ. Carbon nanotube toughened aluminium oxide nanocomposite. Journal of the European Ceramic Society. 2010;**30**:865-873

[15] Jagminas A, Kaciulis S, Klimas V, R'eza A, Mickevicius S, Soltani P. Fabrication of graphene-alumina heterostructured films with nanotube morphology. Journal of Physical Chemistry C. 2016;**120**:9490-9497

[16] Benavente R, Pruna A, Borrell A, Salvador MD, Pullini D, Penaranda-Foix F, et al. Fast route to obtain Al_2O_3-based nanocomposites employing graphene oxide: Synthesis and sintering. Materials Research Bulletin. 2015;**64**:245-251

[17] Liu Y, Ma D, Han X, Bao X, Frandsen W, Wang D, et al. Hydrothermal synthesis of microscale

boehmite and gamma nanoleaves aluminia. Materials Letters. 2008;**62**:1297-1301

[18] Chen YF, Lee CY, Yeng MY, Chiu HT. The effect of calcination temperature on the crystallinity of TiO_2 nanopowders. Journal of Crystal Growth. 2003;**247**:363-370

[19] Zhang Y, Li D, Tan X, Zhang B, Ruan X, Liu H, et al. High quality graphene sheets from graphene oxide by hot-pressing. Carbon. 2013;**54**:143-148

[20] Liu H, Choy KL, Roe M. Enhanced conductivity of reduced graphene oxide decorated with aluminium oxide nanoparticles by oxygen annealing. Nanoscale. 2013;**5**:5725-5731

[21] Wang SX, Zou KX, Qian YX, Deng YF, Zhang L, Chen G.H. et al. Insight to the synergistic effect of N-doping level and pore structure on improving the electrochemical performance of sulfur/N-doped porous carbon cathode for Li-S batteries. Carbon 2019;**144**:745-755

[22] Nan CW, Shen Y, Ma J. Physical properties of composites near percolation. Annual Review of Materials Research. 2010;**40**:131-151

[23] Kim JY, Lee WH, Suk JW, Potts JR, Chou H, Kholmanov IN, et al. Chlorination of reduced graphene oxide enhances the dielectric constant of reduced graphene oxide/polymer composites. Advanced Materials. 2013;**25**:2308-2313

[24] Zaretsky EB. High temperature impact response of 998 alumina. Journal of Applied Physics. 2013;**114**:183518

[25] Yoo E, Kim J, Hosono E, Zhou H-s, Kudo T, Honma I. Large Reversible Li Storage of Graphene Nanosheet Families for Use in Rechargeable Lithium Ion Batteries. Nano Letters. 2008;**8**:2277-2282

[26] Wang L, Zhang G, Zhang X, Shi H, Zeng W, Zhang H, et al. Porous ultrathin carbon nanobubbles formed carbon nanofiber webs for high-performance flexible supercapacitors. Journal of Materials Chemistry A. 2017;**5**:14801-14810

[27] Kang K, Meng YS, Br′eger J, Grey CP, Ceder G. Electrodes with high power and high capacity for rechargeable lithium batteries. Science. 2006;**311**:977-980

[28] Staaf L, Lundgren P, Enoksson P. Present and future supercapacitor carbon electrode materials for improved energy storage used in intelligent wireless sensor systems. Nano Energy. 2014;**9**:128-141

[29] Wang B, Zhang W, Wang L, Wei J, Bai X, Liu J, et al. Three dimensional carbon-bubble foams with hierarchical pores for ultra-long cycling life supercapacitors. Nanotechnology. 2018;**29**:275706

[30] Shopsowitz KE, Hamad WY, MacLachlan MJ. Chiral nematic mesoporous carbon derived from nanocrystalline cellulose. Angewandte Chemie, International Edition. 2011;**50**: 10991-10995

[31] Zhu Y, Murali S, Stoller MD, Ganesh K, Cai W, Ferreira PJ, et al. Carbon-based supercapacitors produced by activation of graphene. Science. 2011;**332**:1537-1541

[32] Sankar S, Lee H, Jung H, Kim A, Ahmed ATA, Inamdar AI, et al. Ultrathin graphene nanosheets derived from rice husks for sustainable supercapacitor electrodes. New Journal of Chemistry. 2017;**41**:13792-13797

[33] Sevilla M, Fuertes AB. Hierarchical microporous/mesoporous carbon nanosheets for high-performance supercapacitors. ACS Nano. 2014;**8**:5069-5078

[34] Sing KS. Reporting physisorption data for gas/solid systems with special reference to the determination of surface area and porosity. Pure and Applied Chemistry. 1985;**57**:603-619

[35] Zhou Q , Chang J, Jiang Y, Wei T, Sheng L, Fan Z. Fast charge rate supercapacitors based on nitrogen-doped aligned carbon nanosheet networks. Electrochimica Acta. 2017;**251**:91-98

[36] Yang G, Niu J, Shao R, Liang J, Dou M, Li Z, et al. A facile molecularly engineered copper (II) phthalocyanine as hole transport material for planar perovskite solar cells with enhanced performance and stability. Nano Energy. 2017;**36**:322-330

[37] Fang Y, Lv Y, Che R, Wu H, Zhang X, Gu D, et al. Two-dimensional mesoporous carbon nanosheets and their derived graphene nanosheets: Synthesis and efficient lithium ion storage. Journal of the American Chemical Society. 2013;**135**:1524-1530

[38] Hou J, Cao C, Idrees F, Ma X. Hierarchical porous nitrogen-doped carbon nanosheets derived from silk for ultrahigh-capacity battery anodes and supercapacitors. ACS Nano. 2015;**9**:2556-2564

Chapter 6

Creation and Evaluation of Atomically Ordered Side- and Facet-Surface Structures of Three-Dimensional Silicon Nano-Architectures

Azusa N. Hattori and Ken Hattori

Abstract

The realization of three-dimensional (3D)-architected nanostructures, that is, the transformation from novel two-dimensional (2D) film-based devices to 3D complex nanodevices, is of crucial importance with the progress of scaling down devices to nanometer order. However, little attention has been devoted to controlling the atomic ordering and structures of side-surfaces on 3D structures, while techniques for controlling and investigating 2D surfaces, namely, surface science, have been established only for planar 2D surfaces. We have established an original methodology that enables atomic orderings and arrangements of surfaces with arbitrary directions to be observed on 3D figured structures by developing diffraction and microscopy techniques. An original technique, namely, directly and quantitatively viewing the side- and facet-surfaces at the atomic scale by reflection high-energy electron diffraction (RHEED) and low-energy electron diffraction (LEED), can be used to determine process parameters in etching. This chapter introduces methods of evaluation by RHEED and LEED based on a reciprocal space map and methods of creating various atomically flat {111} and {100} side-surfaces of 3D Si nano-architectures and tilted {111} facet-surfaces fabricated by lithography dry and wet etching processes, followed by annealing treatment in vacuum.

Keywords: side-surface, facet-surface, atomically flat, Si, lithography, dry etching, wet etching, electron diffraction, reciprocal space map

1. Introduction

The significance of nanostructuring techniques has increased with the progress of scaling down devices to nanometer order in accordance with Moore's law [1]. The realization of three-dimensional (3D)-architected nanostructures, that is, the transformation from novel two-dimensional (2D) film-based planar devices to 3D complex and multifunctional nanodevices, is of crucial importance to future electronic applications [2, 3], the so-called More Moore. So far, various nanofabrication techniques have been proposed and developed with different levels of success. In many cases, materials grown on substrates are affected by substrate structural characteristics such as shape, roughness, and dimensionality. Three-dimensional patterned substrates prepared by a conventional lithographic method can be used to form 3D nanostructures. Although considerable attention has been devoted to controlling the size, shape, and positioning in research on 3D patterning, little attention has been focused on the atomic ordering of arbitrarily oriented surfaces on 3D patterned substrates. The realization of perfect surfaces on 3D structures is required to produce high-quality samples. Since material growth starts on a surface, the surface condition clearly determines the structural and physical properties of the grown material.

Although techniques for studying 2D surfaces, namely, surface science, have been intensively developed and established, little attention has been devoted to controlling the atomic ordering and structures of side-surfaces on 3D architectures. Fabrication techniques for 3D nanoscale structures that are promising for 3D integrated circuits have been individually developed from surface science.

The subject of study should be changed from 2D planar surfaces to 3D assembly surfaces to enable atomically ordered nanofabrication on vertical side-surfaces and/or tilted facet-surfaces in 3D space (**Figure 1**). For this purpose, a simple and accurate structure evaluation technique is required. Currently, scanning electron microscopy (SEM) is widely used for the observation of 3D nanostructures. However, SEM cannot be used to evaluate structures involving atomic ordering. On the other hand, transmission electron microscopy (TEM) is a powerful technique for examining the atomic structure of 3D nanomaterials, but it is destructive and not convenient. An alternative technique is a diffraction method. Reflection high-energy electron diffraction (RHEED) enables the examination of surface properties such as atomic ordering, surface roughness (flatness), and surface homogeneity [4]. Low-energy electron diffraction (LEED) is also a conventional and nondestructive surface observation technique [5]. To apply these techniques to 3D structured samples with various oriented surfaces, instead of 2D planar samples, an appropriate alignment of the incident electron beam considering the configuration in 3D space is indispensable [6–9].

In this chapter, we first show how to obtain and evaluate RHEED and LEED patterns from 3D structured samples by explaining the principle of diffraction. Then, we demonstrate creating and evaluating atomically ordered side- and facet-surface structures of 3D silicon nano-architectures. Finally, we discuss novel structures that have been constructed on 3D patterned Si to form 3D interconnected structures and their physical properties.

2. Theory (methodology)

2.1 Reciprocal space structures of 1D, 2D, and 3D crystals

Electron diffraction is one of the most powerful tools for investigating crystalline structures, particularly nanomaterials and material surfaces owing to their larger atomic scattering cross sections σ than those obtained with X-ray diffraction,

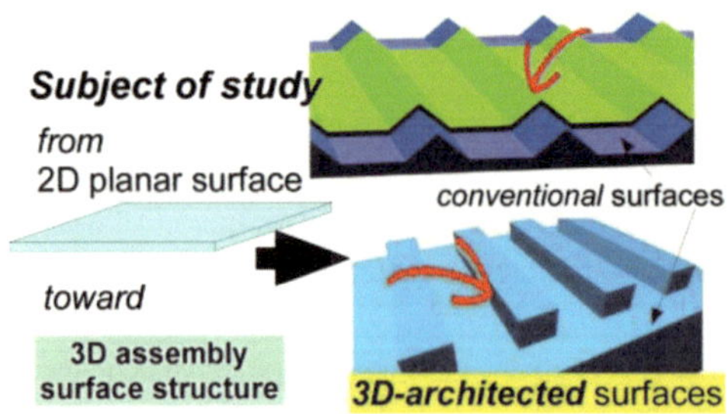

Figure 1.
Concept of our approach toward realizing atomically ordered 3D structures.

for instance, $\sigma \sim 4$ Mb for a 15 keV electron and 0.002 Mb for a Cu K ($\sim$8 keV) X-ray to a Si atom [10, 11]. According to the kinematic diffraction theory, a diffraction pattern can be understood as a reciprocal structure pattern reflecting a crystalline structure (**Figure 2**) in a reciprocal space map (RSM). By comparing an experimental reciprocal structure pattern in an RSM with patterns of candidate crystalline structures, we can analyze the characteristics of crystalline materials, such as dimensionality, atomic structure, orientation, size, and strain [4, 12–16].

For instance, an ideal one-dimensional (1D) material with lattice constant a in the x direction in real space (**Figure 2(a)**) shows a characteristic reciprocal structure pattern consisting of reciprocal lattice planes with a reciprocal lattice unit length of $a^* = 2\pi/a$ in the x direction (**Figure 2(b)**). This can be explained by the Bragg interference condition $a\cos\theta_{out} - a\cos\theta_{in} = n\lambda$, that is, $\left(\vec{k}_{out} - \left(\vec{k}_{in} + na^*\right)\right) \bullet \vec{e}_x = 0$. Here, λ is wavelength, and $\vec{k}_{in}$ and $\vec{k}_{out}$ are incident and outgoing wave-number vectors, respectively, satisfying $k_0 = \left|\vec{k}_{in}\right| = \left|\vec{k}_{out}\right| = 2\pi/\lambda$. Since the intersection of reciprocal lattice planes with an Ewald sphere of radius k_0 produces a diffraction pattern (**Figure 2(c)**), diffraction rings are observed as the diffraction patterns of 1D materials.

A 2D material with lattice constants a and b, that is, unit vectors $\vec{a}$ and $\vec{b}$ (**Figure 2(d)**), shows reciprocal lattice rods with reciprocal lattice unit vectors of $\vec{a}^*$ and $\vec{b}^*$, corresponding to the intersection of orthogonal reciprocal lattice planes (**Figure 2(e)**). The intersection of reciprocal lattice rods with an Ewald sphere produces a pattern of diffraction spots arrayed on arcs. A reciprocal lattice rod at position $h\vec{a}^* + k\vec{b}^*$ is called an $(h\ k)$ rod, which produces an $(h\ k)$ 2D diffraction

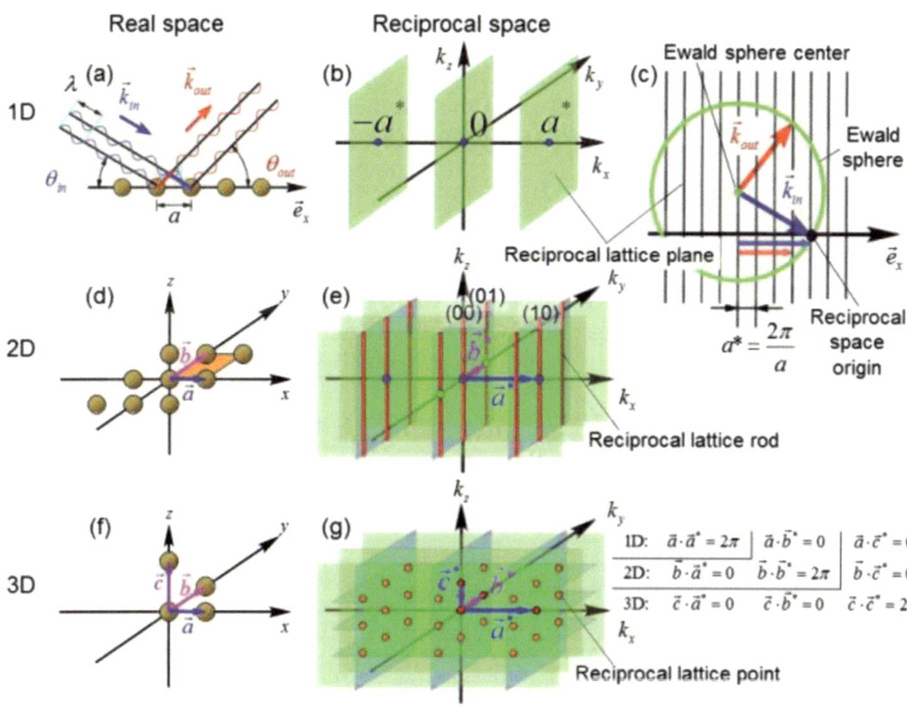

Figure 2.
Schematics of (a)–(c) 1D, (d) and (e) 2D, and (f) and (g) 3D crystals: real space structures (a), (d), and (f) and corresponding reciprocal space structures (b), (c), (e), and (g). 1D, 2D, and 3D crystals show reciprocal lattice planes, reciprocal lattice rods, and reciprocal lattice points, respectively.

spot. A 3D material with unit vectors $\vec{a}$, $\vec{b}$, and $\vec{c}$ (**Figure 2(f)**) shows reciprocal lattice points with unit vectors of $\vec{a}^*$, $\vec{b}^*$, and $\vec{c}^*$ (**Figure 2(g)**). The intersection of reciprocal lattice points with an Ewald sphere produces a pattern of diffraction spots arrayed on a lattice. A reciprocal lattice point at position $h\vec{a}^*$, $k\vec{b}^*$, and $l\vec{c}^*$ produces an $(h\ k\ l)$ 3D diffraction spot.

From diffraction patterns (e.g., diffraction rings or spots) projected on a detection screen, the original reciprocal structure pattern (e.g., reciprocal lattice planes, rods, or points) in a 3D RSM can be regenerated by changing the incident angle of the electron beam to the crystalline material (i.e., the direction of $\vec{k}_{in}$) or the beam energy (i.e., k_0), as illustrated in **Figure 3** for a 3D material with azimuth angle rotation [12–15]. For example, **Figure 4(a)** and **(b)** show a series of transmission electron diffraction patterns of a nanocrystal on a Si(001) substrate as a function of the azimuth angle around the substrate surface normal direction [14]. Indeed, the conversion of diffraction spots on a lattice (indicated by orange arrows) in different diffraction patterns generates reciprocal lattice points in a 3D RSM, as indicated by orange arrows in **Figure 4(c)–(e)**. In this 3D RSM, we can recognize the existence of a certain 3D crystalline structure, in this case the structure is α-$FeSi_2$, among candidate crystal structures [17, 18]; the crystalline orientations are α-$FeSi_2(110)\|$ Si(001) and α-$FeSi_2[001]\|$ Si$\langle 110\rangle$ from the reciprocal structure pattern.

In **Figure 4(a)** and **(b)**, there are also diffraction spots on arcs indicated by red arrows. These spots are often observed under a glancing condition of an electron beam nearly parallel to a substrate surface in RHEED with a primary energy E_p of typically 10–15 keV. The conversion of the spots generates reciprocal lattice rods perpendicular to the substrate surface in a 3D RSM, as indicated by red arrows in **Figure 4(c)–(e)**. The existence of spots on arcs or reciprocal lattice rods implies that the surface perpendicular to the rods is atomically well-ordered; in this case, it implies the formation of a well-defined clean Si(001) substrate surface.

The width of a reciprocal lattice rod is finite (full width at half maximum $\simeq 2\pi/D$) and reflects the crystalline domain size D. Thus, the intersection of the rod with an Ewald sphere is elongated along the rod direction, that is, the surface normal direction. Indeed, in **Figures 4(a)** and **(b)** and **5(a)**, there are the surface spots (red arrows) elongated along the Si[001] substrate surface normal direction.

2.2 Diffraction patterns from side-surfaces of 3D structured sample

One of the advantages of RHEED is its ability to observe atomically ordered surfaces in any direction, as described later in Sections 3.2 and 4. RHEED has been used to investigate crystalline structures on planar substrate surfaces. Recently, for

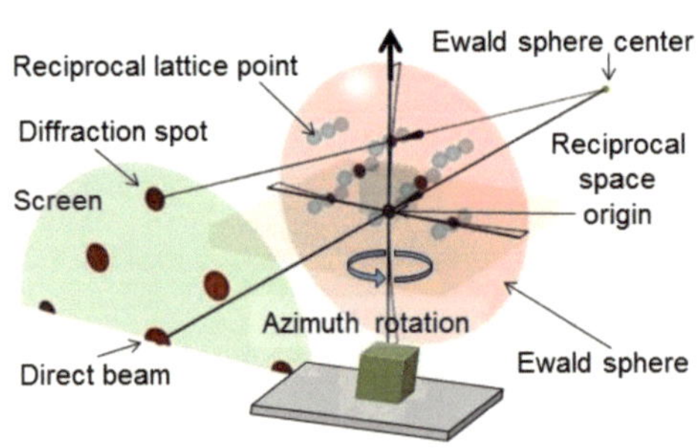

Figure 3.
Schematic of relationship between diffraction spots and reciprocal lattice points intersecting with a partial Ewald sphere in 3D RSM, for 3D crystal.

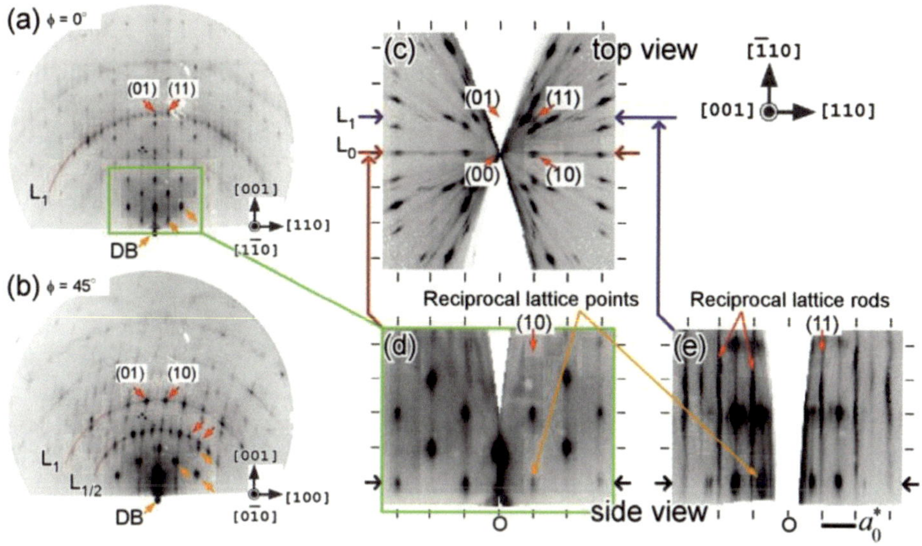

Figure 4.

(a) and (b) RHEED patterns of α-FeSi2(110)[001]|| Si(001)⟨110⟩ at azimuth angles ϕ = 0° and 45°, respectively. Ep is 15.0 keV ($k_0 = 62.7$ Å^{-1}). (c) Top view and (d) and (e) side views of the 3D RSM regenerated from a series of RHEED patterns. In (a) and (b), diffraction spots indicated by orange and red arrows are assigned to reciprocal lattice points of 3D α-FeSi2 nanocrystals and reciprocal lattice rods of a 2D Si(001) substrate surface, respectively. Reciprocal lattice rods [e.g., (00) and (01)] lie on Laue zones (e.g., L0 and L1). DB denotes the direct beam position. $a_0^* = 2\pi/a_0 \approx 2\pi/3.84$ $\left[\text{Å}^{-1}\right]$ is the reciprocal lattice unit length of Si(001)1×1.

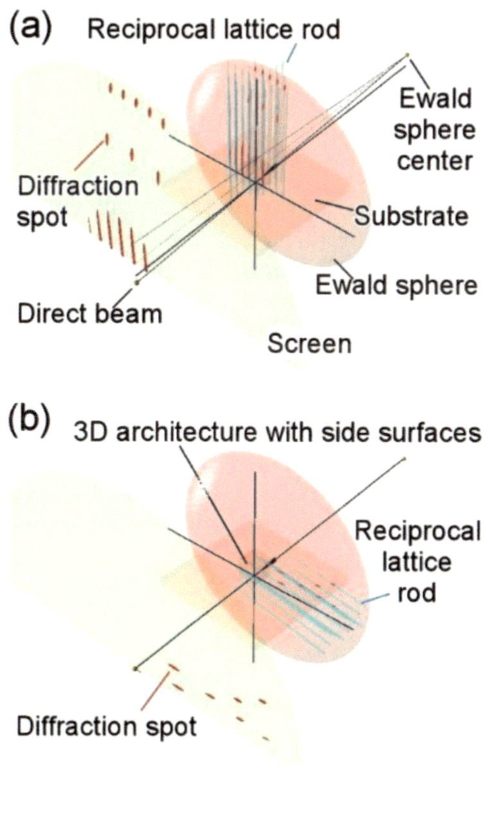

Figure 5.

Schematics of 2D diffraction spots and reciprocal lattice rods of (a) substrate surface and (b) side-surface of 3D fabricated material. In both cases, diffraction spots are elongated in the surface normal direction.

3D fabricated materials, the authors have demonstrated that not only substrate surfaces but also surfaces inclined from or perpendicular to a substrate plane exhibit atomically ordered structures where surface spots elongated along directions inclined from or perpendicular to the substrate normal direction appear in RHEED patterns (**Figure 5(b)**) [6, 7, 9]. We emphasize that a simple surface property of 3D fabricated materials, that is, a surface direction, can be confirmed by the elongation of surface spots.

Low-energy electron diffraction (LEED) with a typical E_p of 50–100 eV focuses the interference caused by the backward scattering of an electron incident to an atom, while RHEED focuses the interference caused by forward scattering. Both LEED and RHEED are sensitive to surface structures. A LEED pattern at the normal incidence corresponds to a top view of reciprocal lattice rods intersecting with an Ewald sphere from the surface normal direction. The diffraction spots move to the (00) spot arising from the (00) rod with increasing E_p [8, 18]. For an inclined surface, the arrangement of diffraction spots in alignment changes that in an arc, the center of which is the (00) spot. The diffraction spots also move to the (00) spot with increasing E_p. Thus, we can confirm the surface direction of 3D fabricated materials by the arrangement and E_p-dependent motion of diffraction spots.

3. Experimental procedure

The authors demonstrated the creation of atomically ordered side- and facet-surfaces owing to the arbitrary selection of planes (**Figure 6**). Atomically reconstructed Si{100}, {110}, and {111} vertical side-surfaces and {111} tilted facet-surfaces were first realized on 3D patterned Si substrates, and the perfection of the 3D surface structures was examined. The {100}2×1, {110}16×2, and {111} 7 ×7 diffraction spots from the 3D surfaces were confirmed by RHEED and LEED.

3.1 Sample fabrication with combination of dry and wet etching processes

The 3D architected Si samples with arbitrary faces discussed here were produced on commercial mirror-polished Si substrates by the following dry and wet etching procedures. All processes were performed at room temperature (RT). First, a line mask pattern was drawn using a photoresist. To produce the vertical side-surfaces (**Figure 7(a)**–**(c)**), Si was etched in an inductive coupled plasma (ICP)-reactive ion etching system (RIE-400iPB, Samco). The process parameters were an ICP source power of 300 W, a bias power of 10 W, and a working pressure of 4 Pa. Mixture gases of 10 sccm SF_6, 5 sccm O_2, and 200 sccm Ar were used in the etching cycle, and 40 sccm C_4F_8, 5 sccm O_2, and 200 sccm Ar were used in the passivation process [6–9]. Depending on dry RIE conditions, various curved structures from trapezoid to triangle shapes were obtained. When we produced tilted facet-surfaces (**Figure 7(d)** and **(e)**), the optimized conditions for a triangle-like shape were a

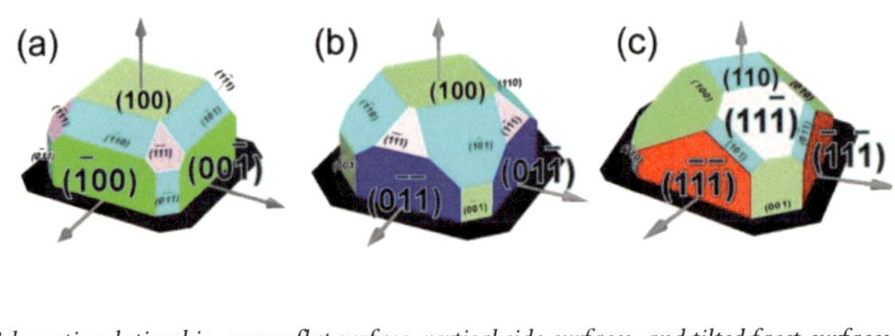

Figure 6.
(a)–(c) Schematic relationship among flat surface, vertical side-surfaces, and tilted facet-surfaces.

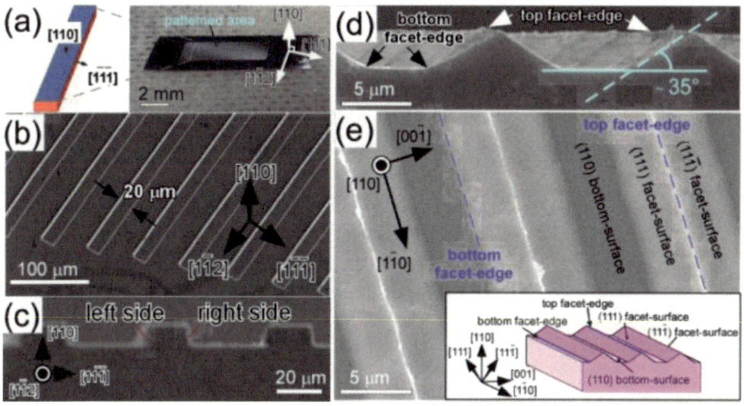

Figure 7.
(a) Photograph of a 3D patterned Si(110) substrate consisting of {111} vertical side-surfaces. (b) Top and (c) cross-sectional SEM images of a patterned area. (d) Cross-sectional and (e) top SEM images of a sample with tilted {111} facet-surfaces.

flow rate of 51 sccm, a pressure of 4.0 Pa, and a bias power of 60 W using an RIE system (RIE-10 NR, Samco) with SF_6 gas [9].

After dry etching, wet etching was performed to reduce the side- and facet-surface roughnesses. The base solution used in the wet process removes the damage and contaminants [19–21], and flash annealing promotes reconstruction through the sublimation of the contaminants [22]. Unoptimized conditions never led to reconstructed patterns on side- or facet-surfaces. In this process, the etching recipe was optimized by considering the plane-dependent etching properties [23]. A {100} sample was dipped in 25 wt% tetramethylammonium hydroxide (TMAH) at 70°C for 3 min. A {110} sample was etched in 25 wt% TMAH with 0.1 vol% surfactant (iso-octylphenoxy polyethoxyethanol) at 70°C for 1 min [7]. Then, the sample was rinsed with pure water, dried by blowing with N_2, and introduced into an ultra-high vacuum (UHV) chamber. The 3D Si sample was degassed and flashed by direct-current heating at ~1150–1200°C at a pressure below 2×10^{-8} Pa.

3.2 RHEED observation of 3D structured samples

RHEED patterns were obtained at RT using an electron beam with an E_p of 15 keV and a diameter of ~0.5 mm. Au advantage of the RHEED technique is the accessibility of various 3D surfaces; that is, the diffraction pattern consists of all diffractions from the substantial crystal surfaces in 3D space. Actually, the selection of the incident e-beam direction enables the identification of 3D surface structures. In this study, the direction of the incident electrons was defined by the glancing angle θ and azimuth angle ϕ for the top and bottom Si surfaces (**Figure 8**). The RHEED patterns were also obtained from the 3D samples with vertical side-surface or tilted facet-surfaces by controlling θ and ϕ. For instance, for 3D Si with a vertical {111} side-surface, θ is the angle between the incident electron beam and the top (110) surface, which was changed from −0.4° to +1.2°. ϕ, which is defined as the angle from $[1\bar{1}2]$ in the in-plane direction, was changed from −4.5° to +4.7°. The RHEED patterns were filtered by a computer to emphasize the spot features in the background. LEED can also be used to observe 3D surface ordering, while 3D accessibility is less than that of RHEED because of the limited range of diffraction conditions. LEED patterns were also observed using a typical E_p of 40–200 eV and a beam size of ~1 mm.

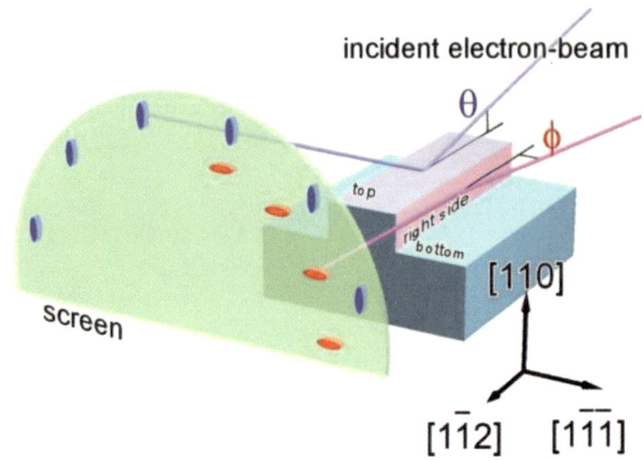

Figure 8.
Schematic of the diffraction from the top- and side-surfaces of a 3D patterned Si(110) substrate. The control of the incident electron beam direction defined by θ and ϕ enables the observable faces in the 3D space to be selected.

4. Surface structures on 3D architected Si sample

4.1 {111} vertical side-surfaces

Figure 9(b)–(d) show typical filtered RHEED patterns obtained from 3D Si with a vertical {111} side-surface (**Figure 7(c)**) various θ and ϕ values. The {111} side-surfaces were produced on a (110) surface (**Figure 6(c)**). The RHEED patterns showed curious characteristics, because half of the patterns were removed, which has not been reported to the best of our knowledge. The appearance of the left-side

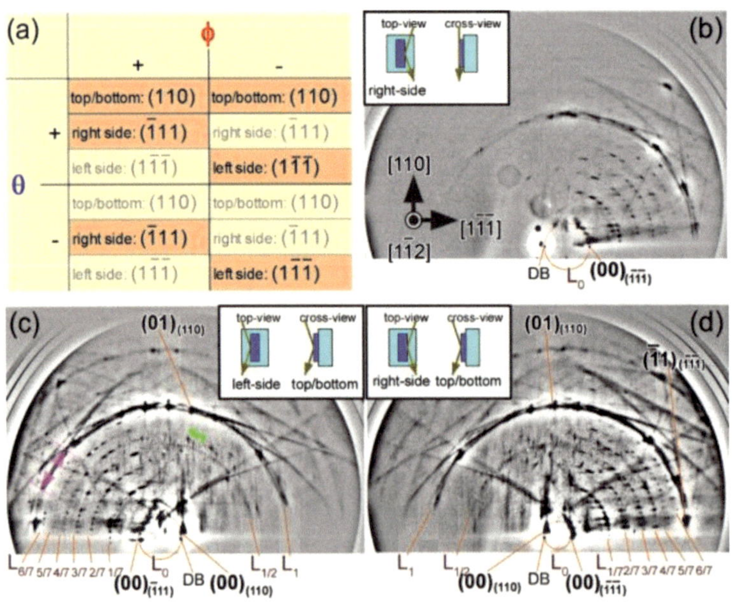

Figure 9.
(a) Dependence of observable conditions (orange) in the RHEED pattern on the polarity of θ and ϕ. (b)–(d) RHEED patterns from 3D Si with {111} vertical side-surfaces observed at (b) θ = 0.0° and ϕ = +1.9°, (c) θ = +0.3° and ϕ = −1.6°, and (d) θ = +0.3° and ϕ = +1.1°. Insets schematically show the relationship between the incident electron beam and the observable surfaces in 3D space.

and right-side Si{111}7×7 patterns depends on ϕ, as shown in **Figure 9(c)** and **(d)**. ϕ is the azimuth angle for the Si(110) top-/bottom-surfaces and simultaneously corresponds to the glancing angle for the Si{111} side-surfaces. On the other hand, θ is the glancing angle for Si(110) and is also the azimuth angle for Si{111}. **Figure 9(a)** summarizes the dependence of the observable and non-observable conditions in the RHEED patterns on the polarity of θ and ϕ. When θ (the glancing angle for Si(110)) decreases, the RHEED pattern from Si(110) disappears, while a pattern from Si{111} is present. Indeed, in **Figure 9(b)** (θ = 0.0° and ϕ = +1.9°), a quarter circle 7×7 pattern with faint 2×16 spot can be seen. The diffraction spots on the left and the right sides (**Figure 7(c)**) are slightly elongated in the horizontal direction. In general, the elongation (streaky) direction corresponds to the surface normal direction [4], as illustrated in **Figure 8**. Thus, these slightly streaky spots indicate the existence of vertical side-surfaces.

Let us analyze the curious RHEED patterns in more detail. In **Figure 9(c)** (θ = +0.3° and ϕ = −1.6°) and **Figure 9(d)** (θ = +0.3° and ϕ = +1.1°), diffraction spots from the direct beam (DB) can be observed in the 1/7th-order Laue zones ($L_{1/7}$-$L_{6/7}$) on the left and right quarter sides, respectively, as well as seven spots within the Kikuchi band width (e.g., indicated by a pink arrow in **Figure 9(c)**). These patterns clearly correspond to Si{111}7×7 reconstruction [4], having shadow edges in the horizontal and vertical directions. **Figure 9(c)** (**Figure 9(d)**) corresponds to Si$(\bar{1}11)$ (Si$(1\bar{1}\bar{1})$) 7×7 diffraction on the surface of the left-side (right-side) wall of the 3D patterned structure. Note that a specular spot $(00)_{(\bar{1}11)}$ ($(00)_{(1\bar{1}\bar{1})}$) from the DB appears on the left (right) side. In addition, strong Kikuchi lines and bands were observed in the side RHEED pattern. These results indicate that atomically flat side-surfaces were achieved on 3D patterned Si(110) by etching and UHV annealing.

One can see the characteristic ϕ and θ dependences of the RHEED patterns (movie) in Supporting Information of Ref [5]. These RHEED patterns clearly show that all the surfaces on 3D Si, that is, the (110) top-/bottom-surfaces, the $(1\bar{1}\bar{1})$ right-side surface, and the $(\bar{1}11)$ left-side surface, have atomically ordered structures.

Figure 10 shows the LEED patterns observed from the figured $(\bar{1}11)$ side-surface. Because in LEED we observe backscattering diffraction, while RHEED reflects forward scattering diffraction, a unique 3D Si sample with a wider side-surface was prepared [9]. A clear 7×7 pattern can be seen in **Figure 10(a)**, where the incident electron beam is along a direction almost normal to the $(\bar{1}11)$ side-surface. Characteristic LEED patterns were observed when the sample was rotated; the electron beam probed both a $(\bar{1}11)$ side-surface and a (110) top-surface. An example with an incident angle of ~37° is shown in **Figure 10(b)**, which reflects surface reconstructions for both the figured $(\bar{1}11)$7×7 side-surface and the pristine (110) 16×2 top-surface [6].

The LEED patterns are in good agreement with those simulated by considering the crystal orientation and electron beam (i.e., the cross sections of the reciprocal lattice rods and an Ewald sphere [22]) shown in the lower-right panels of **Figure 10**. It is possible to identify diffraction spots on Laue zones of both $(\bar{1}11)$ and (110) surfaces in the patterns, demonstrating the observation of atomic crystalline ordering for the 3D surfaces using LEED.

4.2 {100} vertical side-surfaces

Figure 11(a) shows a typical filtered RHEED pattern obtained from 3D Si with {100} side-surfaces. The {100} side-surfaces were created on a commercial mirror-polished Si(100) substrate (**Figure 6(a)**). In **Figure 11(a)** (θ = +0.4° and ϕ = −1.2°), the two overlapping Si{100}2×1 reconstructed diffraction patterns [4, 24] originated

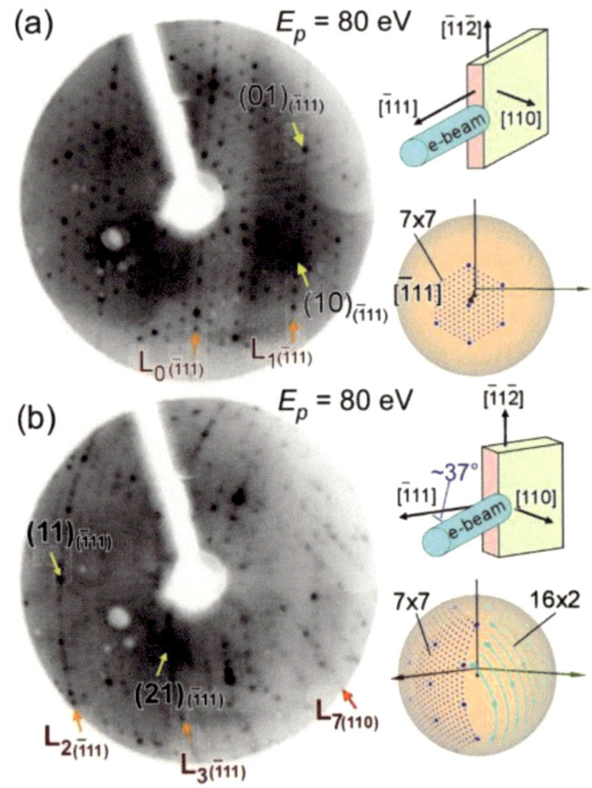

Figure 10.
LEED patterns for 3D Si sample with a wider ($\bar{1}11$) side-surface at different incident angles of the electron beam with Ep = 80 eV: (a) normal incidence to the Si($\bar{1}11$) side-surface and (b) ~37° tilted from the normal direction. The relationship between the incident beam and the sample and the simulated LEED patterns are shown in the upper and lower right, respectively.

from the top/bottom and left-side surfaces. One is a (100) 2×1 pattern with a semicircular shape having a shadow edge $SE_{(100)}$ in the horizontal direction. The other is a (001) 2×1 pattern on the left quarter side with a shadow edge $SE_{(100)}$ in the vertical direction.

We can see that the diffraction spots on the left side are slightly elongated in the horizontal direction, which is a characteristic of a RHEED pattern from a side-surface [6]. The diffractions from the top-/bottom-surfaces disappeared and those from the left-side surface remained when we changed θ from +0.4° to 0.0°. When we set θ = +0.0° and ϕ = −1.2°, only a quarter circle 2×1 pattern with strong Kikuchi lines and bands can been seen, corresponding to the relationship between θ and ϕ, that is, the glancing and azimuth angles for each surface, respectively, similarly to that observed in **Figure 9(a)**. Quarter circle 2×1 patterns were also observed on the right when ϕ was positive.

The left part of **Figure 11(b)** shows the reciprocal lattice of the Si{100}2×1 reconstruction, corresponding to the left-side and top-/bottom-surfaces of the {100} sample. The 2D reciprocal lattice of Si(001) is perpendicular to that of Si(100) with the common crystalline direction of $[0\bar{1}0]$. The Ewald constructions on the (100) top/bottom and (001) left-side surfaces in 3D reciprocal space in our experimental setup, namely, the simulated RHEED patterns at θ = +0.4° and ϕ = −1.2° (**Figure 11(b)** right panel), show good agreement with the observed RHEED pattern (**Figure 11(a)**). These results indicate that atomically flat side-surfaces were achieved on the {100} sample.

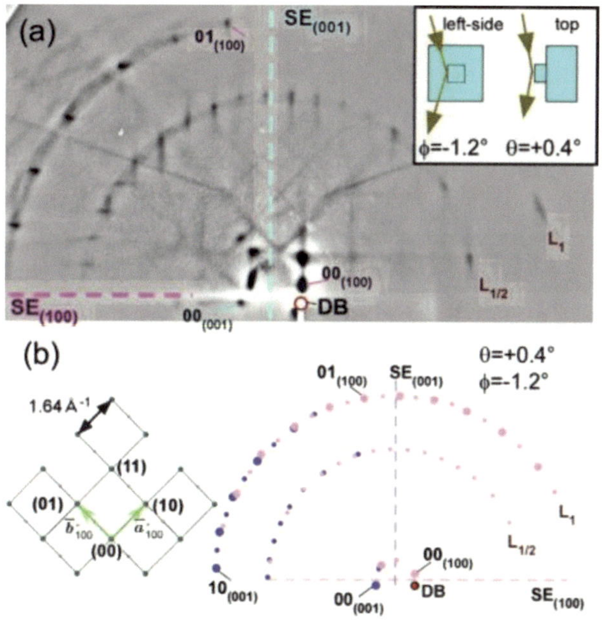

Figure 11.
(a) RHEED pattern from the 3D Si sample with vertical {100} *side-surfaces at* $\theta = +0.4°$ *and* $\phi = -1.2°$*. The inset shows the relationship between the incident electron beam and the 3D surfaces. (b) Schematics of the 2D reciprocal lattices on* {100}2×1 *reconstruction (left) and simulated RHEED pattern from the* {100} *surfaces at* $\theta = +0.4°$ *and* $\phi = -1.2°$*.*

4.3 {110} vertical side-surfaces

Finally, the RHEED pattern from the vertical {110} side-surface is shown in **Figure 12(a)**. Here, a {110} side-surface was produced on a 2D(100) substrate (**Figure 6(b)**). **Figure 12(a)** shows the RHEED pattern from a {110} sample at $\theta = +0.4°$ and $\phi = +1.1°$. A {110}16×2 pattern can be seen in the right quarter, corresponding to the diffraction from the right-side $(0\bar{1}1)$ surface.

The weak spots in the arcs from the DB in the upper right, some of which are indicated by blue arcs, can be assigned to the Laue zone in the $(0\bar{1}1)$ plane ($L_0^{(0\bar{1}1)}$, $L_{1/16}^{(0\bar{1}1)}$, ……, $L_1^{(0\bar{1}1)}$) [24, 25] as shown in **Figure 12(c)**. The horizontal streaks from $L_{1/16}^{(0\bar{1}1)}$ to $L_1^{(0\bar{1}1)}$ represent the diffractions from (00) to (10) at the reciprocal lattice rods. Here, the reconstructed surface unit vectors from Si(110)2×16 are defined as $\vec{a}^*_{s\,110} = 1/2\vec{a}^*_{110}$ and $\vec{b}^*_{s\,110} = 1/16\vec{a}^*_{110} + 1/16\vec{b}^*_{110}$ (domain A, purple points) and $\vec{a}^*_{s\,110} = 1/2\vec{a}^*_{110}$ and $\vec{b}^*_{s\,110} = -1/16\vec{a}^*_{110} + 1/16\vec{b}^*_{110}$ (domain B, green points), as shown in **Figure 12(a)**. In **Figure 12(a)**, domain A is only considered as a $(0\bar{1}1)$ eye guide. The incident electron direction was the [011] direction, and the first Laue zone arc of Si(100) of $L_1^{(100)}$ (**Figure 12(b)**) is the same as $L_1^{(0\bar{1}1)}$ (**Figure 12(d)**).

4.4 {111} tilted facet-surfaces

Figure 13(a) shows a typical filtered RHEED pattern obtained from the {111} facet sample at $\theta_g = +0.7°$ and $\phi = +2.7°$ after flashing in UHV. The RHEED pattern showed notable characteristics, consisting of tilted 7×7 spots (some are marked by yellow circles, and 1/7th-order Laue zones, $L_{1/7}$-$L_{6/7}$, are recognized) and

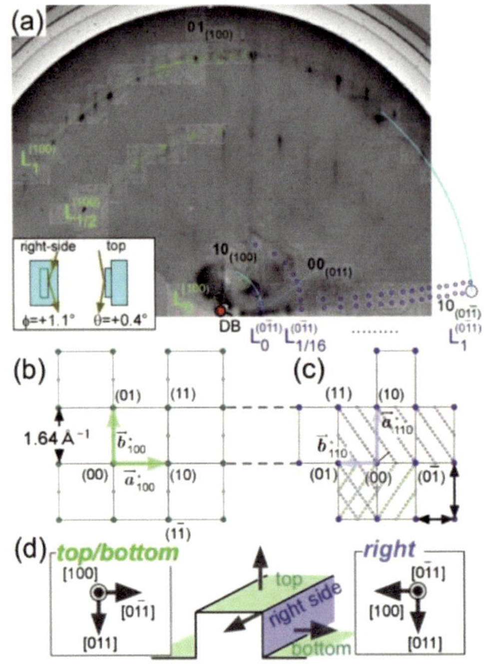

Figure 12.
(a) RHEED pattern from the {110} vertical sample observed at θ = +0.4° and ϕ = +1.1° with eye guides of Laue zone. Schematics of 2D reciprocal lattices on (b) Si(100)2×1 and (c) Si(0$\bar{1}$1)16×2, corresponding to the top/bottom and right-side surfaces, respectively, as shown in (d).

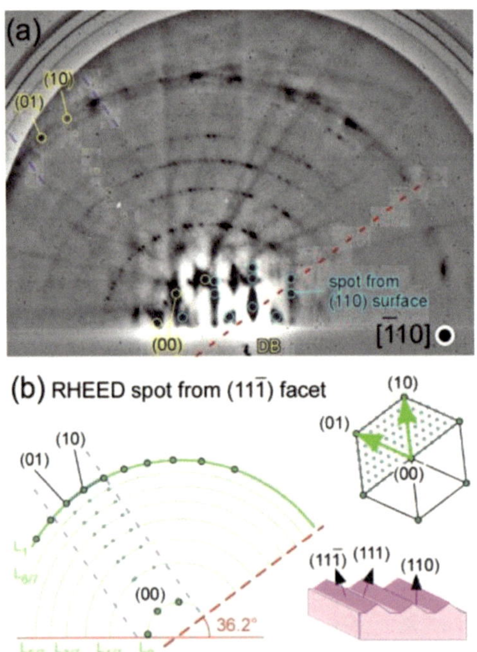

Figure 13.
(a) RHEED pattern for the {111} facet sample at θ = +0.7° and ϕ = +2.7°. The pattern consists of a (11$\bar{1}$)7×7 pattern tilted ~36° in the counterclockwise direction (yellow circles) and a faint 16×2 pattern (cyan circles). (b) Simulated RHEED pattern (left) reflecting geometric relationship (bottom right). The upper right figure represents the corresponding 2D reciprocal lattice normal to the facet direction.

faint horizontal 16×2 spots (cyan circles). The 7×7 pattern was tilted in the counterclockwise direction around DB, and the tilt angle was ~36°, which is consistent with θ_F.

Figure 13(b) shows a simulated RHEED pattern from the (111) facet-surface, corresponding to Ewald sphere cross sections of 2D 7×7 reciprocal lattice rods of the tilted Si(111). The excellent agreement between **Figure 13(a)** and **(b)** indicates the creation of an atomically ordered Si(111)7×7 facet-surface. The fabricated 3D structure shows that the relative glancing and azimuth angles to the tilted Si(111) facet-surface are 2.1 and 1.8°, respectively, under this condition. These RHEED patterns clearly show the existence of three different atomically ordered surfaces on the {111} facet sample, that is, the (110) bottom, (111) facet, and $(11\bar{1})$ facet-surfaces. One can see the characteristic ϕ dependence of the RHEED patterns in Ref [9].

5. Application of 3D architected Si

The creation and observation methods for well-defined surfaces enable the epitaxial growth of an arbitrary geometry, a key technique for nanoconstruction in 3D space [26–31]. Therefore, our established methodology contributes to the realization of well-ordered 3D nanofabrication, where the material stacking direction can be perfectly switched between the out-of-plane and in-plane directions. Novel 3D nanostructures are also expected to help unveil the underlying 3D surface science phenomena. Finally, two demonstrations utilizing a 3D architected Si platform are shown.

5.1 Platform for material growth on 3D surfaces

Atomically well-defined side-surfaces on a substrate can make an enormous contribution to nanofabrication [26–31]. To demonstrate the applicability of material growth on such side-surfaces, we produced Si{111}-Fe and Si{111}-Ag reconstructed side-surface structures on 3D Si with vertical {111}7×7 side-surfaces. Ag and Fe layers with thickness of 1.0 and 0.4 nm were deposited on the $(\bar{1}11)$ left-side and $(1\bar{1}\bar{1})$ right-side surfaces, respectively, and the sample was subsequently annealed at 500°C in UHV.

Figure 14(a) and **(b)** show typical RHEED patterns obtained from the left-side and right-side surfaces, respectively. We can confirm $\sqrt{3} \times \sqrt{3}$ (streaky) spots in L_0 (orange arrows) in **Figure 14(a)**. In **Figure 14(b)**, we can confirm 2×2 (streaky) spots (orange arrows), showing the formation of c-FeSi [18]. Simultaneously, $Si(\bar{1}11)\sqrt{3} \times \sqrt{3}$-Ag [4, 24, 31, 32] was obtained on the left-side surface. These results show that highly developed thin-film formation techniques are applicable for the vertical side-surface of 3D patterned substrates and the material stacking direction can be perfectly switched between the out-of-plane and in-plane directions.

Figure 14(c) and **(d)** show cross-sectional TEM images of the 0.4-nm-thick Fe layer deposited on the $(1\bar{1}\bar{1})$ right-side surface and the 5.0-nm-thick Ag deposited on the $(\bar{1}11)$ left-side surface. We can see four and five MLs of α-Fe [18] on the $(1\bar{1}\bar{1})$ right-side surface (**Figure 14(d)**), where a smooth in-plane heteroepitaxial interface with a length of 50 nm or more was formed between α-Fe and Si. The orientation relations between the α-Fe and the Si right-side surface are $(1\bar{1}\bar{1})_{Fe} \| (1\bar{1}\bar{1})_{Si}$ and $[\bar{1}12]_{Fe} \| [\bar{1}12]_{Si}$, similar to those indicated in previous reports on Fe on a 2D Si(111) surface [18, 33, 34]. A cross-sectional TEM image for Ag deposited on the $(\bar{1}11)$

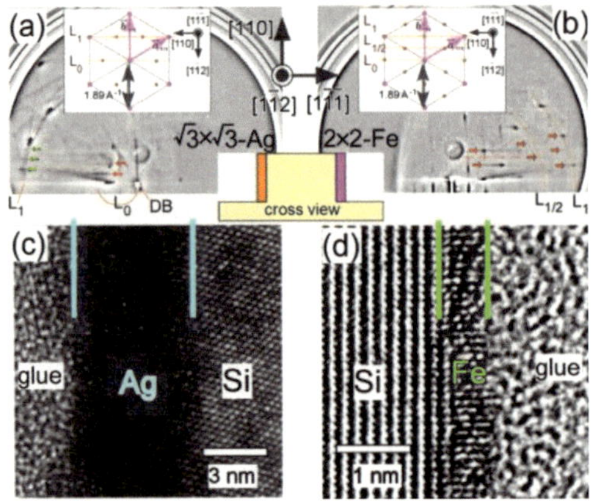

Figure 14.
RHEED patterns obtained from (a) Si($\bar{1}11$)-Ag left-side surface and (b) Si($1\bar{1}\bar{1}$)-Fe right-side surface annealed at 773 K in UHV. Spots in the Lo Laue zone indicated by orange arrows correspond to the $\sqrt{3} \times \sqrt{3}$ reciprocal lattice rods in (a), and spots indicated by orange arrows correspond to the 2×2 superstructure in (b). Cross-sectional TEM images of (c) Ag-deposited and (d) Fe-deposited side-surfaces at RT.

left-side surface (**Figure 14(c)**) showed that fcc-Ag epitaxially grew with an atomically matched interface, Ag($\bar{1}11$) || Si($\bar{1}11$), without any visible defects or dislocations.

Our results clearly show that a coherently grown (ultra) thin film was realized on the vertical side-surface with the growth alternating between the out-of-plane and in-plane directions. The siliciding reaction can be controlled on the side-surfaces. Therefore, highly developed thin-film formation techniques are applicable for the vertical side-surface of 3D patterned substrates, and the material stacking direction can be perfectly switched between the out-of-plane and in-plane directions.

5.2 Novel conductivity on interconnected {100} vertical side-surfaces

Three-dimensional integrated circuits, which contain multiple layers of active devices, have the potential to dramatically enhance chip performance, functionality, and device packing density. Recent 3D structured field-effect transistors (FETs) have surfaces with different orientations; for instance, a fin-type tri-gate structure has one top-surface and two side-surfaces [35, 36].

The electric connection between metal wires on these surfaces—that is, the wiring interconnects at sharp edges of top- and side-surfaces—is one of the issues in the development of 3D devices. Although the conductivity in metal wires on isolated 2D planar or side-surfaces has been well discussed, there are no reports on the metal conductivity of interconnections between 3D surfaces with different orientations, probably owing to the difficulty in measuring the intrinsic conductivity in 3D angular interconnects, which is mainly caused by diffuse scattering on rough surfaces [37–48]. One of the most outstanding factors contributing to the conductivity in the 3D angular interconnects is the facet-edge, which is a boundary of two surfaces with different crystalline orientations. To extract an intrinsic conductive property for metal wires interconnected at facet-edges, atomically flat surfaces in 3D structured substrates are required for the evaluation of the 3D interconnect resistivity by eliminating extra factors, such as roughness. To elucidate the impact of the 3D geometric effect on the conductive property, angularly interconnected Au wires with two configurations, crossing over and parallel to the facet-edges, were produced on atomically flat facet {111} surfaces (**Figure 13**).

Figure 15(a) shows a typical cross-sectional SEM image of a {111} facet sample on which a 10-nm-thick Au layer was deposited. The SEM image indicates that Au

DOI: http://dx.doi.org/10.5772/intechopen.92860

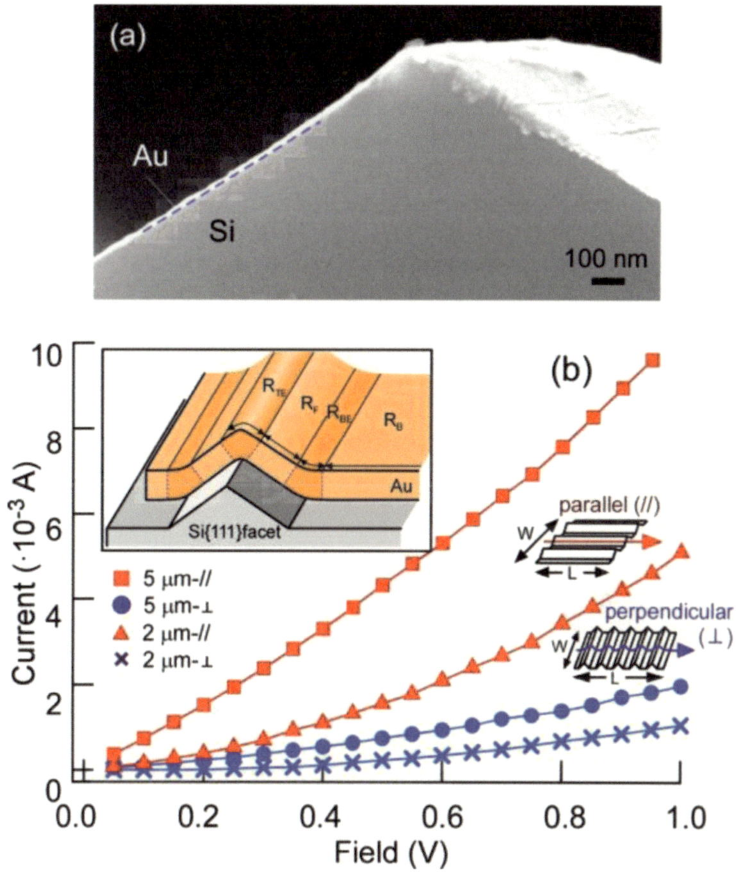

Figure 15.
(a) Typical SEM image for a facet sample on which a 10-nm-thick Au layer was deposited. (b) Current vs. voltage curves of Au wires with the channel area (W×L) of 2×100 and 5×100 μm² at 100 K in the parallel (red) and perpendicular (blue) configurations (insets). The upper left inset shows a schematic of the Au film on the Si{111} facet sample. The Au film is assumed to consist of bottom-surface (B), facet-surface (F), top facet-edge (TE), and bottom facet-edge (BE) regions with resistivity of R_B, R_F, R_{TE}, and R_{BE}, respectively.

was uniformly grown on the Si facet-surface with a smooth interface without any depressions or protrusions. No breaks or discontinuity of the Au wires was observed in SEM images, even at the top facet-edges. **Figure 15(b)** shows current-voltage properties at 100 K for 3D angularly interconnected Au wires of 2 μm width (circles) and 5 μm width (squares) in the parallel (red) and perpendicular (blue) configurations (insets). The length of the Au wires was 100 μm. Both perpendicular wires have an approximately one-order larger resistance (smaller current) than the parallel wires of the same width. For the other wire widths, the resistance in the perpendicular configuration was also larger than that in the parallel configuration.

The simplified resistance ratio, defined as the inverse of the current ratio at a fixed voltage, was 3–10 for the measured wire widths. Note that the dimensions (cross section and length) and crystallinity of the Au film were almost the same in the parallel and perpendicular configurations. Thus, the significant anisotropic resistance in these configurations is ascribed to the properties of the geometric shapes of the Au wires, that is, 3D angular interconnects. Indeed, the perpendicular wires have more facet-edges than the parallel wires; the perpendicular channel crossed over 11 top and 23 bottom facet-edges, while the parallel channel crossed over one top and two bottom facet-edges for a wire width of 5 μm.

Au wires on the 3D facet structure can be separated into four regions: flat bottom-surface, flat facet-surface, convex top facet-edge, and concave bottom facet-edge regions with resistances of R_B, R_F, R_{TE}, and R_{BE}, respectively, as shown in the inset of **Figure 15(b)**. The geometries of crystal grain boundaries in the polycrystalline Au wire may increase the electrical resistance [49]. The aggregation of crystal grain boundaries in the interconnect region, leading to the reduction in effective cross-sectional area, was attributed to the large α ($\approx$30–180), i.e., $R_B = R_F \sim \alpha R_{TE}, \sim \alpha R_{BE}$.

We found that the conductivity passing through the interconnects is sensitive to the alignment of the facet-edges in the electric path, and the series configuration showed a 3–10-fold larger resistance than the parallel configuration, which originated from the increased resistivity across the facet-edges by a factor of ~30–180, as calculated from the circuit model. This work provides a fundamental understanding of the impact of the 3D angular interconnects of a metal wire on electric transport and guidelines for the comprehensive investigation of the intrinsic interconnect transport properties on 3D structures, which is expected to produce critical benefits in the semiconductor industry.

6. Conclusion

This chapter demonstrated the creation and evaluation of atomically ordered side-surfaces and inclined facet-surfaces in nanofabricated 3D Si architectures toward the realization of sequentially 3D integrated and stacked devices for More Moore [3]. Atomically flat and well-ordered 3D structured surfaces play an important role in creating high-performance films on arbitrarily oriented 3D surfaces, similar to films grown on ordinary planar substrates. We emphasized that electron diffraction techniques such as RHEED and LEED are convenient for evaluating atomically ordered 3D surfaces, while conventional SEM has no atomic resolution. The points for recognizing 3D surfaces in diffraction patterns were explained along the basic concept of diffraction in reciprocal space; the intersection of reciprocal lattice rods with an Ewald sphere led to diffraction spots on arcs, and the spots elongate in the surface normal direction.

By diffraction, we evaluated several systems with side- or facet-surfaces of Si{111} or {100} on Si(110) or Si(001) substrates, which were fabricated by lithography dry and wet etching processes, followed by annealing in vacuum. Metal deposition on the well-ordered 3D surfaces (followed by annealing) also produced well-ordered films of nanometer thickness fabricating with arbitrary orientations. This is one of the fundamental techniques for defect-free material construction on 3D architectures. The electrical conductivity of metal wires on atomically flat 3D facet-surfaces crossing facet-edges was also measured toward the design of nanoscale steric wiring. We consider that these techniques for creating and evaluating 3D surfaces are promising for the realization of future 3D architecture devices.

Acknowledgements

The authors appreciate Dr. Shohei Takemoto, Dr. Haoyu Yang, Prof. Hiroshi Daimon, and Prof. Hidekazu Tanaka for their contribution to these studies. We also thank Ms. Saeko Tonda, Ms. Michiko Sakuma, Mr. Shoichi Sakakihara, and Mr. Takeshi Ishibashi for their helpful support in the fabrication of 3D Si samples and Ms. Liliany N. Pamasi and Mr. Ken Maetani for their support in the diffraction measurement of Si samples. This work was partially supported by Adaptable and

Seamless Technology Transfer Program through target-driven R&D (A-STEP) from the Japan Science and Technology Agency (JST) (No. JPMJTM19CM), Japan Society for the Promotion of Science (JSPS) Grant-in-Aid for Scientific Research B (Nos. 18H01871 and 20H02483), the Nanotechnology Platform Project (Nanotechnology Open Facilities in Osaka University, Nos. F-16-OS-0012 and F-16-OS-0016), and the Research Program of "Dynamic Alliance for Open Innovation Bridging Human, Environment and Materials" in "Network Joint Research Center for Materials and Devices" (No. 20203017).

Author details

Azusa N. Hattori[1] and Ken Hattori[2]*

1 Institute of Scientific and Industrial Research, Osaka University, Ibaraki, Osaka, Japan

2 Graduate School of Science and Technology, Nara Institute of Science and Technology, Ikoma, Nara, Japan

*Address all correspondence to: khattori@ms.naist.jp

References

[1] Moore GE. Cramming more components onto integrated circuits. Electronics. 1969;**38**:114. DOI: 10.1109/N-SSC.2006.4785860

[2] Zheng H, Wang J, Lofland SE, Ma Z, Mohaddes-Ardabili L, Zhao T, et al. Multiferroic $BaTiO_3$-$CoFe_2O_4$ nanostructures. Science. 2004;**303**:661. DOI: 10.1126/science.1094207

[3] International Roadmap for Devices and Systems. Available from: https://irds.ieee.org/

[4] Ichimiya A, Cohen PI. Reflection High-Energy Electron Diffraction. Cambridge: Cambridge University Press; 2004. DOI: 10.1017/CBO9780511735097

[5] Van Hove MA, Weinberg WH, Chan C-M. Low-energy electron diffraction: Experiment, theory and surface structure determination. In: Springer Series in Surface Sciences 6. Berlin: Springer; 1986. DOI: 10.1007/978-3-642-82721-1

[6] Hattori AN, Hattori K, Takemoto S, Daimon H, Tanaka H. Creation of atomically flat Si{111}7×7 side-surfaces on a three-dimensionally-architected Si(110) substrate. Surface Science. 2016; **644**:86. DOI: 10.1016/j.susc.2015.09.002

[7] Hattori AN, Takemoto S, Hattori K, Daimon H, Tanaka H. Methods of creating and observing atomically reconstructed vertical Si{100}, {110}, and {111} side surfaces. Applied Physics Express. 2016;**9**:085501. DOI: 10.7567/APEX.9.085501

[8] Yang H, Hattori AN, Ohata A, Takemoto S, Hattori K, Daimon H, et al. Direct observation for atomically flat and ordered vertical 111 side-surfaces on three-dimensionally figured Si(110) substrate using scanning tunneling microscopy. Japanese Journal of Applied Physics. 2017;**56**:111301. DOI: 10.7567/jjap.56.111301

[9] Takemoto S, Hattori AN, Hattori K, Tanaka H, Daimon H. Electric transport properties for three-dimensional angular-interconnects of Au wires crossing facet edges of atomically-flat Si{111} surfaces. Japanese Journal of Applied Physics. 2018;**57**:085503. DOI: 10.7567/JJAP.57.090303

[10] Tanuma S, Powell CJ, Penn DR. Calculations of electron inelastic mean free paths (IMFPS). IV. Evaluation of calculated IMFPs and of the predictive IMFP formula TPP-2 for electron energies between 50 and 2000 eV. Surface and Interface Analysis. 1993;**20**:77. DOI: 10.1002/sia.740200112

[11] Thompson AC, Attwood DT, Gullikson EM, Howells MR, Kim K-J, Kirz J, et al. X-Ray Data Booklet. California: Center for X-ray Optics and Advanced Light Source, Lawrence Berkeley National Laboratory; 2001. Available from: https://xdb.lbl.gov/

[12] Romanyuk O, Kataoka K, Matsui F, Hattori K, Daimon H. Structure analysis of thin iron-silicide film from φ-scan RHEED Patterson function. Czechoslovak Journal of Physics. 2006; **56**:267. DOI: 10.1007/s10582-006-0087-5

[13] Abukawa T, Yamazakai T, Yajima K, Yoshimura K. Weissenberg reflection high-energy electron diffraction for surface crystallography. Physical Review Letters. 2006;**97**:245502. DOI: 10.1103/PhysRevLett.97.245502

[14] Someta M, Maetani K, Hattori K, Daimon H. Phase discrimination of iron-silicides on Si(001) surfaces by three-dimensional reciprocal-lattice mapping. Surface Science. 2010;**604**:21. DOI: 10.1016/j.susc.2009.10.010

[15] Hattori H, Oi H, Tanaka K, Kumagai T, Daimon H. Three-dimensional reciprocal-lattice analysis using azimuth-scan reflection high-energy electron diffraction: Determination of complex crystal orientations of Al grains on Si(111) surface. Japanese Journal of Applied Physics. 2012;**51**:055801. DOI: 10.1143/JJAP.51.055801

[16] Takemoto S, Hattori K, Someta M, Hattori AN, Tanaka H, Kurushima K, et al. Shape-fitting analyses of two-dimensional X-ray diffraction spots for strain-distribution evaluation in a β-$FeSi_2$ nano film. Journal of Applied Crystallography. 2019;**52**:732. DOI: 10.1107/S1600576719007234

[17] Nakano H, Maetani K, Hattori K, Daimon H. Variety of iron silicides grown on Si(001) surfaces by solid phase epitaxy: Schematic phase diagram. Surface Science. 2007;**601**: 5088. DOI: 10.1016/j.susc.2007.04.234

[18] Kataoka K, Hattori K, Miyatake Y, Daimon H. Iron silicides grown by solid phase epitaxy on a Si(111) surface: Schematic phase diagram. Physical Review B. 2006;**74**:155406. DOI: 10.1103/PhysRevB.74.155406

[19] Liang E-Z, Huang C-J, Lin C-F. Use of SiO_2 nanoparticles as etch mask to generate Si nanorods by reactive ion etch. Journal of Vacuum Science and Technology B. 2006;**24**:599. DOI: 10.1116/1.2172251

[20] Ng AMC, Dong L, Ho WK, Djurišić AB, Xie MH, Wu HS, et al. Recovery of clean ordered (111) surface of etched silicon. Applied Surface Science. 2013;**282**:156. DOI: 10.1016/j.apsusc.2013.05.092

[21] Wu B, Kumar A, Pamarthy S. High aspect ratio silicon etch: A review. Journal of Applied Physics. 2010;**108**: 051101. DOI: 10.1063/1.3474652

[22] Oura K, Lftshits VG, Saranin A, Zotov AV, Katayama M. Surface Science: An Introduction (Advanced Texts in Physics). Berlin: Springer; 2003. DOI: 10.1007/978-3-662-05179-5

[23] Pal P, Gosalvez MA, Sato K. Silicon micromachining based on surfactant-added tetramethyl ammonium hydroxide: Etching mechanism and advanced applications. Japanese Journal of Applied Physics. 2010;**49**:056702. DOI: 10.1143/JJAP.49.056702

[24] Lifshits VG, Saranin AA, Zotov AV. Surface Phases on Silicon: Preparation, Structures, and Properties. Chichester: Wiley; 1994. ISBN 0-471-94846-2

[25] Yamamoto Y, Ino S, Ichikawa T. Surface reconstruction on a clean Si(110) surface observed by RHEED. Japanese Journal of Applied Physics. 1986;**25**:L331. DOI: 10.1143/JJAP.25.L331

[26] Hattori AN, Fujiwara Y, Fujiwara K, Murakami Y, Shindo D, Tanaka H. Fabrication of three-dimensional epitaxial $(Fe,Zn)_3O_4$ nanowall wire-structures and their transport properties. Applied Physics Express. 2014;**7**:045201. DOI: 10.7567/APEX.7.045201

[27] Fujiwara Y, Hattori AN, Fujiwara K, Tanaka H. Nanowall shaped MgO substrate with flat (100) sidesurface: A new route to three-dimensional functional oxide nanostructured electronics. Japanese Journal of Applied Physics. 2012;**52**:015001. DOI: 10.7567/JJAP.52.015001

[28] Hattori AN, Fujiwara Y, Fujiwara K, Tanaka H. 3D-architected and integrated metal oxides nanostructures and beyond by three-dimensional nanotemplate pulsed-laser deposition. e-Journal of Surface Science and Nanotechnology. 2015;**13**:279. DOI: 10.1380/ejssnt.2015.279

[29] Hattori AN, Fujiwara Y, Fujiwara K, Nguyen TVA, Nakamura T, Ichimiya M,

et al. Identification of Giant Mott phase transition of single electric nanodomain in manganite nanowall wire. Nano Letters. 2015;**15**:4322. DOI: 10.1021/acs.nanolett.5b00264

[30] Tsubota S, Hattori AN, Nakamura T, Azuma Y, Majima Y, Tanaka H. Enhancement of discrete changes in resistance in engineered VO_2 heterointerface nanowall wire. Applied Physics Express. 2017;**10**:115001. DOI: 10.7567/JJAP.57.090303

[31] Rakshit R, Hattori AN, Naitoh Y, Shima H, Akinaga H, Tanaka H. Three-dimensional nanoconfinement supports Verwey transition in Fe_3O_4 nanowire at 10 nm length scale. Nano Letters. 2019; **19**:5003. DOI: 10.1021/acs.nanolett.9b01222

[32] Le Lay G. Physics and electronics of the noble-metal/elemental-semiconductor interface formation: A status report. Surface Science. 1983;**132**: 169. DOI: 10.1016/B978-0-444-86784-1.50019-5

[33] Shimizu S, Sasaki N, Seki S. Epitaxial growth of α-Fe film on Si(111) substrate by low-energy direct ion beam deposition. Japanese Journal of Applied Physics. 1993;**32**:L943. DOI: 10.1143/JJAP.32.L943

[34] Shimizu S, Sasaki N. Crystalline quality of α-Fe films on Si(111) and Ge(111) substrates grown by direct ion beam deposition. Thin Solid Films. 1996; **281-282**:46. DOI: 10.1016/0040-6090(96)08572-0

[35] Hisamoto D, Lee W-C, Kedzierski J, Takeuchi H, Asano K, Kuo C, et al. FinFET-a self-aligned double-gate MOSFET scalable to 20 nm. IEEE Transactions on Electron Devices. 2000; **47**:2320. DOI: 10.1109/16.887014

[36] Colinge J-P. Multiple-gate SOI MOSFETs. Solid State Electronics. 2004; **48**:897. DOI: 10.1016/j.sse.2003.12.020

[37] Krastev ET, Voice LD, Tobin RG. Surface morphology and electric conductivity of epitaxial Cu(100) films grown on H-terminated Si(100). Journal of Applied Physics. 1996;**79**:6865. DOI: 10.1063/1.361508

[38] Luo EZ, Heun S, Kennedy M, Wollschläger J, Henzler M. Surface roughness and conductivity of thin Ag films. Physical Review B. 1994;**49**:4858. DOI: 10.1103/PhysRevB.49.4858

[39] Chawla JS, Gstrein F, O'Brien KP, Clarke JS, Gall D. Electron scattering at surfaces and grain boundaries in Cu thin films and wires. Physical Review B. 2011;**84**:235423. DOI: 10.1103/PhysRevB.84.235423

[40] Mao K, Saraya T, Hiramoto T. Effects of side surface roughness on carrier mobility in tri-gate single silicon nanowire metal-oxide-semiconductor field-effect transistors. Japanese Journal of Applied Physics. 2013;**52**:04CC11. DOI: 10.7567/JJAP.52.04CC11

[41] Elsom KC, Sambles JR. Macroscopic surface roughness and the resistivity of thin metal films. Journal of Physics F: Metal Physics. 1981;**11**:647. DOI: 10.1088/0305-4608/11/3/012

[42] Lee SC, Neumann A, Jiang Y-B, Artyushkova K, Brueck SRJ. Top-down, in-plane GaAs nanowire MOSFETs on an Al_2O_3 buffer with a trigate oxide from focused ion-beam milling and chemical oxidation. Nanotechnology. 2016;**27**:375707. DOI: 10.1088/0957-4484/27/37/375707

[43] Goodnick SM, Ferry DK, Wilmsen XW, Liliental Z, Fathy D, Krivanek OL. Surface roughness at the Si(100)-SiO_2 interface. Physical Review B. 1985;**32**:8171. DOI: 10.1103/PhysRevB.32.8171

[44] Mou C-Y, Hong T-M. Transport in quantum wells in the presence of interface roughness. Physical Review B.

2000;**61**:12612. DOI: 10.1103/PhysRevB.61.12612

[45] Wang J, Polizzi E, Ghosh A, Datta S, Lundstrom M. Theoretical investigation of surface roughness scattering in silicon nanowire transistors. Applied Physics Letters. 2005;**87**:043101. DOI: 10.1063/1.2001158

[46] Buran C, Pala MG, Bescond M, Dubois M, Mouis M. Three-dimensional real-space simulation of surface roughness in silicon nanowire FETs. IEEE Transactions on Electron Devices. 2009;**56**:2186. DOI: 10.1109/TED.2009.2028382

[47] Kim SG, Luisier M, Paul A, Boykin TB, Klimeck G. Full three-dimensional quantum transport simulation of atomistic Interface roughness in silicon nanowire FETs. IEEE Transactions on Electron Devices. 2011;**58**:1371. DOI: 10.1109/TED.2011.2118213

[48] Ryu H. A multi-subband Monte Carlo study on dominance of scattering mechanisms over carrier transport in sub-10-nm Si nanowire FETs. Nanoscale Research Letters. 2016;**11**:36. DOI: 10.1186/s11671-016-1249-4

[49] Lee H-Y, Yi S-M, Lee J-H, Lee H-S, Hyun S, Joo Y-C. Effects of bending fatigue on the electrical resistance in metallic films on flexible substrates. Metals and Materials International. 2010;**16**:947. DOI: 10.1007/s12540-010-1213-2

Section 3

The Wettability and Permeability of Material Surfaces

Chapter 7

Wettability on Different Surfaces

Yeeli Kelvii Kwok

Abstract

Wettability has been explored for 100 years since it is described by Young's equation in 1805. It is all known that hydrophilicity means contact angle (θ), $\theta < 90°$; hydrophobicity means contact angle (θ), $\theta > 90°$. The utilization of both hydrophilic surfaces and hydrophobic surfaces has also been achieved in both academic and practical perspectives. In order to understand the wettability of a droplet distributed on the textured surfaces, the relevant models are reviewed along with understanding the formation of contact angle and how it is affected by the roughness of the textured surface aiming to obtain the required surface without considering whether the original material is hydrophilic or hydrophobic.

Keywords: wettability, droplet, hydrophilic, hydrophobic, surface tension, contact angle, textured surface, Wenzel model, Cassie-Baxter model

1. Introduction

It is well known that when a small droplet of liquid is deposited on the solid surface, it forms a shape with a contact angle to the solid. This phenomenon is firstly described by Young in 1805, and he proposed that surface energy is the interaction between the forces of adhesion and the forces of cohesion which determine whether the wetting occurs or not (i.e., the spreading of a liquid over a surface) [1]. If it does not occur the complete wetting, the liquid in a bead shape will be formed. In the same time, as a function of the surface energies, a contact angle is defined in the system.

When the liquid wets the solid, three different interfacial boundary surfaces, viz., solid-air (sv), solid-liquid (sl), and liquid-air (lv), are involved. The contact angle, which is included between the interfaces of sl and lv, has to reach a certain value to satisfy the equilibrium state of the three interfacial tensions. It is all known that there are two requirements for the equilibrium.

2. Static equilibrium

The first requirement for keeping a balance of the three interfacial tensions in horizontal direction is described by Young's Eq. (1):

$$\gamma_{lv} \cos\theta = \gamma_{sv} - \gamma_{sl} \quad (1)$$

where γ denotes the interfacial tension in the denomination of the force per unit length, or of the energy per unit area, which are equivalent in measuring the surface energy density, and θ is the contact angle at a location where the tangent along an lv

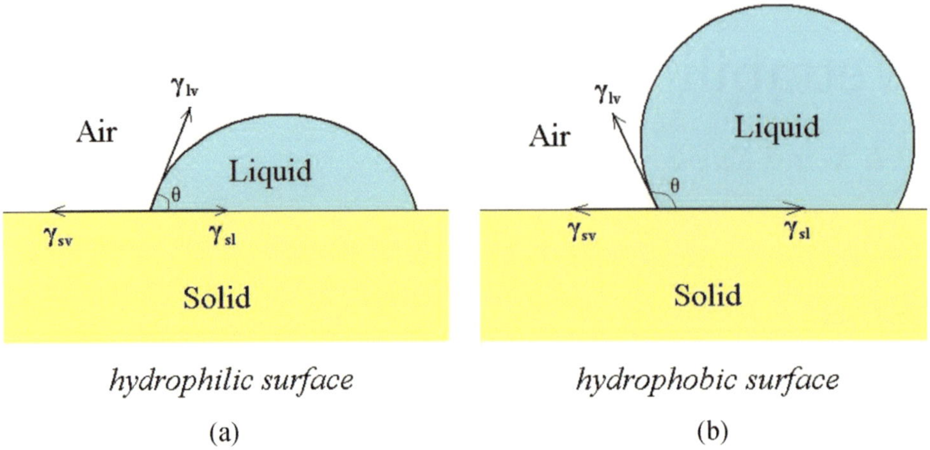

Figure 1.
Contact angle on various surfaces.

interface intersects the solid surface as shown in **Figure 1**. For the surface of solid with high surface energy, $\gamma_{sv} > \gamma_{sl}$, γ_{lv} directs to the side of γ_{sl} and forms a contact angle smaller than 90°. This kind of surface is known to be hydrophilic as shown in **Figure 1a**. For a solid with low surface energy, $\gamma_{sv} < \gamma_{sl}$, γ_{lv} directs to the side of γ_{sv} and forms a contact angle larger than 90° which is known to be hydrophobic as shown in **Figure 1b**.

3. Dynamic equilibrium

Another requirement is the dynamic equilibrium determined by the interface energy which can be calculated from $\gamma \times s$, where s is the area of interface. It should be noted that for a droplet of liquid with certain volume resting on a solid surface, it has the smallest lv interface when the contact angle is 90° (i.e., the droplet is a hemisphere as shown by the blue quarter circle in **Figure 2**); and whether its sl interface spreads (i.e., θ decreases) when $\theta < 90°$ or contract to be more like a sphere (i.e., θ increases) when $\theta > 90°$, the lv interface area increases. Firstly, considering a droplet on a hydrophilic solid surface as shown in **Figure 2a**, the shape of the droplet has not reached equilibrium. With the spreading of the liquid, the area of both the sl interface and the lv interface will increase simultaneously. Because $\gamma_{sv} > \gamma_{sl}$ on hydrophilic surface, the increment of the sl interface area means the conversion from the sv interface to the sl interface. The process involves a release in energy from the sv interface to the sl interface; as a result, the increment of the lv interface area implies a consumption of energy. When the energy changes caused by these two contrary factors are equal, the shape of the droplet will settle and the contact angle will achieve the final value of θ. This energy equilibrium can be described by the following equation:

$$(\gamma_{sv} - \gamma_{sl})ds_{sl} = \gamma_{lv}ds_{lv} \tag{2}$$

where ds_{sl} and ds_{lv} mean a slight variation in the area of sl interface and lv interface, respectively. By combining with Eq. (1), the contact angle can be expressed by:

$$\cos\theta = \frac{ds_{lv}}{ds_{sl}} \tag{3}$$

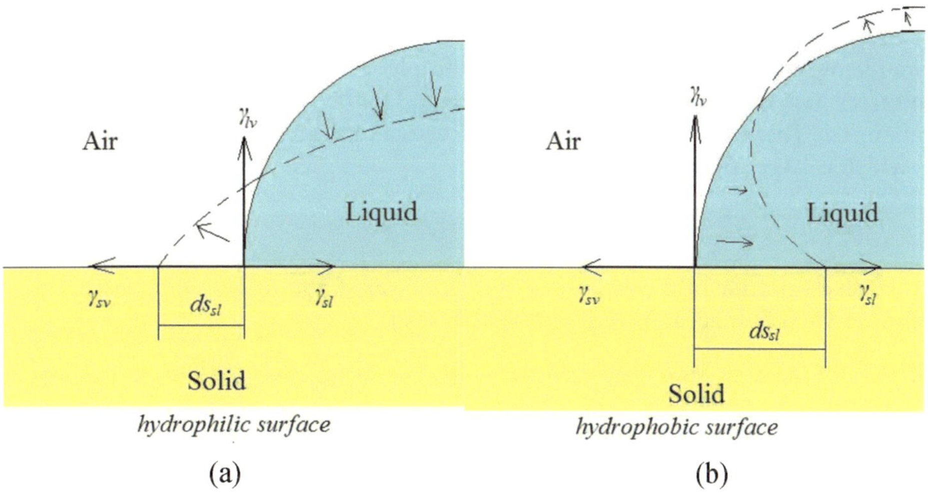

Figure 2.
Drop of liquid on solid surfaces when the equilibrium has not been reached.

It should be noticed that ds_{lv}/ds_{sl} is the area changing rate of the lv interface with the sl interface increasing; it is only determined by the shape of the droplet. Eq. 3 shows the relationship between the contact angle and the profile of the droplet and is independent of materials and surface tension.

For the system applied on a hydrophobic surface as shown in **Figure 2b**, with the effect of the contracting of liquid, the area of the sl interface decreases with increasing lv interface. Because $\gamma_{sv} < \gamma_{sl}$ on hydrophobic surface, the decrement of the sl area involves a release of energy to the increasing lv interface area. When the dynamic equilibrium of energy is reached, Eqs. (2) and (3) can also be applied on this kind of surface.

$$-(\gamma_{sv} - \gamma_{sl})(-ds_{sl}) = \gamma_{lv} ds_{lv} \tag{4}$$

$$\cos\theta = \frac{ds_{lv}}{ds_{sl}}$$

It should be noted that (γ_{sv}-γ_{sl}) and ds_{sl} are negative on hydrophobic surface.

4. Effect of surface roughness on contact angle

It should be noticed that there distinctively exists a difference between the geometric surface and the actual surface and their interface is not ideal as a proposed model in the textbooks. Actually, the surface of any real solid is not a perfect plane. Due to the surface roughness, the real area of the actual surface is larger than the so-called ideal (geometric) surface. Consequently, the surface roughness affects the contact angle and the contact angle distinctively varies with the surface roughness. As a result, in order to keep the equilibrium, the profile of a droplet will vary with the effect of the surface roughness. For studying θ' (new contact angle) distributed on the real rough surface and the effect of its roughness on the relevant wettability, Wenzel and Cassie-Baxter proposed two different models to explain as a key effective factor how solid surfaces with the real geometry features affect the wettability [2–8].

Wenzel model.

According to the model described by Wenzel in 1936 [9], the solid surface completely contacts with liquid under the droplet as shown in **Figure 3**. The *sl* interface area is enlarged to be s'_{sl} which is equal to the "actual surface" by the roughness. There is a ratio of the sl interface area to the geometric surface area, r, which is larger than 1.

$$s'_{sl} = rs_{sl} \tag{5}$$

With a variation of the geometric *sl* interface area, the amount of energy released from it or accumulated in it is increased:

$$(\gamma_{sv} - \gamma_{sl})ds'_{sl} = r(\gamma_{sv} - \gamma_{sl})ds_{sl} \tag{6}$$

In addition, the lv interface is not affected by the surface roughness. So the equilibrium with the new contact angle of θ' can be expressed by:

$$\cos\ \theta' = \frac{r(\gamma_{sv} - \gamma_{sl})}{\gamma_{lv}} \tag{7}$$

Compared with Eq. (1), θ' can be depicted as:

$$\cos\ \theta' = r\cos\theta \tag{8}$$

Taken θ_w and θ_0 to represent θ' and θ, respectively, it is obtained:

$$\cos\theta_w = r\cos\theta_0 \tag{9}$$

where θ_w is the contact angle on the rough surface with Wenzel model and θ_0 is the original contact angle according to the ideal smooth surface. Eq. 9 is the Wenzel equation. It shows that when Wenzel model is applied, $r > 1$, the morphology of the surface always magnifies the underlying wetting properties. θ_w is larger than θ_0 for the hydrophobic material ($\theta_0 > 90°$); and it is smaller than θ for the hydrophilic material ($\theta_0 < 90°$) [10–12].

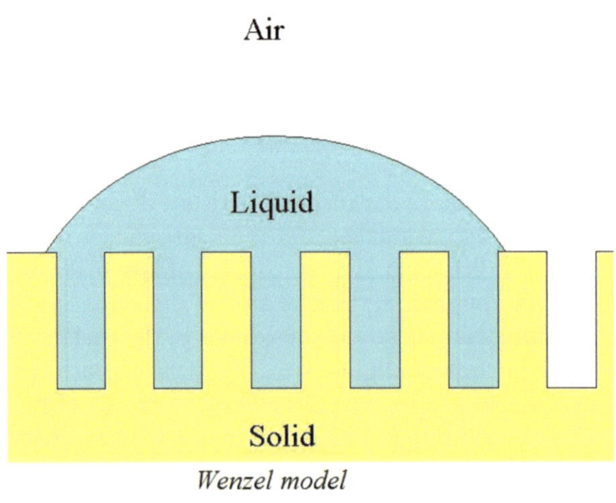

Figure 3.
Schematic of a droplet on the rough surface described by Wenzel.

5. Cassie-Baxter model

In 1944, Cassie applied and explored Wenzel equation on porous materials [13]. According to Cassie-Baxter model, air can be trapped below the drop as shown in **Figure 4**. The area of the *sl* interface is reduced by the surface roughness while a part of that transits to the lv interface in indentations. The ratio of the actual *sl* interface area to the geometric surface area is represented by f, which is smaller than 1 in Cassie-Baxter model.

$$ds'_{sl} = f ds_{sl} \tag{10}$$

$$ds'_{lv} = ds_{lv} + (1-f) ds_{sl} \tag{11}$$

With a variation of the profile of the droplet, the amount of energy transited among the interfaces is changed:

$$(\gamma_{sv} - \gamma_{sl}) ds'_{sl} = f(\gamma_{sv} - \gamma_{sl}) ds_{sl} \tag{12}$$

$$\gamma_{lv} ds'_{lv} = \gamma_{lv} ds_{lv} + (1-f)\gamma_{lv} ds_{sl} \tag{13}$$

The equilibrium with the new contact angle of θ' can be expressed by:

$$\cos\theta' = \frac{f(\gamma_{sv} - \gamma_{sl})}{\gamma_{lv}} - (1-f) \tag{14}$$

Compared with Eq. (1), θ' can be calculated as:

$$\cos\theta' = f(\cos\theta + 1) - 1 \tag{15}$$

Taken θ_c and θ_0 to represent θ' and θ, respectively, it is obtained:

$$\cos\theta_c = f(\cos\theta_0 + 1) - 1 \tag{16}$$

where θ_c is the contact angle on rough surface with Cassie-Baxter model. Eq. (16) is Cassie-Baxter equation. According to Cassie-Baxter model, only the

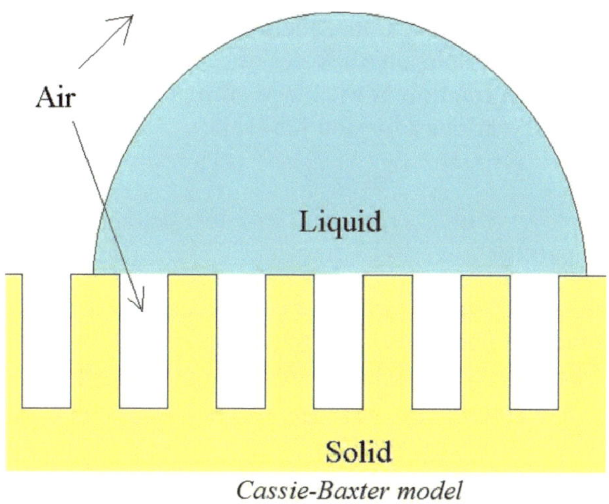

Figure 4.
Schematic of a droplet on the rough surface described by Cassie-Baxter.

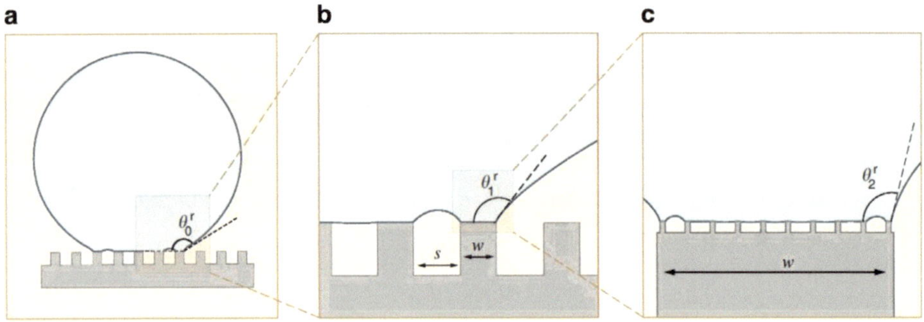

Figure 5.
Schematic of self-similar contact line pinning. (a) A liquid droplet that rests in a Cassie-Baxter state on a hierarchical surface exhibits an apparent receding angle θ_0^r. (b) The apparent contact line of the drop is divided into many smaller first-level contact lines, each at the top of a first-level roughness feature with width w *and spacing* s. *each of these first-level contact lines sits at the base of a first-level capillary bridge, which has a local receding contact angle θ_1^r. (c) The apparent contact line of each second-level capillary bridge is further divided into smaller second-level contact lines, each atop a second-level roughness feature. Each second-level contact line sits at the base of a second-level capillary bridge, which has a local receding contact angle θ_2^r.*

characteristics of hydrophobicity can be enhanced. θ_c is always larger than θ on the rough surface [14–17].

In fact, numerous investigations have been devoted to the wettability on different surfaces, particularly for the surfaces inspired by Nature Mother [18–26]. Paxson et al. [27] fabricated a surface with the hierarchical textures initiated by lotus leaves and revealed the relevant mechanism of the variation or evolution of the adhesion force per unit length of the projected contact line distributed on natural textured surfaces. Results show that the adhesion force varies with the pinned fraction of each level of hierarchy.

Figure 5 shows a droplet sitting on a textured surface in a Cassie-Baxter state. It depicts the real contact line of the droplet, which is changed into many smaller lines. Meanwhile, the contact angle also changes from θ_0^r (the zeroth level) to θ_1^r (the first level of hierarchy) as shown in **Figure 5b**. If the contact line is divided into much smaller lines, viz., the second level of hierarchy, the related contact angle θ_2^r is distinctively different from θ_1^r of the first level of hierarchy as shown in **Figure 5c**. These phenomena will be kept on until a homogeneous wetting interface achieved when reaching a level n. Consequently, the contact angle either increases or decreases by adding multiple length scales of roughness at all smaller levels depending on the pinned fraction of each level of hierarchy, which is critical for designing surfaces with various adhesion [28–33].

6. Conclusion

The droplet on a solid surface will exhibit a certain value of contact angle to achieve the equilibrium of the interfacial tensions. In addition, surface roughness will influence the contact angle, based on Wenzel's and Cassie-Baxter's theories, with the assumption of overhangs. It reveals that the contact angle can be controlled by the intentionally fabricated textured surfaces, and the surface with the fabricated textures can be changed from hydrophilic to hydrophobic, and vice versa, without considering whether the original material is hydrophilic or hydrophobic.

Author details

Yeeli Kelvii Kwok
Department of Mechanical and Biomedical Engineering, City University of Hong Kong, Kowloon, Hong Kong

*Address all correspondence to: yeelikwok@yahoo.com

References

[1] Young T. An essay on the cohesion of fluids. Philosophical Transactions: Royal Society London. 1805;**95**:65-87

[2] Bormashenko E, Bormashenko Y, Stein T, Whyman G, Bormashenko E. Why do pigeon feathers repel water? Hydrophobicity of pennae, Cassie-Baxter wetting hypothesis and Cassie-Wenzel capillarity-induced wetting transition. Journal of Colloid and Interface Science. 2007;**311**(11):212-216

[3] Whyman G, Bormashenko E, Stein T. The rigorous derivation of Young, Cassie-Baxter and Wenzel equations and the analysis of the contact angle hysteresis phenomenon. Chemical Physics Letters. 2008;**450**(4–64): 355-359

[4] Elif Cansoy C, Yildirim Erbil H, Akar O, Akin T. Effect of pattern size and geometry on the use of Cassie-Baxter equation for superhydrophobic surfaces. Colloids and Surfaces A: Physicochemical and Engineering Aspects. 2011;**386**(1–35):116-124

[5] Parry V, Berthomé G, Joud JC. Wetting properties of gas diffusion layers: Application of the Cassie-Baxter and Wenzel equations. Applied Surface Science. 2012;**258**(1515):5619-5627

[6] Lee J, Hwang SH, Yoon SS, Khang DY. Evaporation characteristics of water droplets in Cassie, Wenzel, and mixed states on superhydrophobic pillared Si surface. Colloids and Surfaces A: Physicochemical and Engineering Aspects. 2019;**5625**:304-309

[7] Montes Ruiz-Cabello FJ, Cabrerizo-Vílchez MA, Rodríguez-Valverde MA. Evaluation of the solid-liquid contact area fraction of drops deposited on rough surfaces beyond the Wenzel regime. Colloids and Surfaces A: Physicochemical and Engineering Aspects. 2019;**5685**:455-460

[8] Bormashenko E. Progress in understanding wetting transitions on rough surfaces. Advances in Colloid and Interface Science. 2015;**222**:92-103

[9] Wenzel RN. Resistance of solid surfaces to wetting by water. Journal of Industrial and Engineering Chemistry. 1936;**28**:988-994

[10] Bormashenko E. General equation describing wetting of rough surfaces. Journal of Colloid and Interface Science. 2011;**360**(11):317-319

[11] Khilifi D, Foudhil W, Harmand S, Jabrallah SB. Evaporation of a sessile oil drop in the Wenzel-like regime. International Journal of Thermal Sciences. 2020;**151**:106236

[12] Ceyhan U, Tiktaş A, Özdoğan M. Pinning and depinning of Wenzel-state droplets around inclined steps. Colloid and Interface Science Communications. 2020;**35**:100238

[13] Cassie ABD. Wettability of porous surfaces. Transactions of the Faraday Society. 1944;**40**:546-551

[14] Milne AJB, Amirfazli A. The Cassie equation: How it is meant to be used. Advances in Colloid and Interface Science. 2012;**170**(1–215):48-55

[15] Cengiz U, Elif CC. Applicability of Cassie-Baxter equation for superhydrophobic fluoropolymer-silica composite films. Applied Surface Science. 2015;**33530**:99-106

[16] Aziz H, Amrei MM, Dotivala A, Tang C, Vahedi TH. Modeling Cassie droplets on superhydrophobic coatings with orthogonal fibrous structures. Colloids and Surfaces A: Physicochemical and Engineering Aspects. 2017;**5121**:61-70

[17] Rivera RM, Koltsov A, Lazcano BA, Douce JF. Wettability in water/iron ore

powder systems: To the universality of the Cassie model. International Journal of Mineral Processing. 2017;**16210**:36-47

[18] Malvadkar NA, Hancock MJ, Sekeroglu K, Dressick WJ, Demirel MC. An engineered anisotropic nanofilm with unidirectional wetting properties. Nature Materials. 2010;**9**:1023-1028

[19] Raj R, Adera S, Enright R, Wang EN. High-resolution liquid patterns via three-dimensional droplet shape control. Nature Communications. 2014;**5**:4975

[20] Kreder MJ, Alvarenga J, Kim P, Aizenberg J. Design of anti-icing surfaces: Smooth, textured or slippery? Nature Reviews Materials. 2016;**1**:15003

[21] Schneider L, Laustsen M, Mandsberg N, Taboryski R. The influence of structure heights and opening angles of micro- and nanocones on the macroscopic surface wetting properties. Scientific Reports. 2016;**6**: 21400

[22] Ujjain SK, Roy PK, Kumar S, Singha S, Khare K. Uniting superhydrophobic, superoleophobic and lubricant infused slippery behavior on copper oxide nano-structured substrates. Scientific Reports. 2016;**6**: 35524

[23] Lo YH, Yang CY, Chang HK, Hung WC, Chen PY. Bioinspired diatomite membrane with selective superwettability for oil/water separation. Scientific Reports. 2017;**7**:1426

[24] Ivanchenko P, Delgado-López JM, Iafisco M, Gómez-Morales J, Tampieri A, Martra G, et al. On the surface effects of citrates on nano-apatites: Evidence of a decreased hydrophilicity. Scientific Reports. 2017; **7**:8901

[25] Lee Y, Matsushima N, Yada S, Nita S, Kodama T, Amberg G, et al. Revealing how topography of surface microstructures alters capillary spreading. Scientific Reports. 2019;**9**:7787

[26] Wong TS, Kang SH, Tang SKY, Smythe EJ, Hatton BD, Grinthal A, et al. Bioinspired self-repairing slippery surfaces with pressure-stable omniphobicity. Nature. 2011;**477**: 443-447

[27] Paxson AT, Varanasi KK. Self-similarity of contact line depinning from textured surfaces. Nature Communications. 2013;**4**:1492

[28] Kubiak KJ, Wilson MCT, Mathia TG, Carval P. Wettability versus roughness of engineering surfaces. Wear. 2011;**271**(3–43):523-528

[29] Sarkar A, Kietzig A-M. General equation of wettability: A tool to calculate the contact angle for a rough surface. Chemical Physics Letters. 2013; **57414**:106-111

[30] Bormashenko E. Apparent contact angles for reactive wetting of smooth, rough, and heterogeneous surfaces calculated from the variational principles. Journal of Colloid and Interface Science. 2019;**5371**:597-603

[31] Drelich JW. Contact angles: From past mistakes to new developments through liquid-solid adhesion measurements. Advances in Colloid and Interface Science. 2019;**267**:1-14

[32] Hosseini S, Savaloni H, Shahraki MG. Influence of surface morphology and nano-structure on hydrophobicity: A molecular dynamics approach. Applied Surface Science. 2019;**48515**:536-546

[33] Azimi A, Rohrs C, He P. Hydrodynamics-dominated wetting phenomena on hybrid superhydrophobic surfaces. Journal of Colloid and Interface Science. 2020; **5627**:444-452

Chapter 8

Smart Surfaces with Tunable Wettability

Meenaxi Sharma and Krishnacharya Khare

Abstract

Modification of surface wettability (ranging from complete wetting to complete non-wetting) of various surfaces is often required in many applications. Conventionally, it is done using a coating of suitable materials as per the requirement. In this approach, the old coating needs to be replaced every time by a new appropriate one. Alternatively, smart responsive surfaces can show tunable wettability with external stimulus. Electric field, temperature, light, pH, mechanical strain, etc. can be effectively used as external stimuli, and a suitable coating can be incorporated, which responses to the respective stimulus. These surfaces can be used to tune the surface wettability to any extent based on the magnitude of the stimulus. The primary role of the external stimulus is to vary the liquid-solid interfacial energy, which subsequently changes the surface wettability. The biggest advantage of this approach is that the surface wettability can be reversibly tuned. Each of the techniques mentioned above has many advantages along with certain limitations, and the combination of advantages and limitations helps users to choose the right technique for their work. Many recent studies have used this approach to quantify the tuning of the surface wettability and have also demonstrated its potential in various applications.

Keywords: wetting, surface energy, tunable wetting, responsive surfaces

1. Introduction

There are numerous examples in our daily life where we encounter liquid-solid interaction, for example, wall paints, cookwares, cleaning textiles, raindrops on glass surfaces of automobiles, houses, solar panels, to name a few. In all these examples, drops of different liquids interact with given solid surfaces. The macroscopic behavior of liquids on such solid surfaces is actually governed by the microscopic molecular interactions between the liquid and solid molecules, which in scientific terms is described as the wetting behavior of solid surfaces. **Figure 1** shows the schematics of liquid drops on homogeneous and heterogeneous surfaces describing their final wetting behavior. **Figure 1a** shows Young's wetting state of a liquid drop sitting on a homogeneous surface where the surface wettability is described in terms of Young's contact angle (θ_Y), which can be derived by balancing various interfacial forces (γ's) as given by Eq. (1) where S, L, and V represent solid, liquid, and vapor phases respectively [1].

$$\cos\theta_Y = \frac{\gamma_{SV} - \gamma_{SL}}{\gamma_{LV}} \tag{1}$$

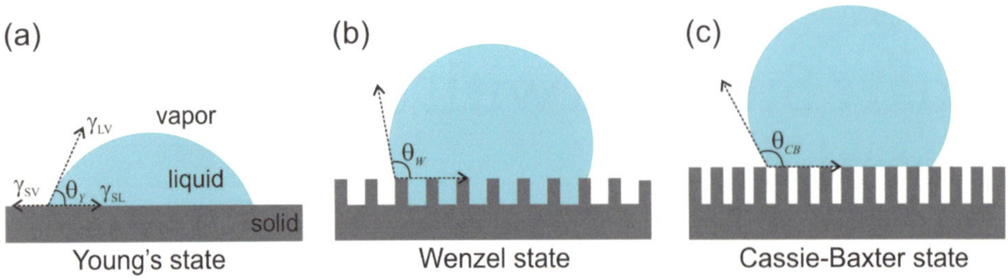

Figure 1.
Schematics of different wetting states of a liquid drop on a homogeneous surface: (a) Young's state and heterogeneous surface; (b) Wenzel and (c) Cassie-Baxter states.

On heterogeneous surfaces, as shown in **Figure 1b, c**, the final wetting behavior is different compared to homogeneous surfaces and is described by Wenzel and Cassie-Baxter contact angles as given by Eq. (2) [2, 3]:

$$\cos\theta_W = r\cos\theta_Y \text{ and } \cos\theta_{CB} = f\cos\theta_Y + (f - 1) \qquad (2)$$

where r is the roughness factor, and f is the area fraction of the solid-liquid interface.

In many cases, maximum interaction of liquid-solid is required, for example, washing textiles, wall paints etc., whereas there are other cases where minimum liquid-solid interaction is desirable, for example, solar panel, nonstick cookware, and glass windshields. Therefore, it becomes essential to manipulate liquid-solid interaction or wetting behavior as per the requirement. Alternative to the conventional approach, where the wetting behavior is manipulated by varying the surface topography or chemistry, continuously and reversibly varying the wettability based on some external energy source has recently gained huge popularity. This method has many advantages over the conventional method due to its nondestructive nature. Various researchers and engineers have investigated many such external energy sources or stimuli (e.g., electric field, temperature, radiation, mechanical strain, pH, magnetic field, etc.), which can be efficiently used to manipulate the wetting behavior of given solid surfaces continuously as well as reversibly. In this chapter, we provide a comprehensive summary of most of these techniques with the fundamental principle behind them as well as their potential applications.

2. Smart responsive surfaces

Based on the nature of external stimuli used to manipulate the wettability of a given surface, smart responsive surfaces can be divided into many categories, which are discussed in the following sections. Each of the smart surfaces has some advantages and disadvantaged over others, which is discussed in depth along with their fundamental mechanism, state-of-the-art status, and potential commercial applications.

2.1 Electric field-responsive surfaces

An electric field as an external stimulus is considered to be one of the fast and convenient ways to switch the surface wettability by virtue of its ability to control the surface chemistry in a few seconds or less. This is one of the most versatile ways to manipulate the liquids on solid as well as liquid surfaces without applying a responsive coating on the surface. However, to perform the electrowetting experiment, it is mandatory to have both the substrate and the liquid to be conducting. The electrowetting phenomenon, which is based on the electrocapillarity principle, was first explained in detail by Lippmann in 1875 [4, 5]. Traditional electrowetting was started around 1973 where the electrocapillarity technique was applied to a three-phase system in which a drop of the aqueous electrolyte solution was in contact with a mercury substrate [6]. Later in 1993, developments in this field were commenced by Berge et al. who gave the idea of introducing a thin insulating layer to avoid the direct contact between the conducting drop and the electrode in order to get rid of the electrolysis of water, which is also known as the electrowetting on dielectrics [7]. The electrowetting phenomenon is known to be the dependence of the contact angle on the applied potential between the conducting drop and the conducting substrate. In the electrowetting process, with the applied voltage, a reduction was observed in the interfacial tension of the solid-liquid interface, which further leads to decrease the observed contact angle without affecting the chemical composition of the substrate (as shown in **Figure 2A**).

$$\gamma_{SL}(V) = \gamma_{SL}(0) - \frac{\varepsilon_0 \varepsilon_d}{2d} V^2 \tag{3}$$

where ε_0 is the permittivity of vacuum, ε_d is the dielectric constant of the insulating material, d is the thickness of the insulating layer, and V is the applied voltage.

Berge proposed the electrowetting equation by energy minimization method to get the relationship between the contact angle and the applied voltage, also known

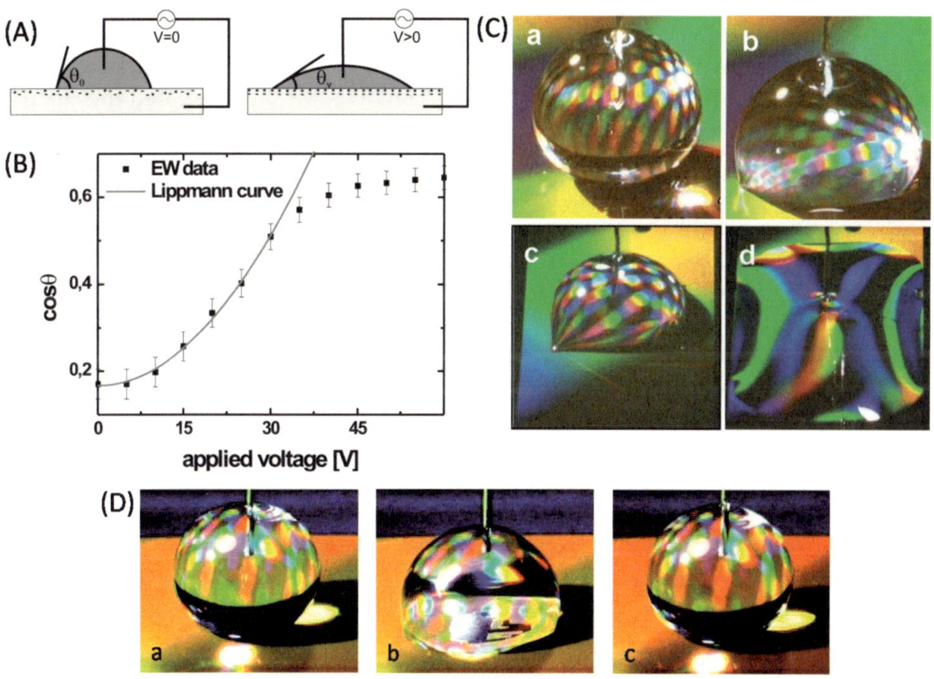

Figure 2.
(A) Sketch of the electrowetting setup on a dielectric (EWOD). The droplet and the conducting Si substrate form a capacitor with the dielectric SiOx layer. (B) Electrowetting curve for the system consisting of triangular grooves used in the experiments reproduced with permission from [8]. (C) Four frames from the video recording demonstrating electrically induced transitions between different wetting states of a liquid droplet on the nanostructured substrate reproduced with permission from [9]. (D) Demonstration of electrically induced reversible transitions between different wetting states of a liquid on a nanostructured substrate, reproduced with permission from [10].

as the famous Young-Lipmann equation [7]. In the absence of trapped charges, the equation can be derived as:

$$\cos\theta(V) = \cos\theta(0) + \frac{\varepsilon_0 \varepsilon_d}{2\gamma_{LV} d} V^2 \quad (4)$$

where θ (0) and θ (V) represent the contact angle without applied voltage and at a finite voltage *V*, respectively.

However, if the trapped charges are present, a finite voltage V_T of the trapped charges should get introduced in the above equation, and the modified equation can be written as

$$\cos\theta(V) = \cos\theta(0) + \frac{\varepsilon_0 \varepsilon_d}{2\gamma_{LV} d} (V - V_T)^2 \quad (5)$$

where V_T is the voltage to consider the effect of trapped charges. According to this equation, the contact angle continuously decreases with increase in the applied voltage. However, beyond the threshold voltage, the contact angle is found to be independent of the applied voltage, and this phenomenon is called the contact angle saturation (as shown in **Figure 2B**). It is difficult to achieve a large change and reversible initial contact angle with the applied voltage. In practice, the reversible electrowetting can only be achieved for small voltage change due to contact angle saturation phenomenon at high voltages (as shown in **Figure 2B**).

Inspired by nature, surfaces with micro/nanostructures give rise to superhydrophobic behavior with a larger water contact angle. On the contrary, electrowetting reduces the initial contact angle by the applied voltage. Combining superhydrophobicity with electrowetting would lead to effectively enhance the range of wettability change. Electrowetting on micro/nanostructured surfaces has acquired a lot of attention in the research community from various aspects, from fundamentals to applications [9, 11–15]. For the first time, Krupenkin et al. demonstrated the electric field-controlled dynamic wetting behavior of liquids on nanostructured surfaces [9]. They had shown that electrowetting was a tool to dynamically control the wetting behavior of liquids by covering a wide range of wetting states from superhydrophobic (non-wetting) to almost complete wetting as shown in **Figure 2C**. When no voltage was applied, drop took the shape of a spherical ball with a contact angle close to 180° (**Figure 2C—(a)**). With an increase in the applied voltage to 22 V, it underwent a transition to the immobile state with a decrease in the contact angle (**Figure 2C—(b, c)**). With a further increase in the voltage to 50 V, the contact angle decreased, making a transition to the completely wetting state (**Figure 2C—(d)**). Later, they extended their work to focus on the electric field-induced reversible wetting transition between the non-wetting state (Cassie-Baxter) and the partial wetting state (Wenzel) on a nanostructured surface [10]. They have shown the controlled reversible wetting transition between a rolling drop and a completely immobile drop on a nanostructured surface. To reverse the transition from the immobile state to the rolling drop, they transported a short pulse of electric current through the nanostructured substrate. One of the major issues associated with the electrowetting experiment is the large contact angle hysteresis, that is, difference between the contact angle values after the completion of one voltage cycle, that is, after the voltage has reached to 0 V.

In the past decade, many research groups have dedicatedly worked on the minimization of contact angle hysteresis during electrowetting on smooth as well rough surfaces [16–18]. To achieve the reversible electrowetting, a lubricating fluid can be used to cover the surface structures to provide a smooth layer of lubricating fluid on top of the dry substrate to reduce the surface hysteresis to a large extent by minimizing the undesirable energy loss [19–21]. Hao et al. performed electrowetting on liquid-infused film (EWOLF) to get complete reversible wetting behavior. They have shown that infusing the lubricating fluid in porous membranes, and viscous energy dissipation could be sufficiently enhanced with suppression of drop oscillation, resulting in fast response without losing the reversible behavior [20]. Upon applying a voltage of 500 V, the contact angle decreased to 50° on the lubricant-infused membranes showing a wettability change of 53° (**Figure 3a**).

Careful observation reveals that wetting ridges are formed at the oil-water interface due to the deformation of the lubricant caused by the capillary pressure (**Figure 3b**). Due to the presence of a smooth liquid-liquid interface, drop freely moves without any pinning and turns back to its initial wetting configuration indicating complete reversible behavior (**Figure 3c**). However, for the superhydrophobic PTFE membranes, due to large electrowetting hysteresis, the wetting transition was irreversible. Measurement of apparent contact after every voltage on-off cycle also confirms the electrowetting reversibility for liquid-infused membranes (**Figure 3d**). Electrowetting hysteresis on the liquid-infused membranes (~3°) is much smaller compared to the superhydrophobic membranes (~40°). Later Bormashenko et al. also reported the low voltage electrowetting on EWOLF utilizing lubricated honeycomb polymer surfaces with very low contact angle hysteresis [22]. Due to fast response, reversible wetting behavior, and low energy consumption, electrowetting has gained a lot of attention among various researchers for its applications in emerging fields such as liquid lens [23, 24], optical display [25], liquid patterning, and controlled movement of liquid drops in narrow channels. One of the most important applications realizing electric field-induced tunable

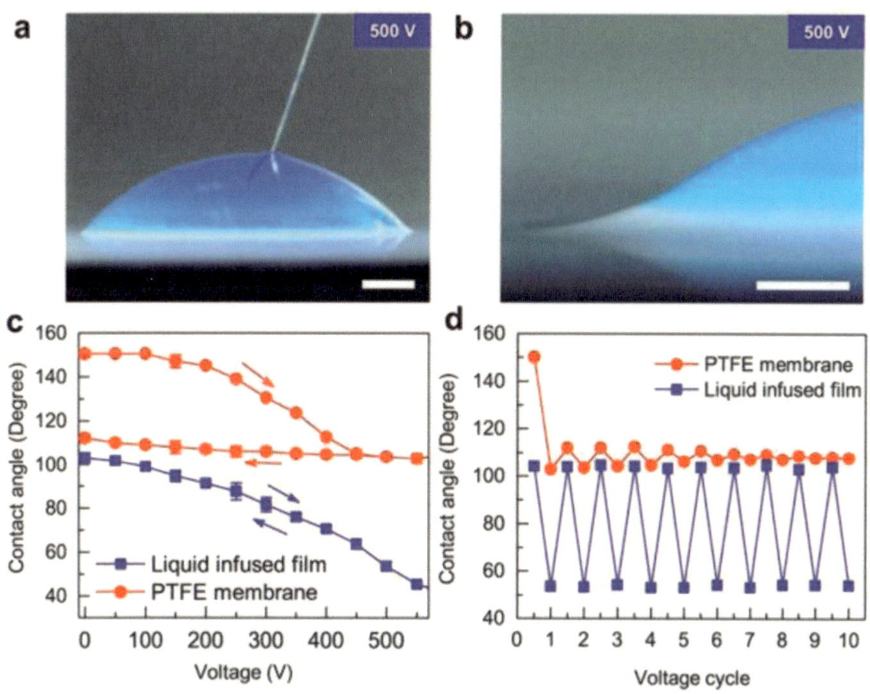

Figure 3.
Wetting properties of liquid-infused film and electrowetting response. (a) Optical image of a stained droplet in EWOLF subject to an actuation voltage of 500 V. (b) the formation of the wetting ridge as a result of oil motion at the liquid–liquid interface. (c) Characterization of the variations of apparent CA in EWOD and EWOLF. (d) The variation of droplet apparent CA subject to electrowetting cycles. Reproduced with permission from [20].

wettability is an optical zoom lens [23, 24]. Kuiper has demonstrated a liquid-based variable foal lens based on the meniscus between two immiscible liquids. The application of voltage would result in the accumulation of charges near the solid-liquid interface, which would further lead to change the shape of the meniscus from convex to concave, as shown in **Figure 4** [24].

2.2 Temperature-responsive surfaces

Temperature-responsive or thermoresponsive polymers often evince a conformational change of polymer chains in response to the temperature change, and they are widely used to develop thermoresponsive surfaces [26]. Poly-isopropylacrylamide (PNIPAAm), the most common temperature-responsive polymer, shows different molecular arrangement at temperatures below and above the lower critical solution temperature (LCST) of about 32°. Below LCST, the intermolecular hydrogen bonding between the amino groups and carbonyl groups as well as water leads to a loosely coiled molecular arrangement resulting in hydrophilic behavior. While above LCST, the intramolecular hydrogen bonding leads to form a compact and collapsed arrangement of PNIPAAm chains, hence making it difficult for carbonyl and amino groups to interact with the water molecules and results in the hydrophobic behavior. PNIPAAm films can be easily grafted on both smooth as well as rough substrates using surface-initiated atom transfer radical polymerization (ATRP) technique. However, on smooth substrates modified with PNIPAAm coating, wettability change was limited as demonstrated by Sun et al. that wettability could only be varied from 63 to 93° by changing the temperature below and above LCST [27]. By introducing the roughness, thermoresponsive wettability of the surface was greatly enhanced, and a large reversible wettability switch from superhydrophobic (~149.3°) to superhydrophilic (~0°) was achieved [27–31] (as shown in **Figure 5a, b**). This behavior is due to the competition between intermolecular and intramolecular hydrogen bonding below and above LCST

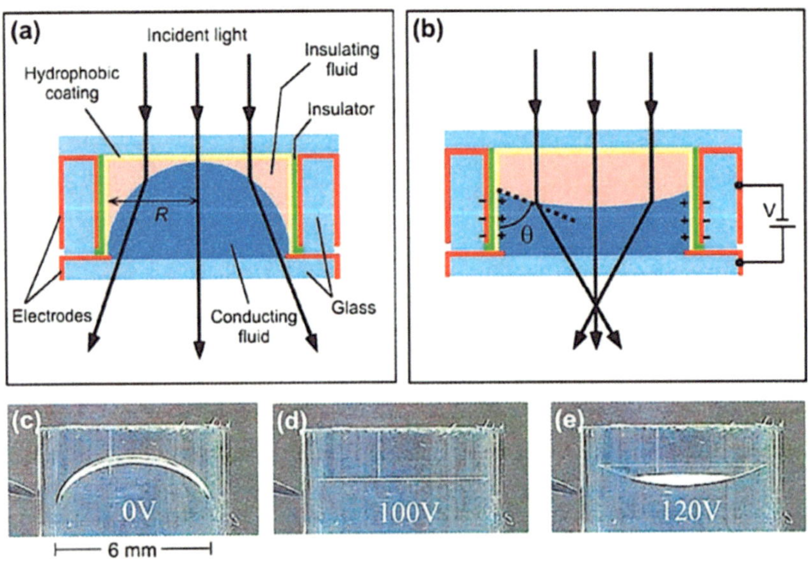

Figure 4.
(a) Schematic cross section of a liquid-based variable lens in a cylindrical glass housing (b) when a voltage is applied, charges accumulate in the wall electrode, and opposite charges collect near the solid/liquid interface in the conducting liquid. (c)–(e) Video frames of a 6-mm-diameter lens taken at voltages of 0, 100, and 120 V. Reproduced with permission from [24].

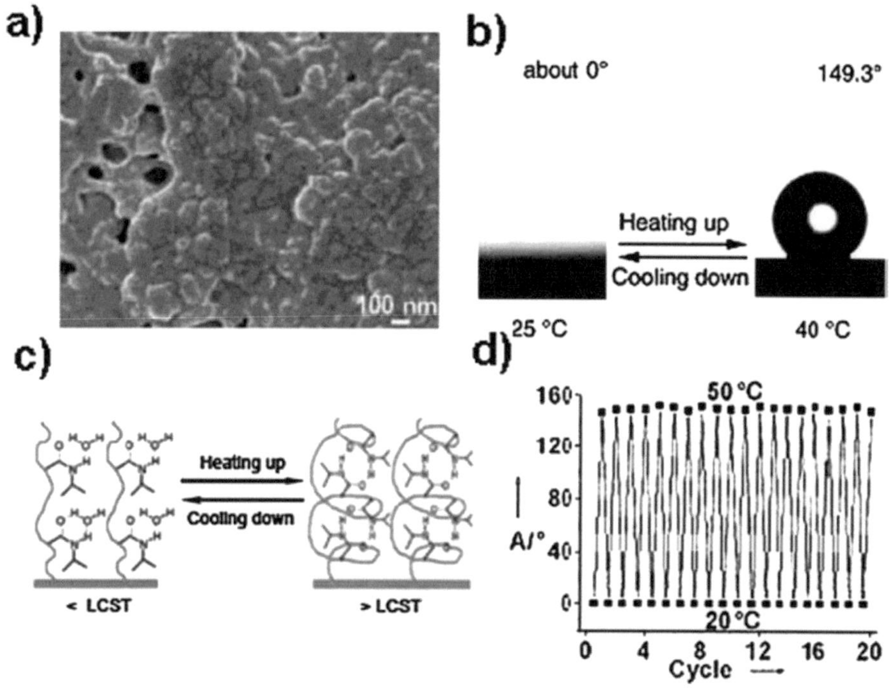

Figure 5.
(a) SEM image of the nanostructures on a rough substrate modified with PNIPAAm. (b) Water drop profile for the responsive surface at 25C and 40°C. (c) Diagram of reversible formation of intermolecular hydrogen bonding between PNIPAAm chains and water molecules (left) and intramolecular hydrogen bonding between C=O and N–H groups in PNIPAAm chains (right) below and above the LCST. (d) Contact angles at two different temperatures 20 and 40°C for PNIPAAm-modified rough substrate. Reproduced with permission from [29].

(**Figure 5C**). The introduction of small surface roughness with a thermoresponsive coating dramatically improves the range of wettability. It makes it possible to get a switch of the wettability between the superhydrophobic and superhydrophilic states in a narrow temperature range. In the past decades, various other investigations reveal the progress in the direction of thermoresponsive surfaces to get the large wettability switch by combining surface features with PNIPAAm coating [28, 32].

Fu et al. reported a versatile approach to demonstrate the dynamical change of surface wettability by preparing nano-porous aluminum surfaces utilizing nanostructured surfaces modified with PNIPAAm coating [33]. Furthermore, a rough copper mesh film with hierarchical micro and nanostructures modified with PNIPAAm coating was used for temperature controlled water permeation [28]. For instance, micro/nanostructured composite films of PNIPAAm and polystyrene with controllable thermoresponsive wettability, which could switch between superhydrophobic and superhydrophilic was prepared by electrospinning technique [34]. Later by combining two thermoresponsive polymers poly(NIPAAm-co-NIPMAM), precise control over wettability switch from gradual to sudden could be achieved by precisely controlling the transition temperature [30]. Furthermore, low-cost block polymer brush containing poly(N-isopropylmethacrylamide)-block-poly(N-isopropylacrylamide) (PNIPMAM-b-PNIPAAm) was fabricated on the microstructured silicon substrate using ATRP technique, and multistage thermoresponsive wettability was observed. ATRP is an excellent method to construct block polymer brushes with different LCST on the surface. For instance, LCST of PNIPAAm and PNIPMAM is about 32 and 44° respectively. At temperature below 32 (LCST of PNIPAAm) PNIPMAM-b-PNIPAAm chains show loosely coiled molecular

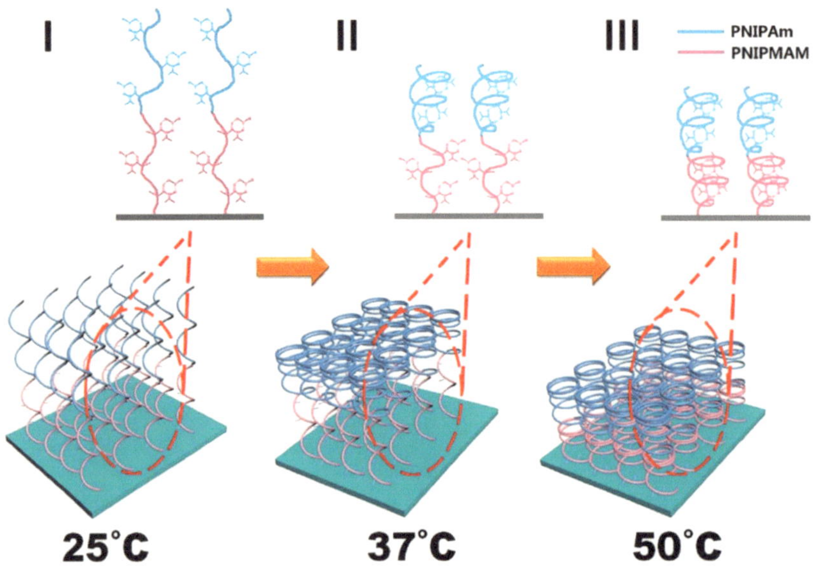

Figure 6.
Proposed principle of the constriction of PNIPMAM-b-PNIPAAm brushes with the increasing of temperature. Reproduced with permission from [32].

arrangement and thus show hydrophilic behavior with a contact angle of 20.9 ± 2.8°; (as shown in **Figure 6—I**); when the temperature is raised beyond 32° but below 44°, PNIPAAm undergo phase change because it has crossed its LCST, which lead to form a collapsed and compact conformation of upper PNIPAAm chains induced by intramolecular hydrogen bonding and show less hydrophilic behavior with a contact angle of about 85.2 ± 2.7° (**Figure 6—II**). Further increasing the temperature above 44° (LCST of PNIPAAm), the molecular arrangement of PNIPAAm chains would change, which would further collapse the polymer chains making the surface more hydrophobic with a contact angle of about 108.6 ± 2.1° (**Figure 6—III**) [32]. Meanwhile, various other investigations have dealt with the fabrication of tunable wettability surfaces based on the polymer brush coating on cotton fabrics [35, 36].

Jiang et al. fabricated thermoresponsive co-polymer, PHFBMA-b-(PGMA-g-PNIPAM), by controlled free radical polymerization, which was used to fabricate a tunable wettability cotton fabric by dip-coating it into the thermoresponsive micelle. Dried cotton fabric was smooth with high fluorine content and showed hydrophobic behavior at low temperatures while at high temperatures; the surface of the cotton fabric was rough with low fluorine content and showed good hydrophilic behavior [36]. In addition to thermoresponsive polymers, which have been used extensively to fabricate the thermoresponsive surfaces, thermoresponsive inorganic oxides can also be used to get reversible thermoresponsive wetting transition [37, 38]. Due to the fast switching wettability in a narrow temperature range, thermoresponsive surfaces offer promising applications in areas including the thermally driven movement of liquid, oil–water separation, and switchable adhesion on the surface [39–44].

2.3 Photoresponsive surfaces

Surface wettability can be intelligently controlled using a variety of photoresponsive inorganic and organic oxides and polymers that are based on two main features: the switch of bi-stable states and change of surface free energy under light (electromagnetic radiation) stimulus. Surface wettability can be reversibly switched to a highly wetting state under ultraviolet (UV) illumination, and the original state is recovered after the surface is placed for a longer time period in dark and/or heated at elevated temperature. Among various photoresponsive materials, inorganic oxides (wide bandgap semiconducting material) are widely used in various applications owing to their excellent chemical and mechanical stability, low cost, and outstanding optoelectronic properties. Among the different photosensitive inorganic materials, titanium dioxide (TiO_2) and zinc oxide (ZnO) are the most common semiconductors widely used for their excellent UV absorption characteristics as their bandgap corresponds to the UV energy. The mechanism involving photogeneration of electron and holes and absorption of water is responsible for the change in the wettability. Photoresponsive wettability change of titanium dioxide was first reported by Wang et al. [45] and later discussed in detail by Pant et al. [46]. They have shown that the wetting behavior on a surface coated with a polycrystalline thin film of TiO_2 could be reversible switched between the hydrophilic and hydrophobic states under UV irradiation and dark storage. Respective hydrophilic behavior is due to the conversion of Ti^{4+} state to Ti^{3+} state under UV illumination. For practical applications, it is highly desirable to get the surfaces with wettability switching in a large range from superhydrophobic to superhydrophilic states, which can be obtained by combining the surface roughness with a photoresponsive smart material coating of desired surface chemistry. Introduction of fine nano roughness enhances the hydrophilic and hydrophobic performances. By introducing nanoscale roughness, Tadanaga et al. demonstrated the reversible wetting transition by UV irradiation on a superhydrophobic surface with three layers: flowerlike Al_2O_3, thin TiO_2 gel, and fluoroalkylsilane [47]. Later Feng et al. reported similar reversible wettability switch between superhydrophobic and superhydrophilic states of ZnO nanorod films, as shown in **Figure 7a–d** [48].

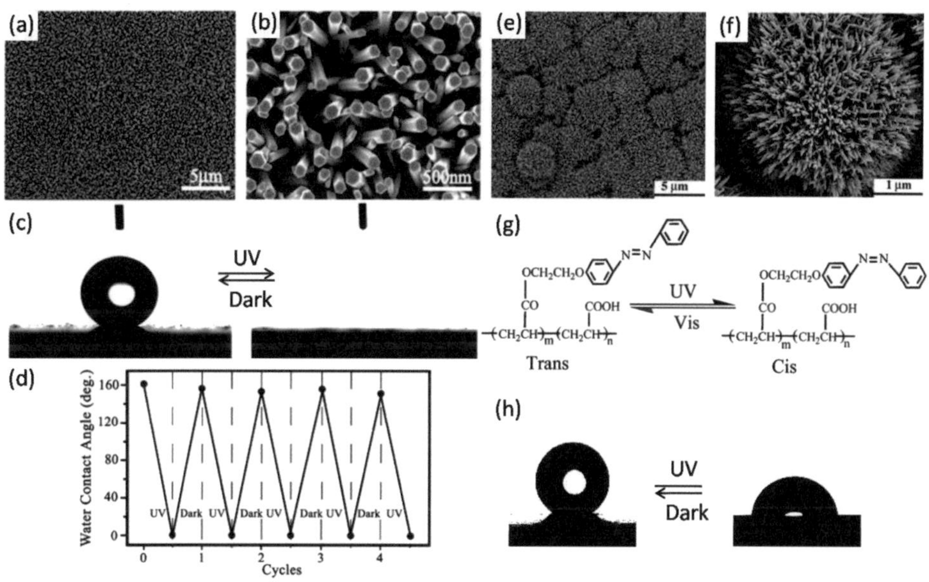

Figure 7.
(a, b) FE-SEM top-images of the as-prepared ZnO nanorod films at low and high magnifications, respectively. (c) Photographs of water droplet shape on the aligned ZnO nanorod films before (left) and after (right) UV illumination, reproduced with permission from [48]. (d) Reversible superhydrophobic to superhydrophilic transition of the as-prepared films under the alternation of UV irradiation and dark storage. (e) Low-magnification FE-SEM image of a TiO_2 nanorod film deposited on a glass wafer; (f) morphology of a single papilla at high magnification, reproduced with permission from [49]. (g) The trans and cis structures of azobenzene upon UV and Vis irradiation. (h) The shapes of water drop on photoresponsive monolayer with a patterned substrate of 40-mm pillar spacing upon UV and Vis irradiation. Reproduced with permission from [50].

They reported that UV irradiation would lead to generate electron-hole pairs and form oxygen vacancies on the surface. Competition between water and oxygen will decide which component will absorb the oxygen vacancies on the surface. Surface hydrophilicity is a cause of the adsorption of hydroxyl groups on the surface; however this state is energetically unstable. Therefore, the hydroxyl groups adsorbed on the surface are replaced by the oxygen groups (which is thermodynamically more stable) after the UV-irradiated surfaces are placed in dark and subsequently, the surface reverts to its original state and the wettability switches from superhydrophobic to superhydrophilic. Further, by introducing micro and nanoscale hierarchical surface structures, Feng et al. showed the reversible wetting switching between superhydrophilic and superhydrophobic states on TiO_2 nanorod films by cooperating micro and nano hierarchical surface structures together with a photosensitive material coating and the wettability change was quite similar to as reported in their previous work [49]. Morphology of hierarchical surface structures with micro and nano features can be clearly visualized from the SEM images shown in **Figure 7e, f**. Due to intrinsic photocatalytic property and reversible wettability switch, TiO_2 has also attracted much attention for its various applications, such as antifogging and self-cleaning [51–55]. Another simple approach to produce rough TiO_2 films was based on CF_4 plasma etching technique, which was adopted by Zhang et al. to produce superhydrophilic and superhydrophobic patterns by UV irradiation using a photomask [56]. However, rough surfaces often lead to scattering of light due to the level of the roughness being larger than the wavelength of visible light. This problem can be addressed by reducing the level of roughness to a scale comparable to the wavelength of the visible light [57]. In addition to TiO_2, other inorganic photoresponsive wide bandgap semiconductor materials such as ZnO, SnO_2, V_2O_5, WO_3 etc. are also used to get the reversible wettability switch between two extreme states [48, 56, 58, 59]. To achieve different functions such as transparency and conductivity, SnO_2 conducting nanorods were prepared, showing 60% transmittance in the visible range, which can meet the demand of smart microfluidic devices [58]. In addition, tungsten oxide films prepared by the electrochemical deposition method demonstrate dual responsive wettability switch and photochromatic characteristics [60].

In addition to inorganic materials, various organic compounds/polymers with stimuli-responsive properties and a reversible photoinduced transformation between two states have been widely used. The various photoresponsive functional groups include azobenzenes, pyrimidines, spiropyran, cinnamates etc. Responsive surface wettability change is associated with the change in the surface free energy caused by the change in chemical composition upon UV irradiation. Among various organic compounds, azobenzene is the most promising photoresponsive material, whose molecular structure reversible transits between trans and cis states under visible and UV illumination [61]. Azobenzene layer with a trans isomer state possesses a larger water contact angle, which is governed by the small dipole moment and low surface energy, while cis state exhibits small contact angle due to high dipole moment and large surface energy. Reversible wettability change from superhydrophobic to superhydrophilic is associated with the corresponding change of state from trans isomer to cis isomer under UV and visible illumination (**Figure 7g**). Using a simple electrostatic self-assembly technique, Jiang et al. showed photo switched wettability on organic azobenzene monolayer with a large reversible change of contact angle (**Figure 7h**) [50]. Since then, various other research groups have worked on the preparation of tunable wetting surfaces based on the photoresponsive organic materials and subsequent manipulation of the wettability via controlling surface properties by irradiating UV light [62–64]. Because of the unique properties of these photoresponsive polymers, they are widely used in various applications such as antifogging, self-cleaning, adhesion control, liquid printing, and oil–water separation [52, 53, 65–67]. Photoresponsive surfaces have also gained much attention in biomedical areas utilizing the light-responsive controlled release system where a guest molecule could be released from the surface in a controlled manner by adjusting the wetting behavior of the surface.

Chen et al. realized a light-responsive release system by controlling the wetting behavior of mesoporous (MS) silica surface-functionalized with an optimal ratio of spiropyran mixed with fluorosilane [67]. They reported that, under UV irradiation (365 nm), the surface could be wetted by water due to the conversion of spiropyran molecule from closed-form to open-form, which resulted in the switch of wettability from hydrophobic to hydrophilic state and release of cargo molecules from the pores (**Figure 8**).

2.4 Mechanically tunable surfaces

As it is mentioned earlier, the surface wettability is mainly governed by chemical composition and structures of the surface. Altering one of these factors would lead to change the surface wetting properties. Most of the external stimulus changes the chemical composition of the substrate; however, changing the surface structures also produces a change in the surface wettability. Various research groups have demonstrated the mechanical stress-responsive tunable wettability using elastic or gel materials [68–75]. Poly(tetrafluoroethylene) PTFE is the most common

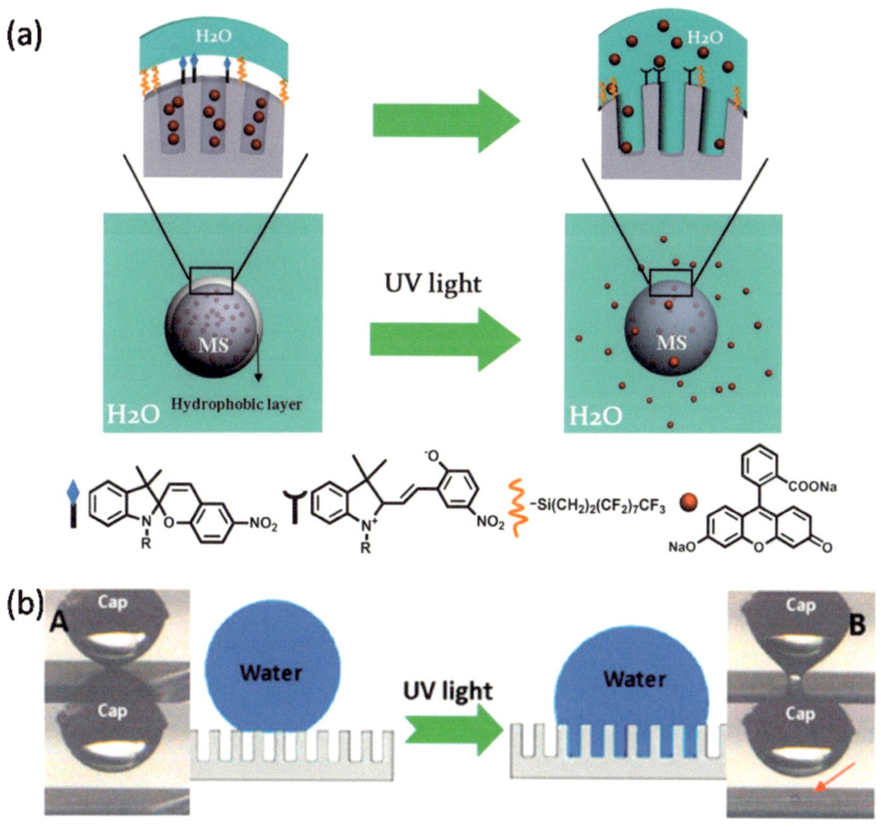

Figure 8.
(a) The schematic of the light-responsive release system. After irradiation with 365 nm UV light, the surface became wet due to the conformational conversion of spiropyran from the "closed" form to the "open" form (b). The mechanism illustrating the wetting process of surface. Reproduced with permission from [67].

hydrophobic material, and Teflon tape, made with PTFE crystals, has been widely used to tune the wetting behavior by applying mechanical strain. Zhang et al. demonstrated a change in the density of PTFE crystal by axial extension [68]. They also reported that with an extension of the PTFE crystals to 190%, water contact angle could be tuned between hydrophobic (118°) and superhydrophobic (165°) states. In their next work, using this approach of biaxial extension and unloading, they have shown reversible wettability switch between superhydrophilic and superhydrophobic behavior of the polyamide films with triangular structures [69]. Initially the polyamide film with the triangular structures showed superhydrophobic behavior with a water contact angle of ~151.2°. When the film is extended to 120%, water penetrates through space among the fiber resulting in the complete wetting situation. While unloading, the polyamide film returned to the superhydrophobic state due to the recovery of the surface structures. In addition, poly(dimethylsiloxane) (PDMS) microwrinkle structures also demonstrate tunable wetting behavior upon stretching. Wrinkling approach is very useful as its wavelength and amplitude can be actively tuned to produce various well-defined structures.

Lin et al. have shown the fabrication of superhydrophobic surface, featured with micro and nanoscale roughness, using nanoparticle coating on PDMS wrinkled structures [71]. For the first time, they have shown the fabrication of a superhydrophobic surface by combining microstructured PDMS wrinkles with nanoparticles. **Figure 9** is a schematic illustration of the fabrication of PDMS elastic film with dual scale roughness. Surface topography can be reversibly tuned between complete relaxed state (wrinkle formation) and the complete stretched state (only nano roughness), that is, reversible switching between the dual scale roughness and nanoscale roughness (as shown in **Figure 9f–k**). Using both theoretical and

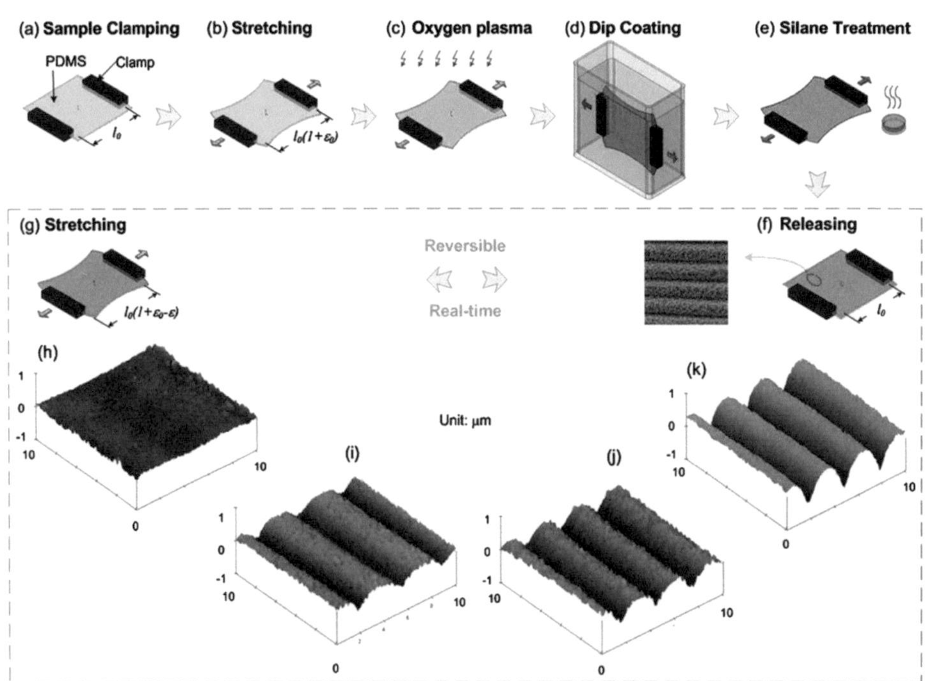

Figure 9.
A schematic illustration of the fabrication of a PDMS film with dual-scale roughness (a–f) and real-time, reversible tunability of its surface topography by mechanical strain (f–k). Inset shows the optical images of a water drop on the surface with dual-scale roughness (right) and nanoscale roughness (left). Reproduced with permission from [71].

experimental approaches, they concluded that only micron-scale roughness is not able to attain Cassie-Baxter state. By regulating the surface topography by applying different mechanical strain, they have shown switching between different wetting states from Cassie-Baxter (dual scale roughness) to Wenzel (both micron wrinkle and nanoparticle-coated elastic film). Chung et al. have shown that the drop shape strongly gets affected by the geometrical anisotropy of the sinusoidal grooves as on the isotropic surface contact angle were uniquely defined; however this was not the case for anisotropic surfaces on which drop showed two different contact angles in the direction parallel and perpendicular to the grooves [70].

2.5 pH-responsive surfaces

In recent years, pH-responsive wetting has gained a lot of attention because of their emerging applications in various fields such as drug delivery and biosensors [76]. It is of particular interest where it is required to change the wetting behavior of various acidic and basic liquids. Xu et al. have shown a novel method to prepare pH-responsive surfaces containing block copolymer thin films of poly(styrene-b-acrylic acid) (PS-b-PAA) and pH-responsive nanostructures composed of cylindrical domains [77]. PS-b-PAA polymer films show different surface morphologies for three different pH regimes. These polymers films swell more rapidly when immersed in high-pH solutions compared to low-pH solutions. They also observed that with the increase in the pH from 2.6 to 9.1, the water contact angle decreased by 30°. The decrease in the contact angle is due to the rear molecular arrangement of PAA chains in response to increase in pH, which results in increasing hydrophilicity [77, 78]. They concluded that wettability can be regulated by controlling the molecular arrangement of PAA chains in response to pH stimulus.

Uhlmann et al. introduced the idea of surface functionalization for the smart coatings using stimuli-responsive binary polymer brushes containing polymer chains of two different polymers using "grafting from" and "grafting to" approach [79]. The concept of reversible switching can be understood based on the reaction of polymers with different solvents, where a polymer brush of hydrophilic and hydrophobic polymers is treated with nonselective and selective solvents for both the polymers. In a good solvent, due to the dominance of the interchain repulsion, polymer chains show stretched conformation while in a bad solvent, due to strong repulsion between solvent and polymer, chains show coiled and collapsed conformation. This can also be interpreted as; brush structure shows hydrophilic behavior when treated with a selective solvent for polymer A while shows hydrophobic behavior when treated with a selective solvent for polymer B. However, in a nonselective solvent, both polymers show laterally segregate structures. Later Lu et al. reported a system of layer by layer (LbL) hydrogels, composed of amphiphilic polymers, which can undergo a reversible transition in response to pH stimulus. To shed light on the exact wetting state of hydrophobic PaAALbL-coated patterned surface, they measured the CAH (**Figure 10A**) and imaged the behavior of a water drop for different pH values (**Figure 10B–E**). For pH < 6, contact angle hysteresis was too large, close to 120° and the drop was sticky to the surface, and did not fall even if the substrate was turned vertically down. Such stickiness behavior with high CAH is realized for the drop in the Wenzel state, and the observed behavior can be accounted for the enhanced contact line pinning by the surface microstructures [80]. Numerous other researchers have also investigated pH-responsive tunable wetting behavior from its fundamental understanding of switchable wettability [78, 81–83].

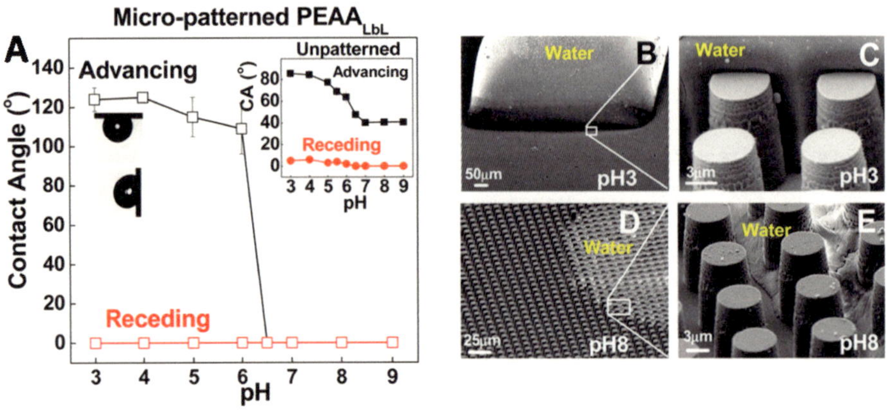

Figure 10.
(A) Advancing and receding CAs of PEAALbL on a micro-patterned substrate with a square array of cylindrical pillar structures. Optical images show water droplets sticking to the substrate at pH 3, suggesting strong pinning. Inset in the upper right corner represents advancing and receding CAs of water droplets on the unpatterned substrate. (B–E) Cryo-SEM images of a water droplet on a PEAALbL-coated micropillar-patterned substrate at (B, C) pH 3 and (D, E) pH 8. Reproduced with permission from [80].

2.6 Other stimuli

Some of the recent work also demonstrates using magnetic field and solvents as external stimuli to manipulate the wettability for given fluids. Smart surfaces that respond to magnetic field have been demonstrated by numerous research groups in the last decade [84, 85]. Grigoryev et al. fabricated a microstructured surface with reentrant geometry composed of Ni micronails, which shows a reversible transition from superomniphobic to omniphilic wetting state in response to the external magnetic field [85]. Cheng et al. also showed the reversible wetting transitions of the microdroplet consisting of superparamagnetic Fe_3O_4 nanoparticles [84]. They have shown that the wettability can be reversibly switched between the Cassie state and the Wenzel state on the microstructured silicon substrate and the transition between two wetting states can be controlled by both concentration of nanoparticles and the intensity of the magnetic field. On the other hand, solvent as an external stimulus is also feasible to tune the wetting behavior as solvent-responsive surfaces are affected by the surrounding medium and their wettability change is governed by the change in their interfacial energy caused by the rearrangement of the molecular chains in response to a solvent [81, 86–88]. Minko et al. demonstrated a novel route to fabricate two-level structured polymer brushes, and the surface wettability could be reversibly controlled by exposing the polymer brush surfaces to the solvent, which is selective to one of the polymers. Surface morphology and surface properties change when exposed to different solvents, which is caused by the interchange between vertical and linear phase segregation of the polymers [81].

3. Conclusions

To conclude, smart surfaces with tunable wettability based on various external stimulus prove to be useful candidates in different applications where the continuous and reversible tuning of surface wettability is important. These surfaces exploit the additional energy gained from the external stimulus to change their wetting behavior as per the requirement. It has been demonstrated that the surface wettability can be varied from hydrophobic to hydrophilic on smooth surfaces and from superhydrophobic to superhydrophobic on patterned surfaces. Many such external stimuli are available to alter the surface wettability because each technique has few advantages and disadvantages over others. As described in different sections, almost complete understanding about most of the techniques has been established. Of course, there are some limiting cases where the scientific understanding is not completely clear and is an active area of research currently. Hence, depending upon the feasibility and available resources, users can choose a suitable stimulus for their application.

Acknowledgements

The authors acknowledge fruitful discussion with Reeta Pant and Subhash Singha, which was particularly useful for the section "**Photoresponsive surfaces**".

Conflict of interest

The authors declare no conflict of interest.

Author details

Meenaxi Sharma and Krishnacharya Khare*
Department of Physics, Indian Institute of Technology Kanpur, Kanpur, India

*Address all correspondence to: kcharya@iitk.ac.in

References

[1] Young T. An essay on the cohesion of fluids. Philosophical Transactions. Royal Society of London. 1805;**95**:65-87

[2] Wenzel RN. Resistance of solid surfaces to wetting by water. Industrial and Engineering Chemistry. 1936;**28**(8):988-994

[3] Cassie ABD, Baxter S. Wettability of porous surfaces. Transactions of the Faraday Society. 1944;**40**:546-551

[4] Gabriel L. Relations entre les phénomènes électriques et capillaires. Annales de Chimie Physique. 1875;**5**:494

[5] Mugele F, Baret J-C. Electrowetting: From basics to applications. Journal of Physics. Condensed Matter. 2005;**17**:R705-RR74

[6] Nakamura Y, Matsumoto M, Nishizawa K, Kamada KI, Watanabe A. Studies on secondary electrocapillary effects: II. The electrocapillary phenomena in thin liquid film. Journal of Colloid and Interface Science. 1977;**59**(2):201-210

[7] Berge B. Electrocapilarity and wetting of insulator film by water. Comptes Rendus de l'Academie des Sciences Serie II. 1993;**317**:157-163

[8] Khare K, Herminghaus S, Baret J-C, Law BM, Brinkmann M, Seemann R. Switching liquid morphologies on linear grooves. Langmuir. 2007;**23**(26):12997-13006

[9] Krupenkin TN, Taylor JA, Schneider TM, Yang S, et al. Langmuir. 2004;**20**(10):3824

[10] Krupenkin TN, Taylor JA, Wang EN, Kolodner P, Hodes M, Salamon TR. Reversible wetting–dewetting transitions on electrically tunable Superhydrophobic nanostructured surfaces. Langmuir. 2007;**23**(18):9128-9133

[11] Bormashenko E, Pogreb R, Balter S, Aurbach D. Electrically controlled membranes exploiting Cassie-Wenzel wetting transitions. Scientific Reports. 2013;**3**(1):3028

[12] McHale G, Brown CV, Sampara N. Voltage-induced spreading and superspreading of liquids. Nature Communications. 2013;**4**(1):1605

[13] Lokanathan M, Sharma H, Shabaka M, Mohanty K, Bahadur VJC, Physicochemical SA, et al. Comparing electrowettability and surfactants as tools for wettability enhancement on a hydrophobic surface. Colloids and Surfaces, A: Physicochemical and Engineering Aspects. 2020;**585**:124155

[14] Mazuruk K, MP V. Electric field effect on the contact angle for non-wetting drops. Journal of Physics: Condensed Matter. 2019;**31**(39):395002

[15] Teng P, Tian D, Fu H, Wang SJMCF. Recent progress of electrowetting for droplet manipulation: From wetting to superwetting systems. Materials Chemistry Frontiers. 2020;**4**(1): 140-154

[16] Li F, Mugele F. How to make sticky surfaces slippery: Contact angle hysteresis in electrowetting with alternating voltage. Applied Physics Letters. 2008;**92**(24):244108

[17] Maillard M, Legrand J, Berge B. Two liquids wetting and low hysteresis electrowetting on dielectric applications. Langmuir. 2009;**25**(11):6162-6167

[18] de Ruiter R, Semprebon C, van Gorcum M, Duits MHG, Brinkmann M, Mugele F. Stability limits of capillary bridges: How contact angle hysteresis affects morphology transitions of liquid microstructures. Physical Review Letters. 2015;**114**(23):234501

[19] Bormashenko E, Pogreb R, Bormashenko Y, Grynyov R, Gendelman O. Low voltage reversible electrowetting exploiting lubricated polymer honeycomb substrates. Applied Physics Letters. 2014;**104**(17):171601

[20] Hao C, Liu Y, Chen X, He Y, Li Q, Li KY, et al. Electrowetting on liquid-infused film (EWOLF): Complete reversibility and controlled droplet oscillation suppression for fast optical imaging. Scientific Reports;**4**:2014, 6846

[21] Barman J, Pant R, Nagarajan AK, Khare K. Electrowetting on dielectrics on lubricating fluid-infused smooth/rough surfaces with negligible hysteresis. Journal of Adhesion Science and Technology. 2017;**31**(2):159-170

[22] Bormashenko E, Pogreb R, Bormashenko Y, Aharoni H, Shulzinger E, Grinev R, et al. Progress in low voltage reversible electrowetting with lubricated polymer honeycomb substrates. RSC Advances. 2015;**5**(41):32491-32496

[23] Berge B, Peseux J. Variable focal lens controlled by an external voltage: An application of electrowetting. European Physical Journal E: Soft Matter and Biological Physics. 2000;**3**(2):159-163

[24] Kuiper S, Hendriks BHW. Variable-focus liquid lens for miniature cameras. Applied Physics Letters. 2004;**85**(7):1128

[25] Hayes RA, Feenstra BJ. Video-speed electronic paper based on electrowetting. Nature. 2003;**425**(6956):383

[26] Wischerhoff E, Zacher T, Laschewsky A, Rekaï ED. Direct observation of the lower critical solution temperature of surface-attached thermo-responsive hydrogels by surface plasmon resonance. Angewandte Chemie, International Edition. 2000;**39**(24):4602-4604

[27] Sun T, Wang G, Feng L, Liu B, Ma Y, Jiang L, et al. Reversible switching between superhydrophilicity and superhydrophobicity. Angewandte Chemie, International Edition. 2004;**43**(3):357-360

[28] Song W, Xia F, Bai Y, Liu F, Sun T, Jiang L. Controllable water permeation on a poly(N-isopropylacrylamide)-modified nanostructured copper mesh film. Langmuir. 2007;**23**(1):327-331

[29] Xia F, Zhu Y, Feng L, Jiang L. Smart responsive surfaces switching reversibly between super-hydrophobicity and super-hydrophilicity. Soft Matter. 2009;**5**(2):275-281

[30] Zhao T, Nie F-Q, Jiang L. Precise control of wettability from LCST tunable surface. Journal of Materials Chemistry. 2010;**20**(11):2176-2181

[31] Konosu Y, Matsumoto H, Tsuboi K, Minagawa M, Tanioka A. Enhancing the effect of the nanofiber network structure on thermoresponsive wettability switching. Langmuir. 2011;**27**(24):14716-14720

[32] Song W, Li H, Wang C, Yang B. Design of multi-stage thermal responsive wettable surface. Advanced Materials Interfaces. 2014;**1**(6):1400009

[33] Fu Q, Rama Rao GV, Basame SB, Keller DJ, Artyushkova K, Fulghum JE, et al. Reversible control of free energy and topography of nanostructured surfaces. Journal of the American Chemical Society. 2004;**126**(29):8904-8905

[34] Wang N, Zhao Y, Jiang L. Low-cost, thermoresponsive wettability of surfaces: Poly(N-isopropylacrylamide)/polystyrene composite films prepared by electrospinning. Macromolecular Rapid Communications. 2008;**29**:485-489

[35] Jiang B, Zhang L, Liao B, Pang H. Self-assembly of well-defined thermo-responsive fluoropolymer and its application in tunable wettability surface. Polymer. 2014;**55**(21):5350-5357

[36] Jiang B, Pang H. Synthesis and self-assembly of thermoresponsive block-graft fluoropolymer as well as its tunable wettability surface. Journal of Polymer Science, Part A: Polymer Chemistry. 2016;**54**(7):992-1002

[37] Yang J, Zhang Z, Men X, Xu X, Zhu X. Thermo-responsive surface wettability on a pristine carbon nanotube film. Carbon. 2011;**49**(1):19-23

[38] Zhang M, Yao G, Cheng Y, Xu Y, Yang L, Lv J, et al. Temperature-dependent differences in wettability and photocatalysis of TiO_2 nanotube arrays thin films. Applied Surface Science. 2015;**356**:546-552

[39] Li C, Guo R, Jiang X, Hu S, Li L, Cao X, et al. Reversible switching of water-droplet mobility on a superhydrophobic surface based on a phase transition of a side-chain liquid-crystal polymer. Advanced Materials. 2009;**21**(42):4254-4258

[40] Wang T, Chen H, Liu K, Wang S, Xue P, Yu Y, et al. Janus Si micropillar arrays with thermal-responsive anisotropic wettability for manipulation of microfluid motions. ACS Applied Materials & Interfaces. 2015;**7**(1):376-382

[41] Wang Y, Lai C, Hu H, Liu Y, Fei B, Xin JH. Temperature-responsive nanofibers for controllable oil/water separation. RSC Advances. 2015;**5**(63):51078-51085

[42] Li J-J, Zhu L-T, Luo Z-H. Electrospun fibrous membrane with enhanced swithchable oil/water wettability for oily water separation. Chemical Engineering Journal. 2016;**287**:474-481

[43] Frysali MA, Anastasiadis SHJL. Temperature-and/or pH-responsive surfaces with controllable wettability: From parahydrophobicity to superhydrophilicity. Langmuir. 2017;**33**(36):9106-9114

[44] Li X, Jiang Y, Jiang Z, Li Y, Wen C, Lian JJASS. Reversible wettability transition between superhydrophilicity and superhydrophobicity through alternate heating-reheating cycle on laser-ablated brass surface. Applied Surface Science. 2019;**492**:349-361

[45] Wang R, Hashimoto K, Fujishima A, Chikuni M, Kojima E, Kitamura A, et al. Light-induced amphiphilic surfaces. Nature. 1997;**388**(6641):431-432

[46] Pant R, Singha S, Bandyopadhyay A, Khare K. Investigation of static and dynamic wetting transitions of UV responsive tunable wetting surfaces. Applied Surface Science. 2014;**292**:777-781

[47] Tadanaga K, Morinaga J, Matsuda A, Minami T. Superhydrophobic–superhydrophilic micropatterning on flowerlike alumina coating film by the sol–gel method. Chemistry of Materials. 2000;**12**(3):590-592

[48] Feng X, Feng L, Jin M, Zhai J, Jiang L, Zhu D. Reversible super-hydrophobicity to super-hydrophilicity transition of aligned ZnO nanorod films. Journal of the American Chemical Society. 2004;**126**(1):62-63

[49] Feng X, Zhai J, Jiang L. The fabrication and switchable superhydrophobicity of TiO_2 nanorod films. Angewandte Chemie, International Edition. 2005;**44**(32):5115-5118

[50] Jiang W, Wang G, He Y, Wang X, An Y, Song Y, et al. Photo-switched wettability on an electrostatic self-assembly azobenzene monolayer. Chemical Communications. 2005;**28**:3550-3552

[51] Zhang X-T, Sato O, Taguchi M, Einaga Y, Murakami T, Fujishima A. Self-cleaning particle coating with antireflection properties. Chemistry of Materials. 2005;**17**(3):696-700

[52] Cebeci FÇ, Wu Z, Zhai L, Cohen RE, Rubner MF. Nanoporosity-driven superhydrophilicity: A means to create multifunctional antifogging coatings. Langmuir. 2006;**22**(6):2856-2862

[53] Zhang L, Dillert R, Bahnemann D, Vormoor M. Photo-induced hydrophilicity and self-cleaning: Models and reality. Energy & Environmental Science. 2012;**5**(6):7491-7507

[54] Nau M, Seelinger D, Biesalski M. Independent two way switching of the wetting behavior of cellulose-derived nanoparticle surface coatings by light and by temperature. Advanced Materials Interfaces. 2019;**6**(17):1900378

[55] Shami Z, Holakooei P. Durable light-driven three-dimensional smart switchable superwetting nanotextile as a green scaled-up oil–water separation technology. ACS Omega. 2020;**5**(10):4962-4972

[56] Zhang X, Jin M, Liu Z, Tryk DA, Nishimoto S, Murakami T, et al. Superhydrophobic TiO_2 surfaces: Preparation, photocatalytic wettability conversion, and superhydrophobic–superhydrophilic patterning. Journal of Physical Chemistry C. 2007;**111**(39):14521-14529

[57] Zhang X, Kono H, Liu Z, Nishimoto S, Tryk DA, Murakami T, et al. A transparent and photo-patternable superhydrophobic film. Chemical Communications. 2007;**46**:4949-4951

[58] Zhu W, Feng X, Feng L, Jiang L. UV-manipulated wettability between superhydrophobicity and superhydrophilicity on a transparent and conductive SnO_2 nanorod film. Chemical Communications. 2006;**26**:2753-2755

[59] Wang S, Song Y, Jiang L. Photoresponsive surfaces with controllable wettability. Journal of Photochemistry and Photobiology, C: Photochemistry Reviews. 2007;**8**(1):18-29

[60] Wang S, Feng X, Yao J, Jiang L. Controlling wettability and photochromism in a dual-responsive tungsten oxide film. Angewandte Chemie, International Edition. 2006;**45**(8):1264-1267

[61] Pan S, Ni M, Mu B, Li Q, Hu X-Y, Lin C, et al. Well-defined pillararene-based azobenzene liquid crystalline photoresponsive materials and their thin films with photomodulated surfaces. Advanced Functional Materials. 2015;**25**(23):3571-3580

[62] Ge H, Wang G, He Y, Wang X, Song Y, Jiang L, et al. Photoswitched wettability on inverse opal modified by a self-assembled azobenzene monolayer. ChemPhysChem. 2006;**7**(3):575-578

[63] Blasco E, Piñol M, Oriol L, Schmidt BVKJ, Welle A, Trouillet V, et al. Photochemical generation of light responsive surfaces. Advanced Functional Materials. 2013;**23**(32):4011-4019

[64] Takase K, Hyodo K, Morimoto M, Kojima Y, Mayama H, Yokojima S, et al. Photoinduced reversible formation of a superhydrophilic surface by crystal growth of diarylethene. Chemical Communications. 2016;**52**(42):6885-6887

[65] Lim HS, Han JT, Kwak D, Jin M, Cho K. Photoreversibly switchable superhydrophobic surface with erasable and rewritable pattern. Journal of the American Chemical Society. 2006;**128**(45):14458-14459

[66] Tian D, Zhang X, Tian Y, Wu Y, Wang X, Zhai J, et al. Photo-induced water–oil separation based on switchable superhydrophobicity–superhydrophilicity and underwater superoleophobicity of the aligned ZnO nanorod array-coated mesh films. Journal of Materials Chemistry. 2012;**22**(37):19652-19657

[67] Chen L, Wang W, Su B, Wen Y, Li C, Zhou Y, et al. A light-responsive release platform by controlling the wetting behavior of hydrophobic surface. ACS Nano. 2014;**8**(1):744-751

[68] Zhang J, Li J, Han Y. Superhydrophobic PTFE surfaces by extension. Macromolecular Rapid Communications. 2004;**25**(11):1105-1108

[69] Zhang J, Lu X, Huang W, Han Y. Reversible superhydrophobicity to superhydrophilicity transition by extending and unloading an elastic polyamide film. Macromolecular Rapid Communications. 2005;**26**(6): 477-480

[70] Chung JY, Youngblood JP, Stafford CM. Anisotropic wetting on tunable micro-wrinkled surfaces. Soft Matter. 2007;**3**(9):1163-1169

[71] Lin P-C, Yang S. Mechanically switchable wetting on wrinkled elastomers with dual-scale roughness. Soft Matter. 2009;**5**(5):1011-1018

[72] Yang S, Khare K, Lin P-C. Harnessing surface wrinkle patterns in soft matter. Advanced Functional Materials. 2010;**20**(16):2550-2564

[73] Roy PK, Ujjain SK, Dattatreya S, Kumar S, Pant R, Khare KJAPA. Mechanically tunable single-component soft polydimethylsiloxane (PDMS)-based robust and sticky superhydrophobic surfaces. Applied Physics A: Materials Science & Processing. 2019;**125**(8):535

[74] Wang H, Zhang Z, Wang Z, Liang Y, Cui Z, Zhao J, et al. Multistimuli-responsive microstructured superamphiphobic surfaces with large-range, reversible switchable wettability for oil. ACS Applied Materials & Interfaces. 2019;**11**(31):28478-28486

[75] Li Z, Liu Y, Lei M, Su A, Sridhar S, Li Y, et al. Stimuli-responsive gel impregnated surface with switchable lipophilic/oleophobic properties. Soft Matter. 2020;**16**(6):1636-1641

[76] Bai D, Habersberger BM, Jennings GK. pH-responsive copolymer films by surface-catalyzed growth. Journal of the American Chemical Society. 2005;**127**(47):16486-16493

[77] Xu C, Wayland BB, Fryd M, Winey KI, Composto RJ. pH-responsive nanostructures assembled from amphiphilic block copolymers. Macromolecules. 2006;**39**(18):6063-6070

[78] Li J, Chen X, Chang Y-C. Prseparation of end-grafted polymer brushes by nitroxide-mediated free radical polymerization of vaporized vinyl monomers. Langmuir. 2005;**21**(21):9562-9567

[79] Uhlmann P, Ionov L, Houbenov N, Nitschke M, Grundke K, Motornov M, et al. Surface functionalization by smart coatings: Stimuli-responsive binary polymer brushes. Progress in Organic Coating. 2006;**55**(2):168-174

[80] Lu Y, Sarshar MA, Du K, Chou T, Choi C-H, Sukhishvili SA. Large-amplitude, reversible, pH-triggered wetting transitions enabled by layer-by-layer films. ACS Applied Materials & Interfaces. 2013;**5**(23):12617-12623

[81] Minko S, Müller M, Motornov M, Nitschke M, Grundke K, Stamm M. Two-level structured self-adaptive surfaces with reversibly tunable properties. Journal of

the American Chemical Society. 2003;**125**(13):3896-3900

[82] Lee CH, Kang SK, Lim JA, Lim HS, Cho JH. Electrospun smart fabrics that display pH-responsive tunable wettability. Soft Matter. 2012;**8**(40):10238-10240

[83] Wang F, Pi J, Li J-Y, Song F, Feng R, Wang X-L, et al. Highly-efficient separation of oil and water enabled by a silica nanoparticle coating with pH-triggered tunable surface wettability. Journal of Colloid and Interface Science. 2019;**557**:65-75

[84] Cheng Z, Lai H, Zhang N, Sun K, Jiang L. Magnetically induced reversible transition between Cassie and Wenzel states of superparamagnetic microdroplets on highly hydrophobic silicon surface. Journal of Physical Chemistry C. 2012;**116**(35):18796-18802

[85] Grigoryev A, Tokarev I, Kornev KG, Luzinov I, Minko S. Superomniphobic magnetic microtextures with remote wetting control. Journal of the American Chemical Society. 2012;**134**(31):12916-12919

[86] Motornov M, Minko S, Eichhorn K-J, Nitschke M, Simon F, Stamm M. Reversible tuning of wetting behavior of polymer surface with responsive polymer brushes. Langmuir. 2003;**19**(19):8077-8085

[87] Liu Y, Mu L, Liu B, Kong J. Controlled switchable surface. Chemistry–A European Journal. 2005;**11**(9):2622-2631

[88] Wang X, Qing G, Jiang L, Fuchs H, Sun T. Smart surface of water-induced superhydrophobicity. Chemical Communications. 2009;**19**:2658-2660

Chapter 9

Microfluidic Devices: Applications and Role of Surface Wettability in Its Fabrication

Shivanjali Saxena and Rakesh Joshi

Abstract

Microfluidic devices are based upon the behavior of fluids at the microenvironment level. They offer innumerable applications in the field of science and technology. Their scope is not limited to single field and now have applications in various fields such as biomedical, energy, chemicals and environment as well. Their major advantages are low experiment to cost ratio, and fast response time. Surface wettability is one of the factors contributing to the working of microfluidic devices. Surface wettability measurement is a very critical technique to measure the flow of micro fluids in microfluidic applications. In microfluidic devices the detection of small volume change with change in fluid properties is very minor because of the micrometer range. In order to detect this small change in micrometer range, an *in situ* wetting measurement is required. In this chapter, we have discussed about types of taxis, microfluidic devices: an application of taxis, microfluidic applications and role of surface wettability in microfluidic devices.

Keywords: microfluidic, devices, surface wettability, lab-on-chip, applications, biomedical

1. Introduction

The era of microfluidics started in 1980s with the development of silicon etching procedures which were made for microelectronics industry. This paved way for manufacturing of first of its kind devices called Micro Electro Mechanical Systems (MEMS). In these devices mechanical microelements were integrated together on a silicon wafer. In the 1990s, researchers explored applications of these devices in the field of biology, chemistry and biomedical. They used these devices for controlling liquid's movement in micro channels which paved way for microfluidics. Laboratories on chip were developed for incorporating all the major procedures of biology, chemistry or biomedical on single platform. But, this use to come with huge cost and infrastructure for microelectronics industry. In 2000s a new era of microfluidics was started with the development of molding micro channels in polymers. This lead to decrease in cost as well as manufacturing time that caused the boom in the area of microfluidics and motivated researchers of all fields to work using them [1]. In the era of fast pacing science, microfluidic are devices to enhance pace of research and decrease the experimental cost. They run on principle of various types of taxis, majorly chemotaxis. Taxis is the

IntechOpen

movement of particles according to some external guiding agent. This agent can be heat, oxygen, pressure, electric field or chemical, etc. Various types of taxis, their applications and microfluidics are discussed in this chapter. Microfluidic is the technology where movement of the particles is on the basis of microenvironment consisting of viscosity, surface tension and pressure. In microfluidics, micro channels are molded or etched over the silicon, glass or various polymer materials such as PolyDimethylSiloxane. These types of devices are vastly being used in all the fields of research, diagnostics and therapeutics. In microfluidics, the micro channels are formed to attain the desired result which can consist of: mixing, sorting, pumping or controlling the biochemical microenvironment. They have the advantage of decreasing the response time and experimental consumables and overall cost. They have the potential to perform large scale experimentation in small scale. Important factors to be considered for fabrication of microfluidic devices are temperature resistance, superior optical transparency of the material, high hardness, excellent electrical isolation, thermal stability, chemical inertness to many fluids, biocompatibility, and surface wettability. The performance of microfluidic device depends majorly upon etched or molded micro channel's surface properties. Therefore surface modification is an important factor to improve overall performance of microfluidic devices. Surface roughness, surface heterogeneity and solution impurity are the key parameter which affects the wettability of microfluidic device.

2. Introduction to taxis

Molecules are always in motion irrespective of their state. Molecules in solid state have least freedom while in gaseous state have maximum freedom. Freedom of movement of molecules in liquid phase lies between molecules in solid and gaseous phases. Heat and temperature are factors that affect the movement of molecules. Enhance temperature increases the translational movement of molecules. Movement of molecules can be random or directed towards certain stimuli. Random movement is termed as "kinesis" while directed movement towards certain stimuli is termed is "taxis". For taxis, there is sensory component to detect the attractant and motor component to enable the movement towards the stimuli [2]. Taxis are classified on the basis of the stimulus into various categories.

3. Types of taxis

3.1 Aerotaxis

Aerotaxis is the movement of molecules where oxygen acts as stimulant [3]. It has been observed in bacteria and other microorganisms. Active movement of cells is observed along the gradient of oxygen. Aerotaxis plays role in cell survival as optimal concentration of oxygen is required for cell metabolism and growth [4].

3.2 Anemotaxis

Anemotaxis is the movement towards wind. It is observed in drosophila and some terrestrial mammals such as rats which tend to follow the wind. Drosophilla has been observed to move against the air current [5]. Rats were also observed to follow anemotaxis as air current carries information regarding location and odor content [6].

3.3 Barotaxis

Barotaxis is the movement towards stimulus that is pressure. Movement due to hydraulic resistance (resistance to flow as a result of liquid) is also termed as barotaxis. This type of movement is observed in the neutrophils, a type of immune cell. Neutrophils follows path of least hydraulic resistance.

3.4 Chemotaxis

Chemotaxis can be defined as the movement of cells towards the higher chemical concentration gradient [7]. It is directional locomotion of cells and was first described in the bracken fern spermatozoa in 1884 by Pfeffer. Later in 1888, this phenomenon was described by Leber in mammalian leukocytes in response to an injury. Chemotaxis is an important process required for the growth and development of multicellular organism, immune response and cancer metastasis [8].

3.5 Durotaxis

Durotaxis is the movement of cells towards more rigid gradient which is a result of variation in the structural property of the extracellular matrix. This type of motion implies movement towards more stiffness [9]. This type of motion has been observed in various cell types such as human fibroblast cells, mesenchymal cells and cancer cells. Substrate rigidity is the stimulus that initiates the movement in durotaxis [10].

3.6 Electrotaxis/galvanotaxis

Electrotaxis is also termed as galvanotaxis and implies movement guided through electric field or current [11]. Living cells have the tendency to sense and follow direct current electric field. This type of movement is observed in both *in vitro* and *in vivo* conditions although the mechanism behind sensing of electric field by cells remains unclear [11]. Its applications are observed in wound healing and development. Disruption of an epithelial layer in wound leads to generation of an endogenous electric field which guides migration of cells towards the wound for regeneration [12].

3.7 Gravitaxis

Gravitaxis is characterized by directional movement in response to gravity [13]. This type of movement is observed in the motile microorganisms such as euglena where gravity acts as stimulus to select their niche in environment. It can be both positive and negative. Positive gravitaxis implies movement towards water while negative gravitaxis implies movements towards the surface [14]. This type of motion is observed in *Drosophila melanogaster* and around 18 genes have been identified that mediate this gravitational motion in them [15]. To elucidate the mechanism of this type of motion, asymmetric self-propelled particles were studied for this motion. It was observed that shape anisotropy alone is sufficient to induce such type of motion [16].

3.8 Hydrotaxis

Moisture acts as stimulus in hydrotaxis. Movement of cells, animals or plants towards more moisture is termed as positive hydrotaxis and towards less moisture is termed as negative hydrotaxis. Hydrotaxis is observed in the *C. elegans* as they move towards their preferred water content for mating, geographical distribution and reproduction [17]. It is also observed in the cyanobacterium in desert crusts. Cyanobacteria colonies are observed 1.5–2.0 mm deep into the desert crust but when crust surface is saturated with water, cyanobacterium moves towards the surface having higher moisture content [18].

3.9 Magnetotaxis

Magnetotaxis is the movement due to magnetic field. This type of movement is a character of diverse group of gram-negative bacteria that perform their orientation and coordination movements according to earth's magnetic field [19]. They are majorly aquatic and swim along the geomagnetic field lines. These types of bacteria are also termed as magnetotactic bacteria [20]. Supramolecular adaptive nanomoters have been developed that exhibit magnetotactic behavior and their guided motion is observed in the tissue model [21].

3.10 Phototaxis

Phototaxis is the movement towards or away from the light source. This type of movement is characteristic of phototrophic organisms and is also observed in plants. Prokaryotes use type-I sensory rhodopsin photoreceptors for phototaxis and it allows them movement towards steep light gradient. Cyanobacteria can also perform phototaxis but they also can perform it in two-dimension only through gliding on the surface. Eukaryotes have the ability to navigate through light vector in three-dimension in open water [22].

3.11 Rheotaxis

Rheotaxis is the movement in response to water or air current. This type of motion is observed in aquatic animals where their movement occurs in response to water current [23]. When movement is towards oncoming water current, it is termed as positive rheotaxis while movement opposite of oncoming water current is termed negative rheotaxis [24]. This type of motion is observed in zebrafish, Crustaceans and American lobsters [25].

3.12 Thermotaxis

Thermotaxis is the movement towards or away from temperature gradient. In this motion, organism move towards temperature source. Slime molds and nematodes are known to move along shallow temperature gradient [26, 27]. Mammalian sperm is also observed to perform theromtaxis to reach towards the oviduct in the female body [28].

4. Microfluidic devices: an application of Chemotaxis

Microfluidics is the technology based upon behavior of fluids in the microenvironment. Fluids tend to behave very differently in micrometric scale as compared to macro scale. These characteristics of fluids are now been used for various studies based upon taxis. In macroscopic system, pressure, volume and temperature are the key players whereas viscosity, surface tension, high shear rate and geometric effects (high surface to volume ratios, constriction, and bifurcation) are the key drivers of the microfluidic system [29]. Microfluidics is the integration of fluids physically restricted to sub-millimeter dimensions with micro/nanostructures and devices [30]. Microfluidics is an emerging interdisciplinary field consisting of engineering, physics, chemistry, microtechnology, biotechnology and material sciences [31]. The reason for its emergence is miniaturization of operational unit in the microfluidic devices. Miniaturization is preferred as all operations can be packed in small form that can be automated and is portable [32]. Low amount of materials and chemicals are required for development and samples required is also less. Automation enables widespread use of the system without any special training requirements. Easy disposals, low cost, reduction of cross-contamination and fast response time are other benefits of the microfluidic system [33].

The global size of microfluidic devices was USD 13.5 billion in 2019 and is supposed to have a compound annual growth rate of 11.3%. The large market size is due to its multi-application and ease of usability. Basic layout of microfluidic devices consist of incorporated fluid channels in at least one direction. These channels provide high surface to volume ratio which is useful in applications such as biochemical analysis, antimicrobial susceptibility test and heat exchange modules. This field started with applications in chromatography and electrophoresis [34]. With time it has evolve and currently it has vast applications due to development of new fabrication materials and technologies [31]. Its applications include environmental sensing, biomedical applications, drug discovery, drug delivery, micro scale energy systems, artificial organs, micro scale chemical testing and production, micro propulsion, combinatorial synthesis and assays. These applications have been classified under broad categories for discussion in this chapter.

4.1 Biomedical applications

Microfluidics can be used in biomedical field as analytical arrays, gradients, separators, microdiluters, gel structures, droplets, painting cells and devices [35]. In arrays, a set of multiple microchannels is used to study the relationship between different cells with proteins or chemicals within a combinatorial system. This type of system can be used for detection of specific proteins in large number of samples, antibiotic resistance testing, etc. Microfluidics can be used for generation of very steep gradients that cannot be created using other macro techniques [35]. These gradients are useful in the study of macromolecules and cells in response to their varying environment. Biochemical gradients are useful in dictating physiological processes such as proliferation, differentiation and migration. These gradients play an important role in tissue generation as well. They are used for organ on chip techniques also. Phil et al. used drug gradients for activity measurement over CHO cells [36]. Migration and behavior of neutrophils according to protein gradient has also been studied [37]. Chung et al. used growth factor gradient to study the differentiation of human neural stem cells [38].

Microfluidics can be used as diluters where solution is passed through series of controlled dilutions to be used in a specific assay. Ainla et al. have shown use of pulse width flow modulation based designing of microdiluter [39]. They used this microfluidic diluter for analyzing the effect of Ca(2+) concentration over phospholipid bilayer spread onto a SiO2 surface. Microdiluters can also be used as immunoassays for detection of multiple antigens at a same time [40]. Microfluidics can be used in conjunction with gels or microchannels can be made in gels using soft lithography technique. Various types of gels in which microfluidic can be fabricated are agarose, agar and calcium alginate. These types of systems can be used to study complex microenvironment of cells. Takeuchi et al. used microchannels fabricated in agarose to grow *Escherichia coli* in presence of various molecules that can alter their phenotype [41]. Cabodi et al. used alginate based microchannels for study of mass transfer in channels [42]. Complex Microfluidic systems are now being highly researched and commercialized to develop point of care/lab-on-chip (LOC) devices and organ on chip. These devices have high potential as they can provide the customer with the easy of usability, less sample requirement, time and cost efficiency.

Point of care devices are diagnostic measures that are directly used by patients and without requirement of medical staff. A simple paper based microfluidic that can be used as point of care device are known as lateral flow test (LFT). Porous material such as glass fiber, nitrocellulose and cellulose paper can be used for fabrication of LFT. The components on microfluidic LFT device are sample collection pad, a dried conjugate pad followed by a reaction area and an absorbent wicking pad. This is incorporated within a plastic housing and plastic barriers throughout to maintain one dimensional flow. The best example of LFT is dipstick pregnancy test kit. This test works on the principle of an immunoassay. Sample which is urine is applied to the sample pad and rehydrates the goldnanoparticles conjugated detection antibodies [43]. These rehydrates antibodies bind to the target antigen present in the sample. Together they flow to the capture region which consists of control and test line. At the test line, non-labeled antibodies specific for the detection antigen are immobilized. When rehydrated labeled antibodies conjugated with the sample antigen reaches test line, it binds to the non-labeled antibodies specific for that same antigen. This interaction gives visual color change thereby making test line visible in case of positive results. This process is depicted through **Figure 1a** and **b**. The wicking pad in the device performs function of attracting the sample through LFT. After reaction membrane is completely wetted, the capture region functions through capillary action.

Paper microfluidics has also been used to provide point of care diagnostics for non-communicable diseases such as cardiovascular disease and cancer. In this

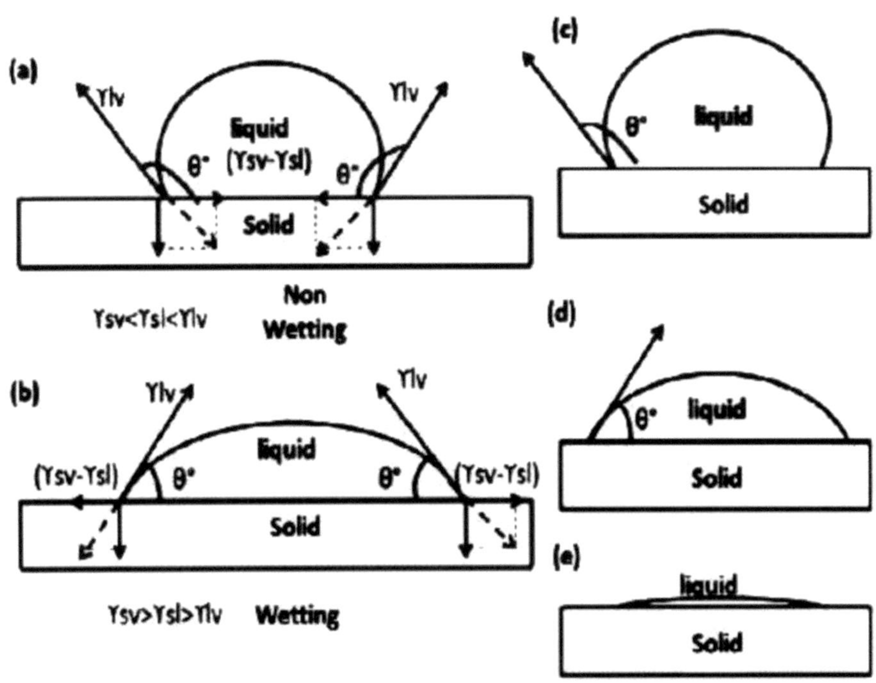

Figure 1.
(a) Non-wetting phenomenon, (b) wetting phenomenon, (c) larger contact angle (non wetting), (d) wetting, (e) angle close to zero complete (wetting).

work, synthetic urinary biomarker is used which is detected through paper microfluidics [44]. These types of devices are also being used for saliva based detection of oral diseases. In the research work by Amy et al. point of care diagnostic device for oral diseased was developed using monolithic disposable cartridge. It was designed in a compact analytical device. This device combined sample pre-treatment procedure of filtering, enrichment and mixing of sample with electrophoretic immunoassays. It can efficiently and quickly measure analyte concentration in the minimally treated and very low volume (20 μl) of saliva sample [45]. Microfluidic devices are also used for digital polymerase chain reaction (PCR) which is a very powerful gene expression analytical tool. Christina et al. showed use of microfluidic based digital PCR for prenatal detection of fetal aneuploidy. Fetal Aneuploidy is the presence of an abnormal number of chromosomes (structures that contain genetic information) in the fetus [46].

Organ on chip is the new class of laboratory models that have advantages of both *in vivo* and *in vitro* models. These chips are microfluidic devices in which tissue of interest is cultured in the favorable microenvironment simulating the actual physiological conditions efficiently. These types of devices can also be used in the field of personalized medicine. For personalized medicine, cells from specific donor patients and healthy patients can be studied under the same environment. Various examples of such devices are lung on a chip [47], atherosclerosis on a chip that made study of physiological functions of an organ and its response to various stimuli feasible [48]. Other examples are bacteria inhabited gut on a chip [49] and blood brain barrier on a chip [50]. This field of organ on chip is emerging rapidly and showcasing various organs' culture and their physiological microenvironment simulation on these microfluidic chips.

4.2 Energy applications

The development in the field of integrated microfluidics was successfully laid by its incorporation with the optical elements such as plasmonic surfaces [51] and waveguides [52]. In 2000s, the development of liquid-crystal switchable gratings, microfluidically tunable photonic crystal fibers and bubble switch laid the foundation of using microfluidics as an essential part of the photonic devices. During mid-2000s a new field of "optofluidics" was evolved from the existing technologies in the field of photonics and microfluidics. Using microchannels and photonic elements, optofluidics has the strength of having precise control over light and fluidics at small scale [53]. Microfluidic systems are being used for development of photocatalytic microreactor. A planar microfluidic reactor was developed by Lei et al. It consisted of the small planar chamber where two $TiO2$ coated slides were used as top cover and bottom substrate. Microstructured UV-cured NOA81 layer was used as the sealant and flow input/output. This reactor has advantages of microfluidics such as easy control of flow, rapid fabrication and large surface/volume ratio. It is the key to more efficient photocatalytic water treatment [54]. $TiO2$ based microreactor has been developed by Matic et al. for photocatalytic applications. This system was fabricated on metal-titanium foil. Titania nanotubes were mechanically engraved in the substrate foil. Using anodization & hydrothermal treatment $TiO2$ anatase film was immobilized over the inner layer of these tubules. An additional $TiO2$ anatase layer was added on top of the film to provide larger photocatalytic area. This microreactor depicted enhanced durability and efficiency [55]. Meng et al. also developed microfluidic based photocatalytic microreactor. They used nanofibrous $TiO2$ through electrospun to develop this photocatalytic microreactor. It depicted enhance efficiency as compared to $TiO2$ film based microreactor [56].

Recently, applications of microfluidics have been developed in the form of microfluidic fuel cells. In these cells, all the systems such as fluid delivery, removal, etc. is confined to the microfluidic channel only. These cells do not require a physical barrier for separation of fuel and oxidant species and therefore they operate in co-laminar flow mode. Whereas, in conventional fuel cell a physical barrier such as proton exchange membrane is required. They can be used to power microsystems, generate on-chip power and in consume electronics as well [31]. Microfluidic fuel cells have attracted huge researchers as they are portable power sources with short startup time and environment friendly nature. Microfluidic fuel cell using laminar air flow had been developed by Eric et al. (**Figure 2a**). It was made through a Y-shaped microchannel consisting of two catalyst covered electrodes on opposite walls. Through these channels, fuel and oxidant merge and flow laminarly parallel between these two electrodes without turbulent mixing. They showed that this type of system can be effectively used to generate microscopic power source for room temperature [57]. There is patented microfluidic fuel cell system for portable energy applications. In this system, microfluidic container, substrate for catalytic composition, a liquid/gas separator, a fuel cell consisting of anode and cathode and electrical connections were all assembled to form this portable energy system [58]. The design of the system and fuel cell components is depicted through **Figure 2b** and **c**, respectively [58]. Luke et al. also developed these microfluidic cells based on microbial fuel that can be used to provide power supply to integrated biosensors. This system was developed in polydimethylsiloxane. Here, two carbon cloth electrodes and proton exchange membrane was used. *Shewanella oneidensis* MR-1 was used in anode chamber as electrogenic bacterial strain and ferricyanide was used in cathode chamber (**Figure 2d**). Maximum current of 2.59 μA was generated using this miniature microbial fuel cell [59]. Svetlana et al. developed a microfluidic cell for energy conversion. They developed hydrogen and oxygen based microfluidic cell using polydimethylsiloxane (PDMS) device. In this

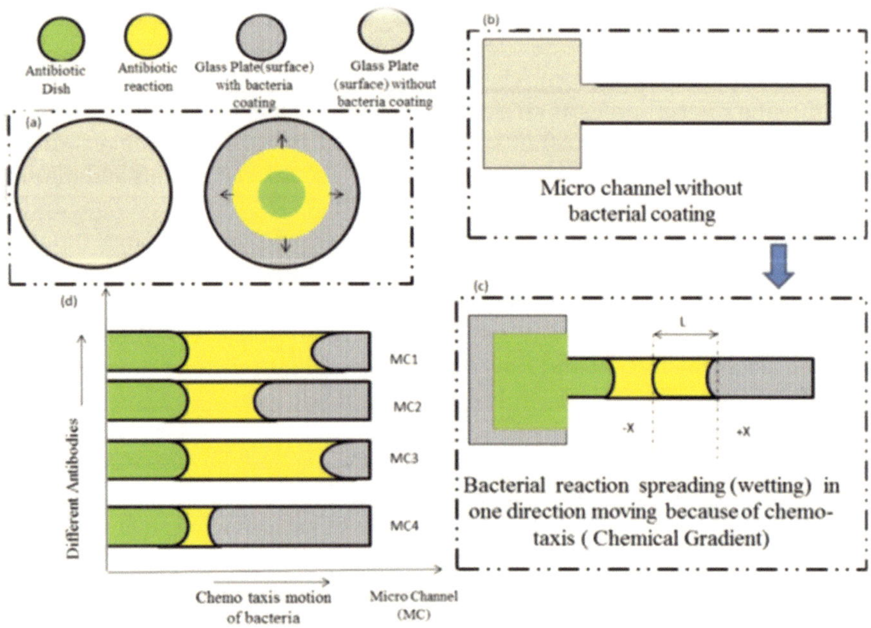

Figure 2.
(a) The reaction of antibiotic dish with bacteria, (b) glass microchannel without bacterial coating, (c) reaction of antibiotic with bacteria in microchannel, and (d) the spreading of chemical reaction in microchannel with different antibodies.

device Pt/quartz electrodes in the form of thin film were embedded into the device. The PDMS microchannel network containing liquid electrolyte was used for immersion of electrode array into it. This also performs the function of thin glass permeable membrane for feeding reactants to the electrodes. This fuel cell operated at room temperature with the maximum power density of 700 μW/cm^2. The overall lifetime of this cell was comparatively higher to the exiting higher surface electrodes based fuel cells [60].

4.3 Chemical and environmental applications

Recent growth in the field of microfluidics has been observed in the field of environmental assessment. Microfluidics is advantageous as multiple processes such as pre-treatment, pre-concentration, separation and detection are incorporated at the same platform. It is used for trace analysis of materials as less risk of contamination is there due to preclusion of sample transportation process. Microfluidics play role in the development of subsurface energy based technologies in the future. Mark et al. developed a microfluidic system based upon high temperature and pressure. Within geo-material micromodels such as rock, cement, clay, etc., direct observations for flow and transport can be made using this system and that too in reservoir conditions. In this micromodel fabrication method, 3D tomography images of real fractures were used as micromodel template. This provided better representation of the pore space and fracture geometries in subsurface formations [61]. Several microfluidic devices can be used for detection and analysis based upon electrochemistry, surface enhanced Raman spectroscopy, chemiluminescence, absorbance and laser-induced fluorescence. These electrochemical and optical based systems can be conjugated on a single micro platform to perform environmental monitoring. These labs on chip systems can be used for real time tracking of pollutants in the environment. Major advantages of these systems are portable compact size, better process control, low-cost production, real-time analysis, low sample consumption and fast response. LOC is used for real-time analysis of pollutants in wastewater. Combining it with the wireless communication, make it a strong tool for modifying data acquisition parameters and data transfer [62].

Microfluidic systems are being used for detection of formaldehyde as well. Formaldehyde is the organic volatile compound found in many household products. It is associated with health risk factors and is also a cause of sick building syndrome. Therefore its detection at real-time in the surroundings is essential for a healthy living. Liu et al. developed a paper based microfluidic system for detection of formaldehyde. Acetoacetanilide reagent is used to implant paper-based chip at reaction site. Concentration of formaldehyde is detected using UV light which induces fluorescence intensity in the dihydropyridine. Dihydropyridine is the complex of formaldehyde with acetoacetanilide. This method was used to detect formaldehyde in the commercial food samples and proved to be an efficient method for detection of formaldehyde concentration [63]. Similarly, Czugala et al. developed a fully integrated microfluidic device to provide wireless and portable analytical platform. This system can be used for detection of nitrite anions in the water. Nitrite anions are one of the water contaminants along with lead, cadmium and nitrate. In this system detection is done through analysis of color intensity of complex formed between nitrite anions and Griess reagent. This color intensity was assessed using low cost Paired Emitter Detector Diode. Biomimetic photo-responsive ionogel microvalve controlled by LED was used for manipulation of on-chip fluid. This system was one of its type that conjugated fully functional microfluidics with photobased valving and photo detection [64]. Microfluidic devices along with porous plugs have also been developed. This device can be used for size based separation of particles including microorganisms and therefore have implications as miniature filter for analysis of water samples. Living radical photopolymerization technique using wide range of polymers was used for fabrication of these devices. Salt-leaching technique was used for placement of porous plug in the microfluidic channels. Pore size of the porous plug in this device was determined using flow field-flow fractionation. It is a new and cost efficient simple tool for water assessment [65]. Research is moving at a fast pace for development and commercialization of such paper based microfluidic devices that can be conjugated with other existing techniques.

5. Role of In situ surface wettability for the development of microfluidic devices

Surface wettability or wetting is the ability of the liquid to maintain contact and interact with the solid surface over which it is flowing. It results from the interaction of intermolecular forces between the molecules of liquid and molecules over the surface of the solid. Surface wettability measurement is a very critical technique to measure the flow of micro fluid in microfluidic applications. In microfluidic devices the detection of small volume change with change in fluid properties is very small because of the micrometer range. In order to detect this small change in micrometer range *in situ* wetting measurement is required. Microfluidic devices offer innumerable application in the field of science and technology. The scopes of these types of devices have been increasing for recent decades. For example, in clinical trials for drug development the amount of antibodies used is very high and cover large cost. On the other hand microfluidic devices reduce amount and cost of antibodies as well as time.

In microfluidic devices the motion of chemical reaction governed by chemo taxis gradient and this gradient is responsible for the motion of droplet. The motion of droplet generally measured with the help of wettability and wettability of droplet depends upon the surface. Wettability has a dynamic impact on the displacement of fluid inside micro fluidic device. The change in displacement of any fluid inside any microfluidic device measured in term of spreading of fluid. The spreading behavior of any flowing liquid measured with its wetting behavior and it is generally measured in term of contact angle. The magnitude of contact angle formed by micro fluid with micro-channel wall has great importance to study the characteristics of micro fluidic device.

Example: Suppose a static fluid is placed at the center of any plate and we apply taxis gradient at the two end of plate. The taxis gradient (magnetic, chemotaxis) is responsible for the displacement of fluid inside. The fluid try to spread both in linear (parallel to gradient axis) and lateral (perpendicular to direction of applied gradient) direction. The two directional spreading of liquid makes difficulty in the quantitative measurement in displaced liquid. In order to overcome this issue microfluidic devices play the important role in various scientific testing applications.

The contact angle measurement is carried out using young's equation is given in Eq. (1). The equation is derived by balancing different interfacial energy in all direction.

$$\cos\theta = \left(\sigma_{sv} - \sigma_{sl}\right) / \sigma_{lv} \tag{1}$$

where θ = contact angle

σ_{sv} = solid/vapor interfacial energy

σ_{sl} = solid/liquid interfacial energy
σ_{lv} = liquid/vapor interfacial energy.

Wettability of fluid over the solid surface is measured in terms of contact angle (θ). The higher value of contact angle leads to lower wettability (low spreading area of displaced fluid) as shown in **Figure 1c**. The contact angle close to 0°, as droplet turns into flat puddle shows complete wetting (highest spreading area) as shown in **Figure 1e**, if angle exceeds zero but is less than 90° as shown in **Figure 1d** shows wetting [66].

In microfluidic devices the fluid displacement takes place only in linear direction because of micro channel cavity. The quantitative measurement of displaced fluid inside micro channel can be made by measuring the dimension of micro channel and displace length of fluid. The measurement of displace volume with little change in taxis gradient improves the overall sensitivity of device. Sensitivity of device is defined as the measurement of small change in the system by varying input parameter.

5.1 Case study 1

The existing process of antibiotic susceptibility measurement uses Petri dish coated with bacteria and divides the Petri dish into required number of segment using marker. An antibiotic dish (different concentrations) is then placed over the bacterial coated Petri dish. The petri dish is then placed over incubator for 24–48 hours. The reaction of bacteria with antibiotic takes place in petri dish and reaction takes place in radial outward direction as shown in **Figure 1a**. The measurement of reaction in radial direction is difficult to quantify in required scale. To overcome this issue a microfluidic device can be used as antibiotic susceptibility testing device. In this type of device, glass slide micro channel is coated with bacterial coating like petri dish as shown in **Figure 1b**. Different antibodies are then placed over bacterial coated micro channel for measuring the spreading of reaction

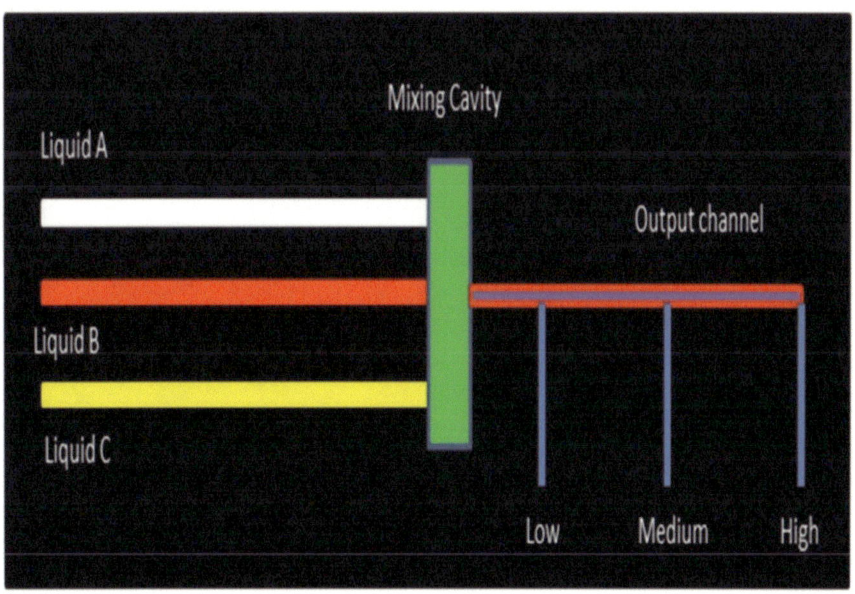

Figure 3.
Microfluidic device for mixing different liquid.

due to chemo taxis in one direction and chemo taxis spreading phenomenon can be quantify using microfluidic chip time lapse microscopy as shown in **Figure 1d**. The spreading of reaction is then measured by the dimension of micro channel as shown in **Figure 1c**.

5.2 Case study 2

In this study, microfluidic device for mixing three liquid is used. In this device, three different liquid A, B and C is used to mix in different concentration and their mixing reaction is measured with the range of output mixing micro channel as shown in **Figure 3**. In this type of device the change in output parameter can be detect significantly my using small volume of liquid droplet. These devices are very useful to measure mixing behavior of two or more liquid for various chemical mixing applications.

6. Factor affecting wettability

The wettability is generally are properties of displaced liquid measured in term of contact angle. The surface morphology, material impurity and porosity are the properties which affect the wettability.

Effect of surface roughness: All smother surfaces look rough in microscopic level. The rough surface of solid specimen affects the wettability of liquid over the solid surface. The contact angle formed with flat surface is called apparent contact angle θ_a and it is consider by considering ideal surface condition. The actual contact angle θ_A is generally higher than that of apparent contact angle θ_a as shown in **Figure 4a,b**. To calculate real surface free energies of liquid actual contact angle is used. Generally hydrophilic surface is considered to be the best surface where lower

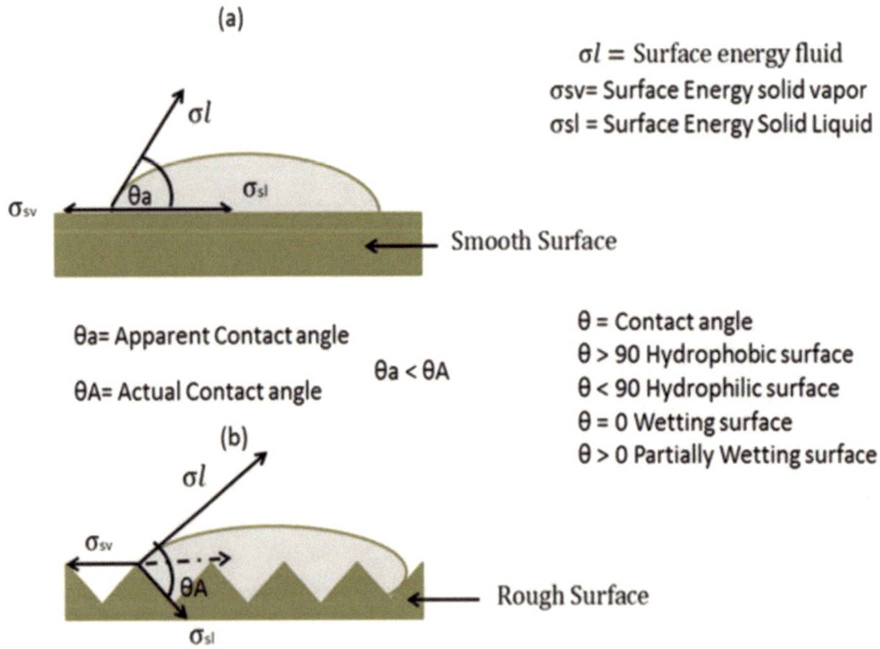

Figure 4.
Effect of surface roughness on the wettability of fluid.

value of contact angle is obtain. The wettability of liquid surface generally increases as we decrease the surface roughness of solid surfaces. The relation between roughness and wettability was explained by Wenzel and stated that if the surface is chemically hydrophobic it will become more hydrophobic when surface roughness is added. According to Wenzel,

$$\cos(\theta_A) = R\cos(\theta_a) \tag{2}$$

R is the surface ratio between actual and projected area of solid surface over which fluid is flowing. For smother surface R = 1 and apparent contact angle becomes equal to actual contact angle. Other than surface roughness, impurity and porosity in solid surface effect wettability.

7. *In situ* wettability measurement in microfluidic devices

In microfluidic device, the displacement of fluid takes place continuously, and it is very difficult to measure wettability (contact angle formed by moving fluid with the wall of micro channel). The Sessile drop method and image analysis techniques are the method used only for measuring the static contact angle of liquid in micro channel device. For biomedical and clinical application the chemotaxis reaction takes place continuously and required continuous monitoring of contact angle *in situ* image capturing system is used. In this technique the position of chemical reaction captured with the help of microscope and high speed camera installed over the viewing point of microscope [67]. The camera records the position of reaction in different time interval and measured the contact angle and contact angle is further used to measure interfacial energy of fluid in microfluidic devices. The schematic of *in situ* image capturing system is as shown in **Figure 5**.

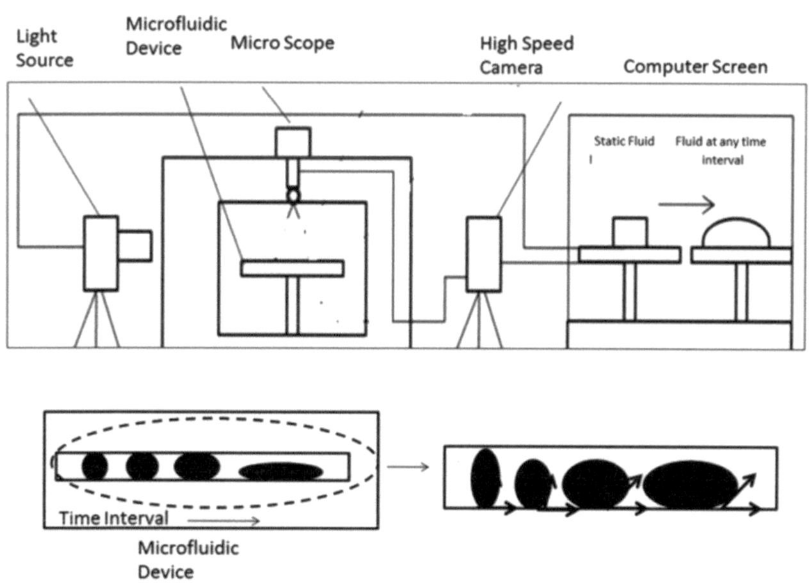

Figure 5.
In situ image capturing system for measuring wettability of microfluid devices.

In this system a microscope is just place at the top of microfluidic device and it captures the motion of chemical reaction change in micro channel. A light source is applied from the side to capture the video with more celerity with the help of high speed camera and store video into computer. The video is than sliced into image in required time interval as shown in lower left corner of figure. The Enlarge version of captured screen is shown in lower right corner of **Figure 5** which shows the measurement of contact angle variation at different time interval.

8. Fabrication process of microfluidic devices

The selection of microfluidic fabrication process is dependent on type of material selection for different microfluidic application. The special grade stainless steel, borosilicate glass, PDMS (polydimethylsiloxane), PMMA(Poly methyl metacrylate) copper, aluminum and Acrylic have been used as solid material for microfluidic device fabrication. Chemical etching, 3D printing, Additive manufacturing, micromachining are the common manufacturing practices for the development of different microfluidic devices [68–70].

9. Conclusion

Microfluidic devices are one of the most widely used devices of twenty-first century. They are being used in almost all the fields including biomedical, energy, chemical, environmental, etc. Microfluidics is the technology based upon various types of taxis, specifically chemotaxis. Surface wettability is an essential factor in the development of microfluidics. Elucidating mechanisms to improve surface wettability will help in the betterment of microfluidic devices. There are still unexplored applications of microfluidics such as in paint industry: to study the mixing and spreading of paints. Initially, microfluidics developed due to advancement in the field of silicon etching and molding of micro channels technique. The further advancement in the fabrication techniques will pave way for development of high-leveled microfluidic devices that will open a new era of research in all the fields.

Acknowledgements

We would like to acknowledge DARVUN India for providing us the opportunity of working for them.

Conflict of interest

The authors declare no conflict of interest.

Author details

Shivanjali Saxena and Rakesh Joshi*
Research Scientist, DARVUN India, Patiala, Punjab, India

*Address all correspondence to: joshi.1@iitj.ac.in

References

[1] Casquillas GV, Houssin T, Durieux L. Microfluidics and microfluidic devices: A review. Elveflow Microfluidic Instruments RSS. 2018

[2] Delves PJ, Roitt IM. Encyclopedia of Immunology. Academic Press; 1998

[3] Mazzag B, Zhulin I, Mogilner A. Model of bacterial sband formation in aerotaxis. Biophysical Journal. 2003;**85**:3558-3574

[4] Taylor BL, Zhulin IB, Johnson MS. Aerotaxis and other energy-sensing behavior in bacteria. Annual Reviews in Microbiology. 1999;**53**:103-128

[5] Kalmus H. Anemotaxis in *Drosophila*. Nature. 1942;**150**:405

[6] Yan S, Graff MM, Bresee CS, Man YB, Hartmann MJ. Whiskers aid anemotaxis in rats. Science Advances. 2016;**2**:e1600716

[7] Lauffenburger DA, Zigmond SH. Chemotactic factor concentration gradients in chemotaxis assay systems. Journal of Immunological Methods. 1981;**40**:45-60

[8] Clark AG, Simon A, Aizel K, Bibette J, Bremond N, Vignjevic DM. 3D cell migration in the presence of chemical gradients using microfluidics. Methods in cell biology: Elsevier. 2018;**147**:133-147

[9] Plotnikov SV, Pasapera AM, Sabass B, Waterman CM. Force fluctuations within focal adhesions mediate ECM-rigidity sensing to guide directed cell migration. Cell. 2012;**151**:1513-1527

[10] DuChez BJ, Doyle AD, Dimitriadis EK, Yamada KM. Durotaxis by human cancer cells. Biophysical Journal. 2019;**116**:670-683

[11] Cortese B, Palama IE, D'Amone S, Gigli G. Influence of electrotaxis on cell behaviour. Integrative Biology. 2014;**6**:817-830

[12] Tai G, Reid B, Cao L, Zhao M. Electrotaxis and wound healing: experimental methods to study electric fields as a directional signal for cell migration. Chem: Springer. 2009:77-97

[13] Adams C, Paul A. Phototaxis and geotaxis of light-adapted zoeae of the golden king crab lithodes aequispinus (anomura: *Lithodidae*) in the laboratory. Journal of Crustacean Biology. 1999;**19**:106-110

[14] Häder D-P, Hemmersbach R. Gravitaxis in *Euglena*. Euglena: Biochemistry, Cell and Molecular Biology. Advances in Experimental Medicine and Biology; Springer. 2017;**979**:237-266

[15] Armstrong J, Texada M, Munjaal R, Baker D, Beckingham K. Gravitaxis in *Drosophila melanogaster*: A forward genetic screen. Genes, Brain, and Behavior. 2006;**5**:222-239

[16] Ten Hagen B, Kümmel F, Wittkowski R, Takagi D, Löwen H, Bechinger C. Gravitaxis of asymmetric self-propelled colloidal particles. Nature Communications. 2014;**5**:4829

[17] Wang W, Qin L-W, Wu T-H, Ge C-L, Wu Y-Q, Zhang Q, et al. cGMP signalling mediates water sensation (hydrosensation) and hydrotaxis in *Caenorhabditis elegans*. Scientific Reports. 2016;**6**:19779

[18] Pringault O, Garcia-Pichel F. Hydrotaxis of cyanobacteria in desert crusts. Microbial Ecology. 2004;**47**: 366-373

[19] Lefèvre CT, Bazylinski DA. Ecology, diversity, and evolution of magnetotactic bacteria. Microbiology

and Molecular Biology Reviews. 2013;**77**:497-526

[20] Parker G. Encyclopedia of Materials: Science and Technology. Elsevier; 2001

[21] Peng F, Tu Y, Men Y, van Hest JC, Wilson DA. Supramolecular adaptive nanomotors with magnetotaxis behavior. Advanced Materials. 2017;**29**:1604996

[22] Jékely G. Evolution of phototaxis. Philosophical Transactions of the Royal Society, B: Biological Sciences. 2009;**364**:2795-2808

[23] Suli A, Watson GM, Rubel EW, Raible DW. Rheotaxis in larval zebrafish is mediated by lateral line mechanosensory hair cells. PLoS One. 2012;**7**

[24] Baxendale S, Whitfield T. Methods to study the development, anatomy, and function of the zebrafish inner ear across the life course. Methods in Cell Biology: Elsevier. 2016:165-209

[25] Queiroga H, Blanton J. Interactions between behaviour and physical forcing in the control of horizontal transport of decapod crustacean larvae. Advances in Marine Biology. 2005;**47**:107-214

[26] Poff KL, Skokut M. Thermotaxis by pseudoplasmodia of *Dictyostelium discoideum*. Proceedings of the National Academy of Sciences. 1977;**74**:2007-2010

[27] Pline M, Diez JA, Dusenbery DB. Extremely sensitive thermotaxis of the nematode *Meloidogyne incognita*. Journal of Nematology. 1988;**20**:605

[28] Bahat A, Eisenbach M. Sperm thermotaxis. Molecular and Cellular Endocrinology. 2006;**252**:115-119

[29] Xue X, Patel MK, Kersaudy-Kerhoas M, Desmulliez MP, Bailey C, Topham D. Analysis of fluid separation in microfluidic T-channels. Applied Mathematical Modelling. 2012;**36**:743-755

[30] Zhang M, Gong X, Wen W. Manipulation of microfluidic droplets by electrorheological fluid. Electrophoresis. 2009;**30**:3116-3123

[31] Lin B. Microfluidics: Technologies and Applications. Springer; 2011

[32] Aryasomayajula A, Bayat P, Rezai P, Selvaganapathy PR. Microfluidic Devices and their Applications. Springer Handbook of Nanotechnology. Berlin, Heidelberg: Springer; 2017. pp. 487-536

[33] Igata E, Arundell M, Morgan H, Cooper J. Interconnected reversible lab-on-a-chip technology. Lab on a Chip. 2002;**2**:65-69

[34] Seeley JV, Micyus NJ, Bandurski SV, Seeley SK, McCurry JD. Microfluidic deans switch for comprehensive two-dimensional gas chromatography. Analytical Chemistry. 2007;**79**:1840-1847

[35] Weibel DB, Whitesides GM. Applications of microfluidics in chemical biology. Current Opinion in Chemical Biology. 2006;**10**:584-591

[36] Pihl J, Sinclair J, Sahlin E, Karlsson M, Petterson F, Olofsson J, et al. Microfluidic gradient-generating device for pharmacological profiling. Analytical Chemistry. 2005;**77**:3897-3903

[37] Irimia D, Liu S-Y, Tharp WG, Samadani A, Toner M, Poznansky MC. Microfluidic system for measuring neutrophil migratory responses to fast switches of chemical gradients. Lab on a Chip. 2006;**6**:191-198

[38] Chung BG, Flanagan LA, Rhee SW, Schwartz PH, Lee AP, Monuki ES, et al.

Human neural stem cell growth and differentiation in a gradient-generating microfluidic device. Lab on a Chip. 2005;**5**:401-406

[39] Ainla A, Gözen I, Orwar O, Jesorka A. A microfluidic diluter based on pulse width flow modulation. Analytical Chemistry. 2009;**81**:5549-5556

[40] Jiang X, Ng JM, Stroock AD, Dertinger SK, Whitesides GM. A miniaturized, parallel, serially diluted immunoassay for analyzing multiple antigens. Journal of the American Chemical Society. 2003;**125**:5294-5295

[41] Takeuchi S, DiLuzio WR, Weibel DB, Whitesides GM. Controlling the shape of filamentous cells of *Escherichia coli*. Nano Letters. 2005;**5**:1819-1823

[42] Cabodi M, Choi NW, Gleghorn JP, Lee CS, Bonassar LJ, Stroock AD. A microfluidic biomaterial. Journal of the American Chemical Society. 2005;**127**:13788-13789

[43] Hu J, Wang L, Li F, Han YL, Lin M, Lu TJ, et al. Oligonucleotide-linked gold nanoparticle aggregates for enhanced sensitivity in lateral flow assays. Lab on a Chip. 2013;**13**:4352-4357

[44] Warren AD, Kwong GA, Wood DK, Lin KY, Bhatia SN. Point-of-care diagnostics for noncommunicable diseases using synthetic urinary biomarkers and paper microfluidics. Proceedings of the National Academy of Sciences. 2014;**111**:3671-3676

[45] Herr AE, Hatch AV, Throckmorton DJ, Tran HM, Brennan JS, Giannobile WV, et al. Microfluidic immunoassays as rapid saliva-based clinical diagnostics. Proceedings of the National Academy of Sciences. 2007;**104**:5268-5273

[46] Fan HC, Blumenfeld YJ, El-Sayed YY, Chueh J, Quake SR. Microfluidic digital PCR enables rapid prenatal diagnosis of fetal aneuploidy. American Journal of Obstetrics and Gynecology. 2009;**200**:543.e1-543.e7

[47] Huh D, Matthews B, Mammoto A, Montoya-Zavala M, Hsin H, Ingber D. A mechanosensitive transcriptional mechanism that controls angiogenesis. Science. 2010;**328**:1662-1668

[48] Westein E, van der Meer AD, Kuijpers MJ, Frimat J-P, van den Berg A, Heemskerk JW. Atherosclerotic geometries exacerbate pathological thrombus formation poststenosis in a von Willebrand factor-dependent manner. Proceedings of the National Academy of Sciences. 2013;**110**:1357-1362

[49] Kim HJ, Huh D, Hamilton G, Ingber DE. Human gut-on-a-chip inhabited by microbial flora that experiences intestinal peristalsis-like motions and flow. Lab on a Chip. 2012;**12**:2165-2174

[50] van Der Helm MW, Van Der Meer AD, Eijkel JC, van den Berg A, Segerink LI. Microfluidic organ-on-chip technology for blood-brain barrier research. Tissue barriers. 2016;**4**:e1142493

[51] Lee HJ, Goodrich TT, Corn RM. SPR imaging measurements of 1-D and 2-D DNA microarrays created from microfluidic channels on gold thin films. Analytical Chemistry. 2001;**73**:5525-5531

[52] Pawlak M, Schick E, Bopp MA, Schneider MJ, Oroszlan P, Ehrat M. Zeptosens' protein microarrays: A novel high performance microarray platform for low abundance protein analysis. Proteomics. 2002;**2**:383-393

[53] Erickson D, Sinton D, Psaltis D. Optofluidics for energy applications. Nature Photonics. 2011;**5**:583-590

[54] Lei L, Wang N, Zhang X, Tai Q, Tsai DP, Chan HL. Optofluidic planar reactors for photocatalytic water treatment using solar energy. Biomicrofluidics. 2010;**4**:043004

[55] Krivec M, Žagar K, Suhadolnik L, Čeh M, DraŽić G. Highly efficient TiO2-based microreactor for photocatalytic applications. ACS Applied Materials & Interfaces. 2013;**5**:9088-9094

[56] Meng Z, Zhang X, Qin J. A high efficiency microfluidic-based photocatalytic microreactor using electrospun nanofibrous TiO2 as a photocatalyst. Nanoscale. 2013;**5**:4687-4690

[57] Choban ER, Markoski LJ, Wieckowski A, Kenis PJ. Microfluidic fuel cell based on laminar flow. Journal of Power Sources. 2004;**128**:54-60

[58] Gervasio D, Zenhausern F. Microfluidic fuel cell system and method for portable energy applications. Google Patents; 2006

[59] Wagner LT, Yang J, Ghobadian S, Montazami R, Hashemi NN. A microfluidic reactor for energy applications. Open Journal of Applied Biosensor. 2012;**1**:21

[60] Mitrovski SM, Elliott LC, Nuzzo RG. Microfluidic devices for energy conversion: Planar integration and performance of a passive, fully immersed H2-O2 fuel cell. Langmuir. 2004;**20**:6974-6976

[61] Porter ML, Jiménez-Martínez J, Martinez R, McCulloch Q, Carey JW, Viswanathan HS. Geo-material microfluidics at reservoir conditions for subsurface energy resource applications. Lab on a Chip. 2015;**15**:4044-4053

[62] Lin Y, Gritsenko D, Feng S, Teh YC, Lu X, Xu J. Detection of heavy metal by paper-based microfluidics. Biosensors & Bioelectronics. 2016;**83**:256-266

[63] Liu C-C, Wang Y-N, Fu L-M, Huang Y-H. Microfluidic paper-based chip platform for formaldehyde concentration detection. Chemical Engineering Journal. 2018;**332**:695-701

[64] Czugala M, Fay C, O'Connor NE, Corcoran B, Benito-Lopez F, Diamond D. Portable integrated microfluidic analytical platform for the monitoring and detection of nitrite. Talanta. 2013;**116**:997-1004

[65] Chon K, Moon J, Kim S, Kim S-D, Cho J. Bio-particle separation using microfluidic porous plug for environmental monitoring. Desalination. 2007;**202**:215-223

[66] Joshi R, Chhibber R. Effect of surface roughness and oxidation conditions on wettability of borosilicate glass sealant. Materials Science Forum: Trans Tech Publ. 2017;**909**:61-66

[67] Joshi R, Chhibber R. Development and interface characterization of unmatched glass-metal joint. Journal of Manufacturing Processes. 2018;**31**:787-800

[68] Ackermann KR, Henkel T, Popp J. Quantitative online detection of low-concentrated drugs via a SERS microfluidic system. ChemPhysChem. 2007;**8**:2665-2670

[69] Bruzzone A, Costa H, Lonardo P, Lucca D. Advances in engineered surfaces for functional performance. CIRP Annals. 2008;**57**:750-769

[70] Ungerböck B, Mistlberger G, Charwat V, Ertl P, Mayr T. Oxygen imaging in microfluidic devices with optical sensors applying color cameras. Procedia Engineering. 2010;**5**:456-459

Chapter 10

Performance Evaluation and Mechanism Study of a Silicone Hydrophobic Polymer for Improving Gas Reservoir Permeability

Jie Zhang, Xu-Yang Yao, Bao-Jun Bai and Wang Ren

Abstract

The permeability of tight gas reservoirs is usually lower than 1 md. When the external fluids from drilling and completion processes invade such reservoirs, formation damage occurs and causes serious damage to oil and gas production. Fluorocarbon surfactants are most often recommended for removing such damage because they have extremely low surface tension, which means that they can change the reservoir wettability from water wet to gas or oil wet. However, they are not normally applied in the field because they are not cost-effective. Besides, some environmental concerns also restrict their application. In this work, we studied the effects of an oligomeric organosilicon surfactant (OSSF) on wettability modification, surface tension reduction, invasion of different fluids, and fluid flow back. It was found that the amount of spontaneous imbibition and remaining water could be reduced by the surfactant as a result of surface tension reduction and wettability alteration. Compared to the distilled water, the concentration of 0.20 wt% OSSF could decrease water saturation of cores by about 4%. At a flow-back pressure of 0.06 and 0.03 MPa after 20 PV displacement, permeability recovery could increase from 8 to 7–93% and 86%, respectively. We also found that the mechanism of OSSF includes the physical obstruction effect, surface tension reduction of external fluids, and wettability alteration of the reservoir generated. Meanwhile, quantum chemical calculations indicated that adsorbent layer of polydimethylsiloxane could decrease the affinity and adhesion of CH_4 and H_2O on the pore surface.

Keywords: low-permeability reservoir, silicone hydrophobic polymer, spontaneous imbibition, antiwater blocking, water saturation, gas permeability improvement

1. Introduction

Water blocking can damage the low permeability of a normally tight gas reservoir due to the increase of water saturation and the reduction of gas phase permeability in the process of drilling, completion, and stimulation. Zhong et al. [1] showed that the permeability damage rate in a gas reservoir could reach 70–90%, and the gas well production could decrease by more than 70% when water-blocking damage occurred. Therefore, it is important to develop an antiwater blocking agent with high efficiency and low cost for the development of a superior low-permeability gas reservoir.

At present, most researchers consider water blocking as caused by capillary thermodynamics and dynamics [2]. The commonly used antiwater blocking agents include lower alcohol content agents (especially methanol), alcohol ethers, silyl ethers, hydrocarbon surfactants [3], and fluorocarbon surfactants, and "... the alcohols react with formation water in the reservoir to form low boiling point azeotrope, which is helpful for gasification and flow back of the injected water" [4]. Nasr et al. indicated that the glycol ether and polyethylene glycol monobutyl ether could remove the water blocking damage and improve the permeability of oil fields in Arab countries [5]. Zhang et al. evaluated the effects of methanol, ethanol, and ethylene glycol on alleviating the water blocking damage of the low-permeability sandstone gas reservoir. The results showed that methanol had the most favorable performance, followed by ethanol and ethylene glycol, respectively [6]. "Bai et al. compared the anti-water blocking effects of methanol and petroleum sulfonate on a low permeability gas reservoir and found methanol to have better performance ... " Surfactants can reduce the surface tension and change the reservoir wettability from water wetting to gas or oil wetting [7]. Bang et al. reduced the water blocking damage and increased the fluid flow in the fracture reservoirs with an alcohol solution containing a fluorocarbon surfactant. They also reduced the water blocking damage of the condensate in gas reservoirs by applying the fluorocarbon surfactant. The results showed that the hydrated silanol groups could adsorb water from pore surfaces via covalent bonding with the alkenyl groups of fluorocarbon surfactants. This made fluorine-containing alkyl directionally arranged, changing the reservoir wettability from water wetting to oil wetting and finally increasing the gas phase permeability [8–10]. Li et al. developed an antiwater blocking agent containing perfluoroalkyl side chains by using a stepwise emulsion polymerization method, which showed low surface tension and interfacial tension. The surfactant can adsorb on the surface of the pore via chemical adsorption and modify the surface to preferential gas wetting [11]. Li et al. used two nonionic fluorocarbon surfactant methanol solutions for water phase displacement experiments and found that it could improve the permeability [12]. Liu et al. synthesized a cationic fluoride Gemini surfactant and recorded ultralow surface tension; moreover, its solution could remarkably reduce the damage of water blocking on low-permeability formations [13]. Liu et al. found that a wettability variation from water wetting to gas wetting was achieved by adding 0.1 wt% fluoride. The core flow tests indicated that both the flow back rate and gas relative permeability were significantly improved [14]. Biosurfactants such as *Sophora japonica*, trehalose lipid, rhamnolipid, peptide, and some polymer surfactants also show good prospects in relieving water blocking damage because they can change the wettability by adding a large amount of active groups adsorbed on the rock surface. Moreover, they are natural, environmentally friendly, and easy to produce in the industry. Zhang prepared a biosurfactant referred to as the stearic acid glucose ester methoside maleic acid diester, which can decrease the interfacial tension and overcome the water blocking effect. It has good thermal stability and chemical stability [15].

Due to the high cost and environment concerns of fluorocarbon surfactants, an oligomeric organosilicon surfactant (OSSF) contained functional groups such as the silicon hydroxyl group. The silicon oxygen chain and silicon methyl were prepared by polycondensation reaction [16]. Gas preferential wettability could be achieved by oriented adsorption of silicon methyl via hydrogen bonding or chemical condensation. It could also decrease the dynamic surface tension of fluids owing to faster interfacial adsorption of small molecules during diffusion adsorption.

Besides, it possesses super-strong surface activity, and a very small amount can decrease the surface tension to less than 25 mN/m, while the cost is less than 1/10 of fluorocarbon surfactants.

In this study, the influence of OSSF on the aqueous phase trapping damage was evaluated through the measurement of spontaneous imbibition distilled water saturation, water blocking damage rate, permeability recovery, and remaining water saturation. The mechanisms were explained using aggregates blocking, directional adsorption, wettability alteration, and quantum chemistry calculations.

2. Experimental approach

2.1 Materials

2.1.1 Surfactants

In this study, OSSF and ABSN were used to alleviate the aqueous phase trapping damage to low-permeability sand formation. Both surfactants are water soluble and transparent liquids and were made in our laboratory. The relative molecular mass of an OSSF is 2000–3000. It includes a hydrophobic group of permethylated siloxane, hydrophilic groups of sulfonic acid groups, and hydroxyls. The polydimethylsilioxane chain enables the surfactant to have water-repelling characteristics and low surface tension. These properties enable its adsorption on the rock via multiple points of attachment and give the surfactant a long-term effectiveness. ABSN is quaternary ammonium salt cationic surfactant, and the hydrophobic tail is dodecyl benzene. **Figure 1(a)** and **(b)** shows the chemical structure of dodecyl benzene sulfonate triethanolamine (ABSN) and oligomeric organosilicon surfactant (OSSF), respectively. The artificial and reservoir cores used in the same experiment have the same composition and similar porosity and permeability.

2.1.2 Cores

The cores used in this study include artificial cores and reservoir cores. They have same compositions and similar porosity and permeability. The artificial cores (obtained from Haian Petroleum Scientific Research Instrument Co., Ltd. in Nantong, China) are mainly composed of quartz sand, which does not contain any water sensitive substance. Reservoir cores were drilled at the Yingxi exploratory area in the Qinghai oilfields. The physical characteristics of these cores are listed in **Table 1**.

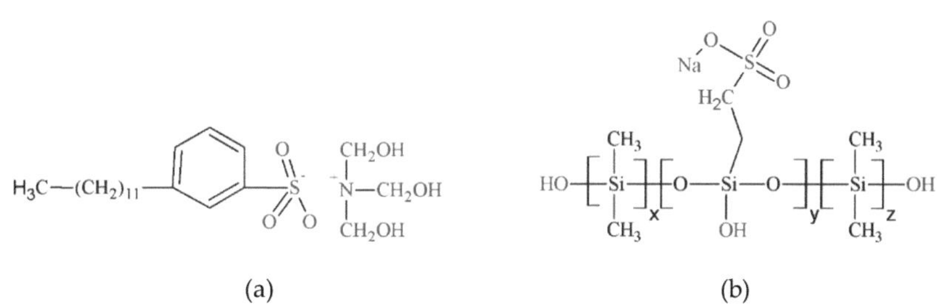

Figure 1.
Chemical structure of (a) ABSN and (b) OSSF.

Type	L/(mm)	D/(mm)	m/(g)	Φ/(%)	K/(10^{-3} μm^2)	Standard error	
						Φ	K
Artificial cores	50.89	24.77	54.83	14.69	28.59	0.66	0.48
Artificial cores	50.75	24.83	54.12	13.91	27.61	0.54	0.61
Artificial cores	50.92	24.81	53.72	15.27	29.86	0.67	0.52
Reservoir cores	51.02	24.95	62.67	0.67	3.27	0.49	0.57
Artificial cores	50.81	24.83	53.15	14.34	26.91	0.83	0.64
Artificial cores	49.95	24.94	54.27	14.76	28.95	0.70	0.67
Artificial cores	50.24	24.86	53.65	14.08	26.69	0.51	0.59

Table 1.
Physical characteristics of cores used in this study.

2.1.3 Fluids

Triply distilled water was used in all experiments. About 4% of sodium bentonite mud was added with 0.2% Na_2CO_3 and then stirred for 2 days. It was used as the basic mud. Sodium bentonite was obtained from Weifang Hua Bentonite Group Co., Ltd. Na_2CO_3 is a commercially pure reagent.

2.2 Experimental method

2.2.1 Determination of contact angle, wettability, and surface energy

Cores were immersed in NaOH solution with a pH of 9 at 150°C for 16 h. Each core was then taken out and dried at 150°C for 4 h. After that, it was cooled down to room temperature. The contact angles among the cores, distilled water, and ethylene glycol were measured with a JC200D3 contact angle analyzer. When the contact angle of water is less than 75°, it is termed "water wetting." When the contact angle of water is greater than 110°, it is considered "oil wetting" [17]. When the contact angle is between 75° and 110°, it is considered "intermediate wetting." This is a condition of gas preferential wetting. The surface energy of cores was calculated using the Owens-Wendt formula [18]. Aging in this paper means rolling under the high temperature of 150°C for 16 h.

2.2.2 Determination of surface tension

The surface tension measurements of NaOH solution with a pH of 9 and basic mud filtrates before and after aging (rolling in the temperature of 150°C oven for 16 h) were performed with an interfacial tensiometer sigma 701 (KSV, Finland) using the dynamic Wilhelmy plate method.

2.2.3 Spontaneous imbibition experiments

Figure 2 shows the experiment's core inside the core holder, and a certain confinement pressure was added. The distilled water was added into the metering tube followed by the OSSF solution. The initial contact volume of the liquid with cores was determined based on the different flow rates. The increased water saturation, defined as spontaneous imbibition volume divided by the core's pore volume, was recorded during the experiment [19]. Core permeability change rates

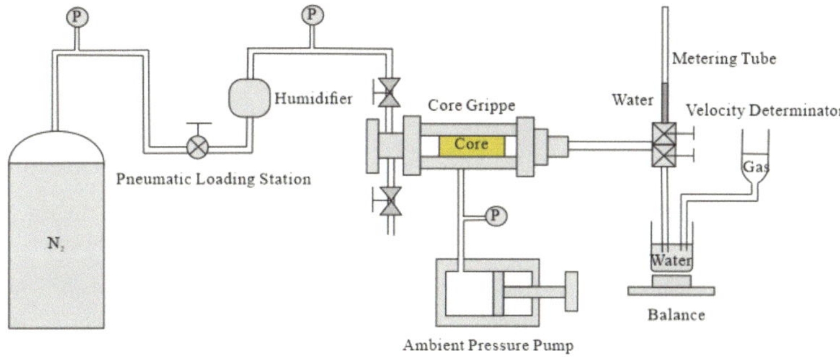

Figure 2.
Schematic of the apparatus used for spontaneous imbibition and reverse displacement.

at different times were defined as the water blocking damage rate. This rate can be calculated [20]:

$$I = \frac{K_0 - K_i}{K_0} \times 100\% \tag{1}$$

where K_O (10^{-3} μm^2) is the gas permeability of artificial cores with irreducible water saturation, and K_i (10^{-3} μm^2) is the gas permeability of artificial cores under different times.

2.2.4 Gas driven flow-back experiment

The cores were gas driven and saturated with flow-back spontaneous imbibition. They were used to weigh the core mass at different flow-back stages and to calculate the residual water saturation. The gas permeability recovery was calculated as follows [21, 22]:

$$D = \frac{K_{npv}}{K_0} \times 100\% \tag{2}$$

where K_O (10^{-3} μm^2) is the gas permeability of artificial cores at irreducible water saturation established by the unsteady state gas drive. K_{nPV} (10^{-3} μm^2) is the gas permeability of artificial cores after they were displaced by gas; these cores represent the gas permeability during different flow-back stages.

3. Results and discussion

3.1 Wettability properties

This section presents an evaluation of the wettability alteration on both artificial cores and reservoir cores using different surfactants in alkaline solutions. Both the distilled water and ethylene glycol experiments were carried out under high temperatures. As shown in **Table 2** (**Figure 3**), water and ethylene glycol spreading on the artificial cores were treated by 0.2 and 0.4 wt% ABSN; thus, the contact angles were too low to measure. The contact angles of 0.2 wt% OSSF treated cores were 110.12° and 27.54°, respectively. Considering the microscopic anisotropy of the cores and the measurement error, we believe that the real contact angles should not vary dramatically from one another. The surface energy of the cores decreased to 20–25 mJ/m^2, which indicated that OSSF could greatly change the wettability from hydrophilic to highly hydrophobic.

As shown in **Table 3 (Figure 4)**, the 0.2 and 0.4 wt% ABSN-treated cores displayed lower water contact angle, and the ethylene glycol droplet completely

Concentration/wt%	Contact angle/°		Surface energy/ (mJ/m^2)	Standard error		
	Distilled water	Ethylene glycol		Distilled water	Ethylene glycol	Surface energy
0.00	Droplet infiltration	Droplet infiltration	—	—	—	
0.20% ABSN	Droplet infiltration	Droplet infiltration	—	—	—	
0.40% ABSN	Droplet infiltration	Droplet infiltration	—	—	—	
0.20% OSSF	110.12	27.54	24.66	2.55	1.80	6.53
0.40% OSSF	110.71	32.14	22.83	2.49	1.84	5.76

Table 2.
Contact angle and surface energy of artificial cores after adsorption equilibrium in ABSN and OSSF solutions (T = 150°C, pH = 9).

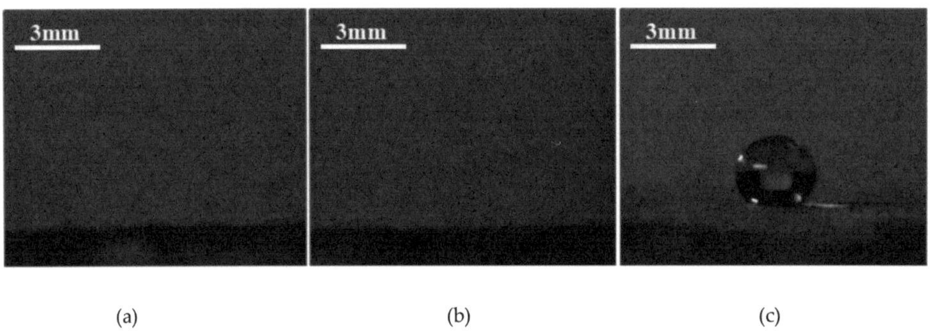

Figure 3.
Contact angle of water on artificial cores (room temperature). (a) Untreated, (b) treated by 0.20 wt% ABSN, and (c) treated by 0.20 wt% OSSF.

Concentration/wt%	Contact angle/ (°)		Surface energy/ (mJ/m^2)	Standard error		
	Distilled water	Ethylene glycol		Distilled water	Ethylene glycol	Surface energy
0.0	13.47	Droplet spreading	—	—	—	
0.20% ABSN	8.75	Droplet spreading	—	—	—	
0.40% ABSN	6.42	Droplet spreading	—	—	—	
0.20% OSSF	104.53	39.19	28.05	1.87	2.14	4.13
0.40% OSSF	107.48	48.50	26.49	1.33	2.39	1.32

Table 3.
Contact angle and surface energy of reservoir cores after adsorption equilibrium in ABSN and OSSF solutions (T = 150°C, pH = 9).

spread on the surface. However, the contact angle of water on the OSS-modified cores still reached 110°, and the surface energy decrease was between 26 and 28 mJ/m^2. These results indicated that water did not spread on the surface of the pore due to the formation of low energy adsorption film treated by OSSF. Since ABSN does not affect wettability alteration, its surface energy is not evaluated here.

3.2 Interfacial properties

To evaluate the surface activities of OSSF and ABSN in a high-temperature basic environment, the surface tension of solutions and fluid filtrates was measured after aging at 150°C for 16 h. In **Figures 5** and **6**, both 0.2–0.4 wt% ABSN and OSSF could decrease the surface tension under high temperature. Meanwhile, OSSF was better at reducing surface tension, meaning that it could better reduce filtrate adsorption and aqueous phase trapping damage in the rock.

3.3 Evaluation of antiwater blocking properties of OSSF

3.3.1 Spontaneous imbibition property of cores

As shown in **Figure 7**, water saturation of cores gradually increased with time. Obvious spontaneous imbibition and diffusion stages could also be seen. The increase in water saturation was dramatic at the spontaneous imbibition stage, but water

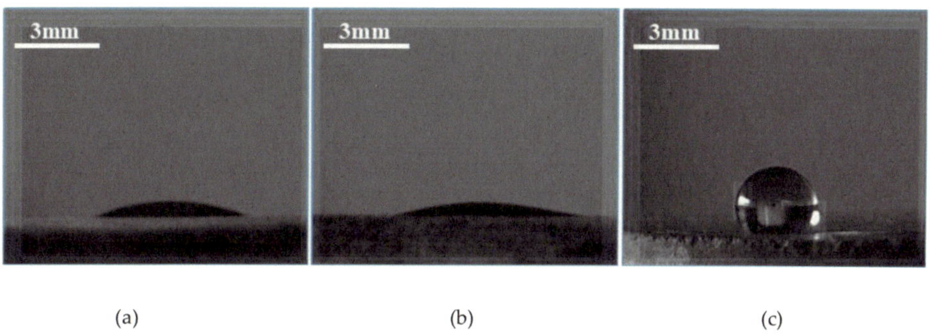

Figure 4.
Contact angle of water on reservoir cores (room temperature). (a) Untreated, (b) treated by 0.20 wt% ABSN, and (c) treated by 0.20 wt% OSSF.

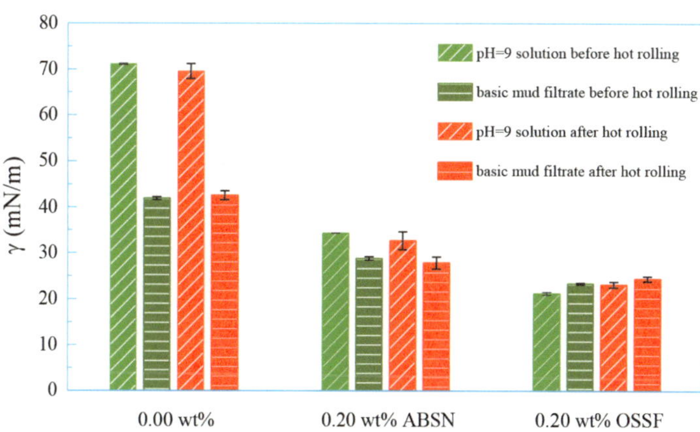

Figure 5.
Surface tension of 0.20 wt% solutions before and after aging (cooling down to room temperature, pH free).

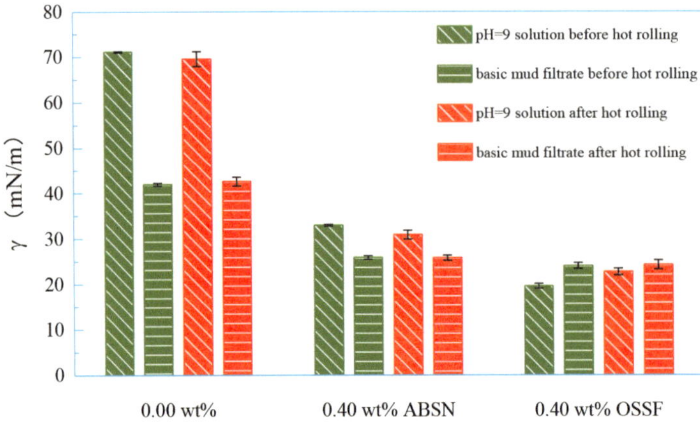

Figure 6.
Surface tension of 0.40 wt% solutions before and after aging (T = 25°C, pH free).

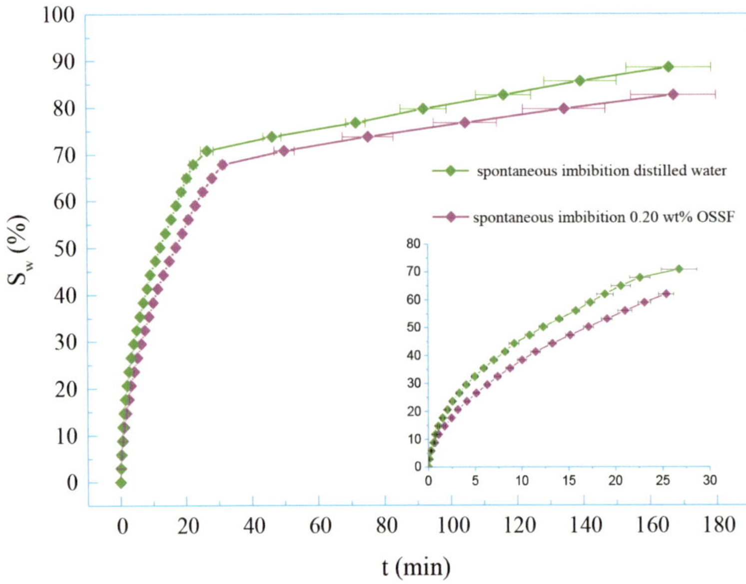

Figure 7.
Water saturation (S_w) of artificial cores versus spontaneous imbibition time (t).

saturation was compromised at the diffusion stage. Effective pores can imbibe the liquid by capillary force, whereas liquid accessing to unconnected pores only relies on diffusion from connected pores, which requires more time. Water saturation decreased from about 71–68% with a 0.20 wt% OSSF in the spontaneous imbibition stage. The spontaneous imbibition time then changed from about 33 to 28 min.

Figure 8 shows the cores treated by 0.2% OSSF solution. Water saturation decreased to 56% and the saturation time decreased to 68 min in the spontaneous imbibition stage. These results indicate that the spontaneous imbibition rate and the amount of external liquid in a reservoir can easily be reduced.

3.3.2 Water blocking damage rate of cores

As shown in **Figure 9**, the water blocking damage rate of distilled water and OSSF solutions increased sharply at the beginning. However, after 20 min, it slowly started to increase. Compared with distilled water, OSSF solution could achieve an additional damage rate reduction of 15%. The results suggest that the water blocking damage rate of the reservoir increased remarkably when the external liquids invaded into reservoir at the beginning. After reaching a certain water saturation, the water blocking damage did not increase significantly.

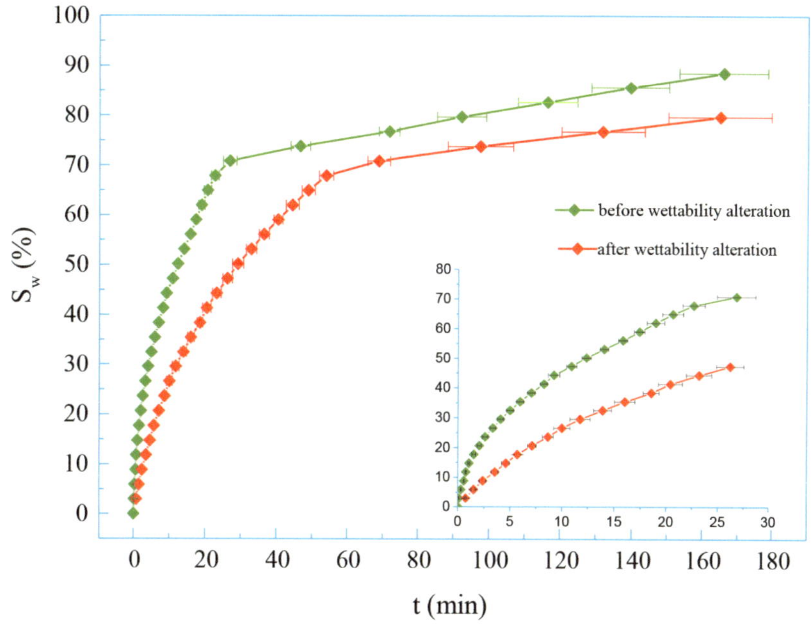

Figure 8.
Water saturation (S_w) of artificial cores treated by 0.2% OSSF solution.

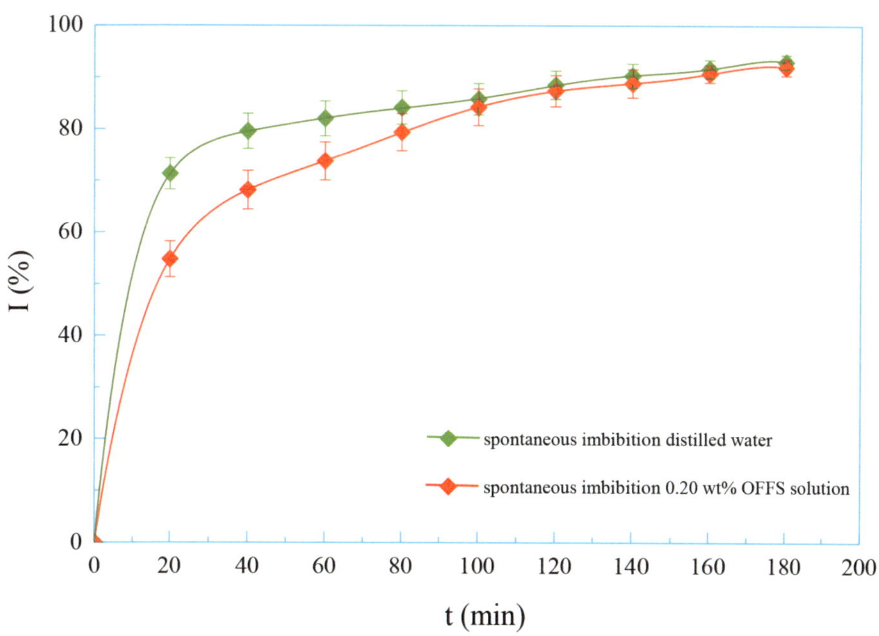

Figure 9.
Water blocking damage rate (I) of artificial cores versus self-absorbed time (t).

3.3.3 Permeability recovery of cores

As shown in **Figure 10**, the permeability recovery of the cores gradually increases with flow-back PV. Meanwhile, the recovery rate gradually decreases. When gas pressure was 0.06 MPa and the flow-back PV is 20, the cores' permeability recovery changes from 7 to 77%. It is evident that the higher the displacing pressure, the higher the permeability recovery. A permeability recovery of 0.20%

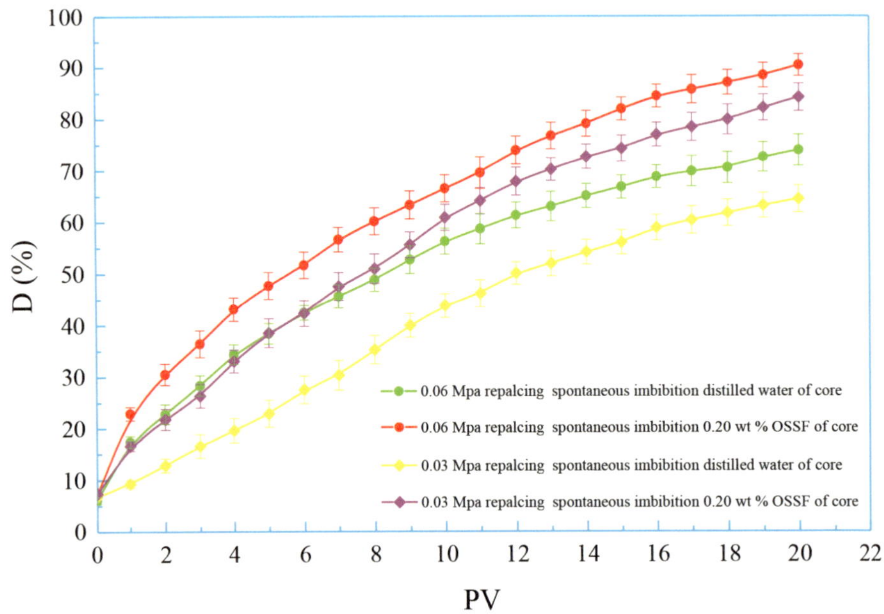

Figure 10.
Pore volumes (PV) versus permeability recovery rate of artificial cores (D).

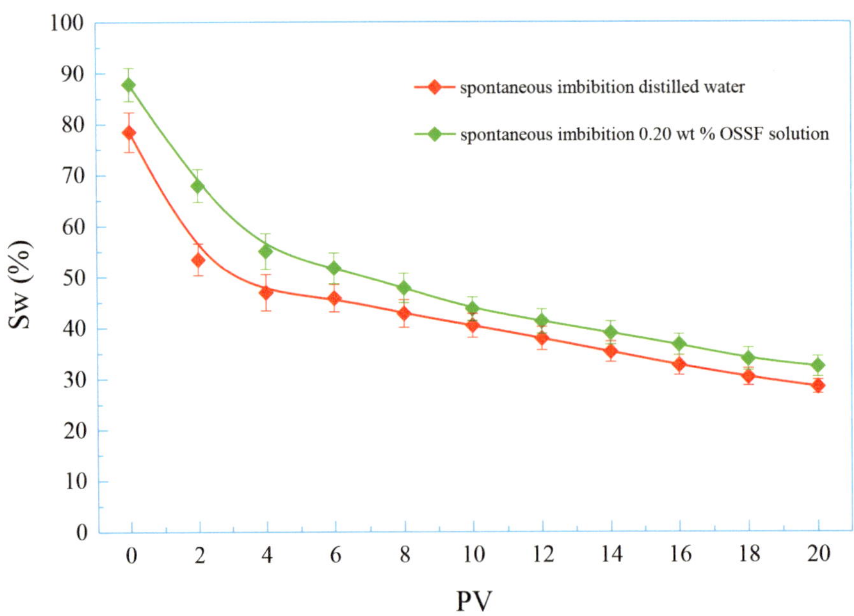

Figure 11.
Gas flow-back volumes versus remaining water saturation (S_w) of artificial cores.

OSSF solution increased from about 8 to 93%; then, with the distilled water, the permeability recovery was 77%. Both increases were at 0.06 MPa and 20 PV. At 0.03 MPa and 20 PV, permeability recovery of 0.20% OSSF solution increased from about 7 to 86%, with the distilled water 67% by contrast. These results illustrate that the increase of flow-back volume and gas pressure can improve the permeability recovery and reduce the water blocking damage of cores. In addition, OSSF is beneficial to the permeability recovery of the reservoirs damaged by external fluids.

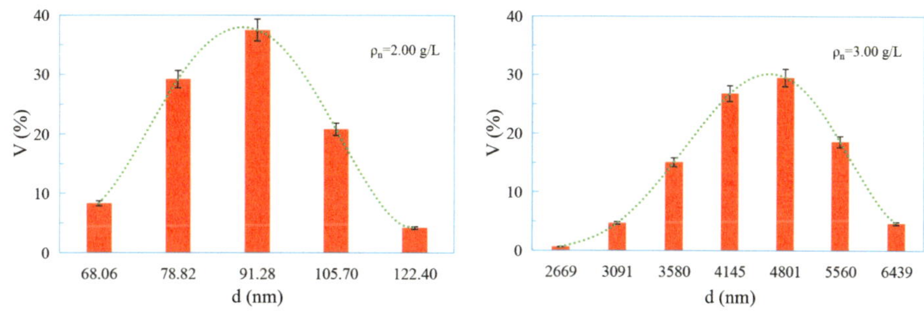

Figure 12.
Size and distribution of OSSF aggregate versus weight percentage (T = 25°C, free pH).

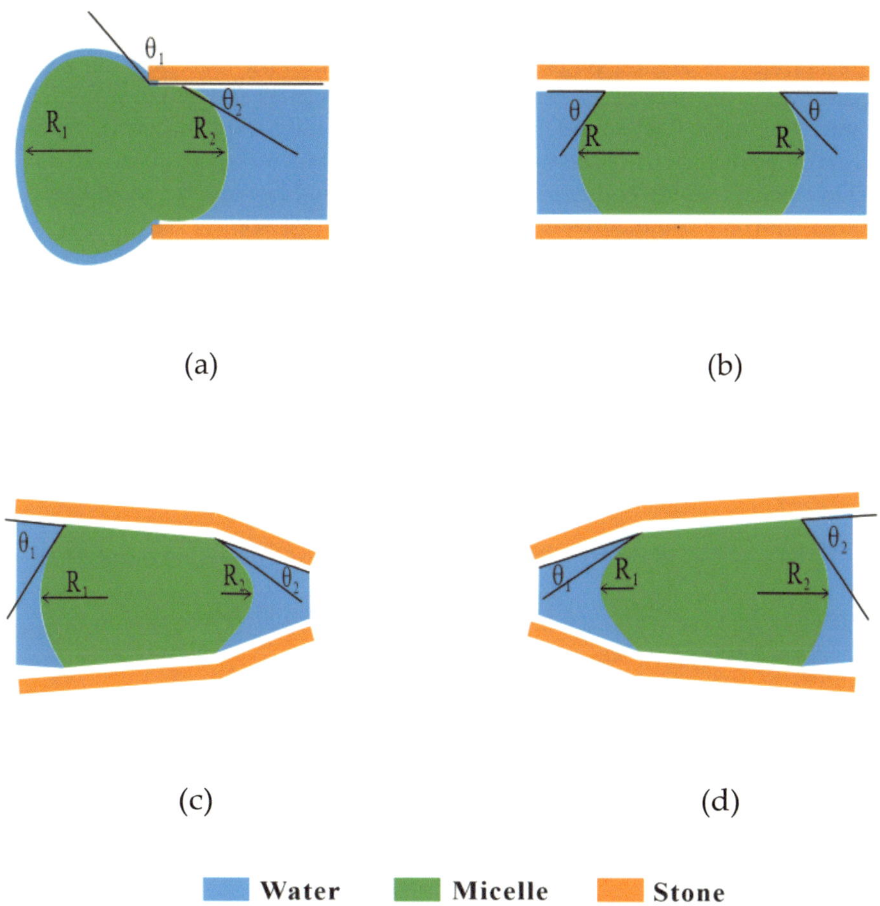

Figure 13.
Flow patterns of OSSF aggregate in the reservoir pores.

3.3.4 Retained water saturation of cores

Accumulation of external fluids in gas reservoir is the principle factor responsible for water blocking damage [23, 24]. As shown in **Figure 11**, the remaining water saturation decreased with PV, but the decreasing rate reduced gradually. Retained water saturation of 0.20% OSSF solution decreased from about 83 to 30%; moreover, with the distilled water, the water saturation decreased from 90 to 34% by contrast. The results reveal that increasing flow-back time can reduce the water saturation and the water blocking damage of the reservoir. Besides, OSSF is also beneficial to the flow back of invasive external fluid in a reservoir.

3.4 Mechanism analysis

3.4.1 Physical blocking

Figure 12 shows that aggregates can be formed when the concentration exceeds its critical micelle concentration (CMC). Adjusting the OSSF concentrations to match reservoir needs with varied pore and throat sizes is necessary when the aggregates invade reservoir with external fluids. This causes friction to form on the surface in which pores or throats and fluid come in contact with each other.

When the aggregates move into the pores or throats of reservoirs, as shown in **Figure 13(b)**, the interfacial tensions on the two contact surfaces are even. When the aggregates flow from the pores to the reservoir throats (**Figure 13(c)**), friction can decrease the flow rate. When aggregates move from the throats to the pores (**Figure 13(d)**), the interfacial tension between aggregates and external fluid becomes smaller. In the process of production, the fluid begins to flow back, the aggregates move back from the throats to the pores, and then, the differences in the interfacial tensions between two contact surfaces could become the driving force for flowing, which could accelerate these fluids to flow back.

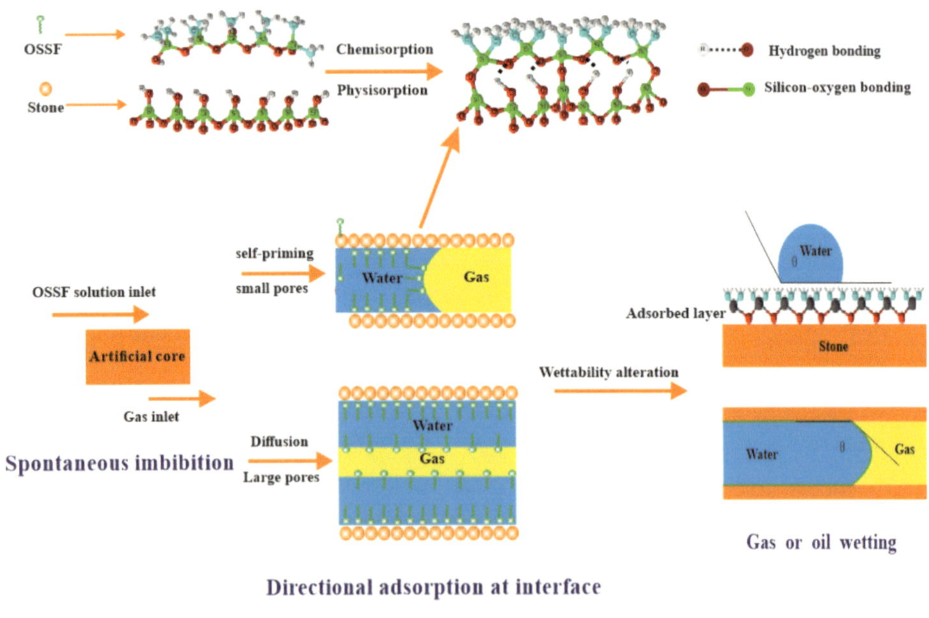

Figure 14.
Principle diagram of OSSF adsorption on water-pore interface.

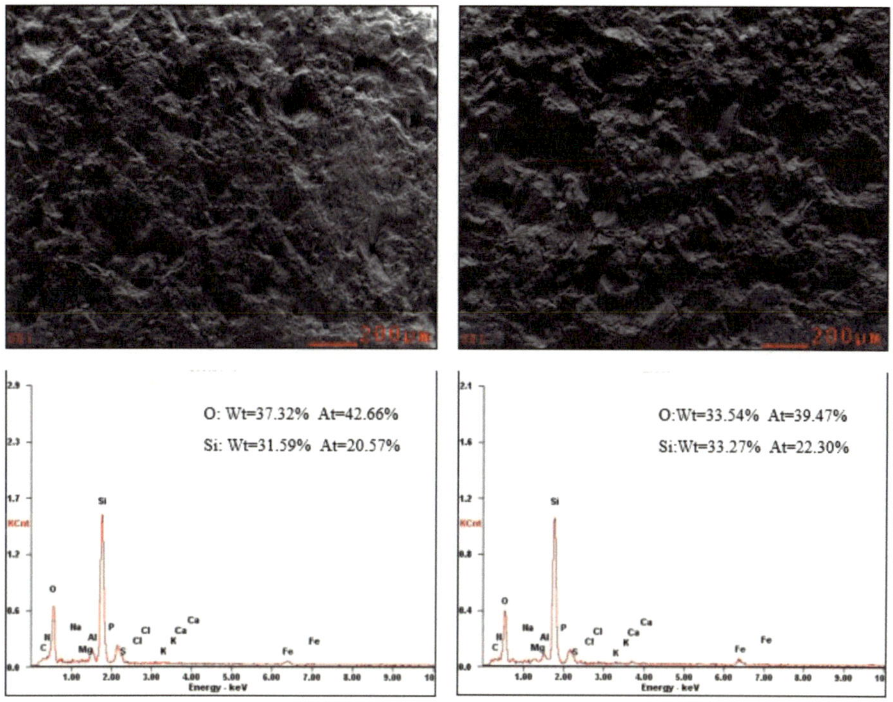

(a) (b)

Figure 15.
SEM image and X-ray spectrum of artificial cores. (a) Untreated and (b) treated by 0.20 wtT OSSF.

3.4.2 Wettability alteration

When the alkaline solution invades into the reservoir, the structural groups of the silicon-oxygen bond (Si-O-Si) on the surface of the silicate minerals can produce silanol groups (Si-OH) by hydrolysis. The density can reach up to 6–7 nm per square. These silanol groups can adsorb the water molecules owing to the hydrogen bonding and the Van Der Waals force upon the water-pore interface. Within both forces, hydrogen bonding plays an important role. It is needed to overcome the adhesion force between rock and water, thereby decreasing the remaining water in the formation [25]. The basic structural unit of the OSSF is polydimethylsiloxane. The silicon-oxygen chain is the polar part of the OSSF composition, and the methyl group represents the nonpolar part.

Under high temperature and catalyst conditions, polarization occurs in the silicon oxygen main chain. This leads to multipoint adsorption of silanol groups (Si-OH) and silicon-oxygen bond (Si-O-Si) through chemical bonding and hydrogen bonding. Meanwhile, the methyl groups directionally rotate and sequentially arrange on the rock surface. This process is shown in **Figure 14** [26]. These directionally adsorbed methyl groups reduce the surface energy of the rock and lead to the hydrophobization on the rock surface, changing the composition of the rock surface. Meanwhile, the capillary force direction of fluids and the physical properties of the gas reservoir change [17].

As shown in **Figure 15**, the cores treated by OSSF have a higher weight and atom Si number as well as a lower weight and atom O number. This is because the cores are composed of silicon and oxygen, with a chemical structural much like that of OSSF. **Figure 16** shows how the OSSF adsorption on the reservoir core surface can

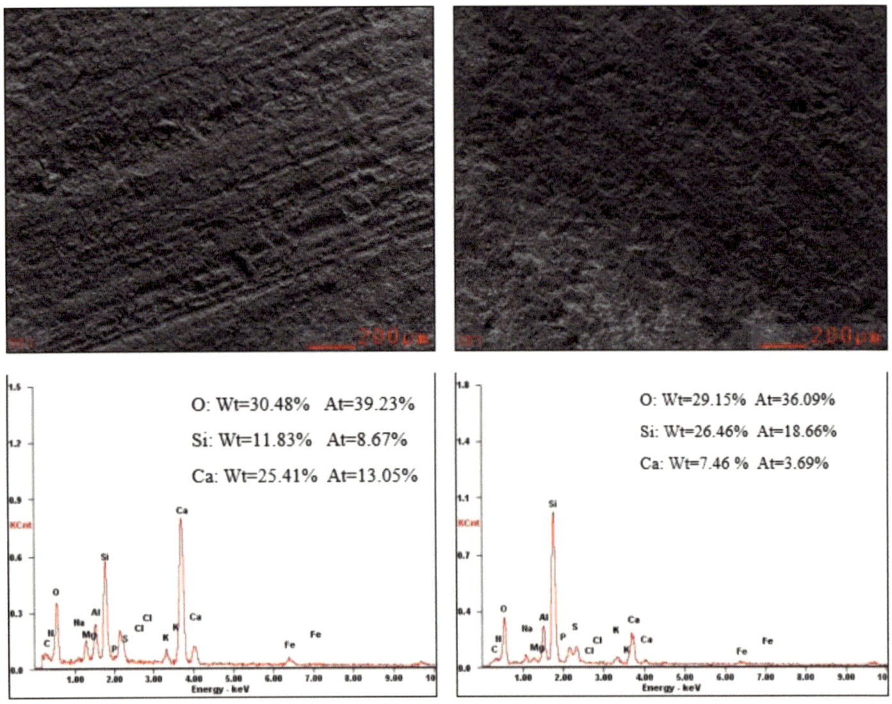

(a) (b)

Figure 16.
SEM image and X-ray spectrum of reservoir cores. (a) Untreated and (b) treated by 0.20 wt% OSSF.

significantly increase the weight and number of atom Si and Ca but slightly decrease that of atom O. Since we use carbonate cores, their surface components change greatly as a result of OSSF adsorption. Both artificial and reservoir core surfaces show more Si atoms. The outermost areas of the core are composed of Si-CH_3, and the cores have become hydrophobic.

3.4.3 Adsorption energy

The weak interaction among the molecules can be calculated accurately by the computational technique based on quantum chemistry. In this paper, (a) silicon-oxygen tetrahedron and (b) dimethyl siloxane were applied to simulate the surface of the untreated cores and OSSF treated cores, respectively. The oxygen atoms at the edge were saturated with hydrogen atoms.

Gaussian 09 W software (Gaussian Inc., Wallingford, CT, USA) is used as the simulation tool, and 6-31G is selected as the basic set. Correction is performed using a basis set superposition error (BSSE). In the adsorption system, the distance between the C atoms of methane and the O atoms of water is the equilibrium distance r_e. The following Eqs. (3) and (4) were used to calculate the adsorption energies (E_e) of water and methane on the core surface [27, 28]:

$$E_e = E_{CH_3-Si//CH_4/Si-O//CH_4} - E_{Si-O/CH_3-Si} - E_{CH_4} \quad (3)$$

$$E_e = E_{CH_3-Si//H_2O/Si-O//H_2O} - E_{Si-O/CH_3-Si} - E_{H_2O} \quad (4)$$

Here, $E_{CH_3-Si//CH_4/Si-O//CH_4}$ represents the system energy in kJ/mol when silicon-oxygen tetrahedron and dimethyl siloxane adsorb methane. $E_{CH_3-Si//H_2O/Si-O//H_2O}$ represents the system energy when silicon-oxygen tetrahedron and dimethyl siloxane adsorb water in kJ/mol. E_{Si-O/CH_3-Si} is the energy of silicon-oxygen tetrahedron or dimethyl siloxane in kJ/mol. E_{H_2O/CH_4} represents the energy of water or methane in kJ/mol.

As shown in **Figure 17** and **Table 4**, the distances of the dimethyl siloxane models between CH_4 and H_2O are larger or similar than that of silicon-oxygen tetrahedron. The adsorption energy of silicon-oxygen tetrahedron and the dimethyl siloxane models to CH_4 is 11.21 and 0.46 kJ/mol, respectively, which indicates that the adsorption of methane on the core surfaces is nonhydrogen-bonding physical adsorption. The adsorption energy of silicon-oxygen tetrahedron and the dimethyl siloxane models to H_2O is 81.78 and 52.18 kJ/mol, respectively, suggesting the adsorption of water is hydrogen-bonding adsorption. Since the adsorption binding energy of CH_4 and H_2O on the silicon-oxygen tetrahedron model is higher than that on the dimethyl siloxane model, the stability of CH_4 and H_2O adsorbed on the pore surface of cores treated by OSSF is lower than that on the untreated cores. At the same flow-back pressure, CH_4 and H_2O adsorbed on the surface of treated cores are often more easily desorbed. Under the same temperature and pressure conditions, after the adsorption of OSSF, the affinity and adhesion of the rock surface to CH_4 and H_2O are reduced, leading to the reduction of the shear stress of CH_4 and H_2O on the pore surfaces of the reservoir, when gas or external fluids flow in the reservoir [29]. This can increase the fluidity of CH_4 and H_2O and finally improve the permeability of CH_4 and H_2O, thereby decreasing water saturation of the reservoir.

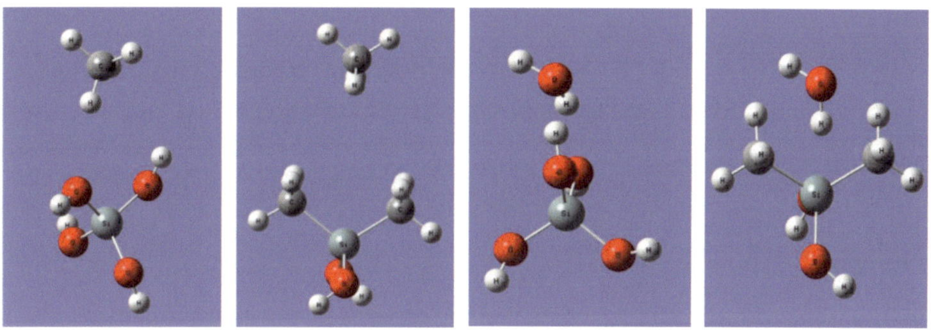

(a) (b) (c) (d)

Figure 17.
Optimized structure of the models absorbed the CH_4 as shown in (a) and (b) and H_2O as shown in (c) and (d) (T = 273.15 K, P = 1 atm).

Adsorption model	Adsorbate	r_e/ (nm)	E_e/ (kJ·mol^{-1})
Silicon-oxygen tetrahedron (a)	CH_4	4.33	−11.21
	H_2O	3.32	−81.77
Dimethylsiloxane (b)	CH_4	5.17	−0.46
	H_2O	3.17	−52.18

Table 4.
Adsorption equilibrium distance (r_e) and adsorption energy (E_e) of the absorbed the CH_4 and H_2O (T = 273.15 K, p = 1 atm).

4. Conclusions

1. An ultralow surface tension could be obtained using OSSF. A 0.20 wt% OSSF solution can alter cores from water wetting to preferential gas wetting.

2. OSSF can remarkably decrease the water saturation and spontaneous imbibition rate of cores, resulting in a reduction in the cores' water blocking damage rate. In this case, the effect of cores' wettability alteration to water saturation and the spontaneous imbibition rate is higher.

3. In the gas driven flow-back experiment, OSSF improved core permeability recovery. The remaining water saturation was also found to be low.

4. The main mechanism for improving core permeability includes the physical obstruction effect, surface tension reduction, and adsorption energy reduction, as well as the capillary force, adhesion, and shear stress of methane and liquid on rock surfaces.

Acknowledgements

This work was supported by the National Major Science and Technology Project of China (Nos. 2016ZX05051, 2016ZX05052, and 2016ZX05030) and Research on Qaidam's Excellent Drilling and Completion Technology (2016E-0109).

Conflicts of interest

The authors declare that there is no conflict of interest regarding the publication of this paper.

Author contributions

The supervisor, Jie Zhang, conducted the experimental work and analyzed the results of the research. All authors contributed to the analysis of results. Jie Zhang wrote the manuscript. Xu-Yang Yao contributed to experiment's design and analysis of the results. Bao-Jun Bai contributed to analysis and discussion of the results, reviewing, and editing of the manuscript.

Author details

Jie Zhang[1,3]*, Xu-Yang Yao[2], Bao-Jun Bai[3] and Wang Ren[1]

1 Drilling Fluids Research Department, CNPC Engineering Technology R&D Company Limited, Beijing, P.R. China

2 Petrochina Xin Jiang Oil Field Company, Karamay, P.R. China

3 Department of Petroleum Engineering, Missouri University of Science and Technology, Rolla, MO, USA

*Address all correspondence to: zhangjiedri@cnpc.com.cn

References

[1] Zhong XY, Huang L, Wang LH. Research progress on water blockage effect in low permeability gas reservoirs. Special Oil & Gas Reserves. 2008;**15**: 12-23

[2] He CZ, Hua MQ. Quantitative study of water blocking mechanism. Drilling Fluid & Completion Fluid. 2000;**3**:4-7

[3] Pham VP, Jo YW, Oh JS, Kim SM, Park JW, Kim SH, et al. Effect of plasma-nitric acid treatment on the electrical conductivity of flexible transparent conductive films. Japanese Journal of Applied Physics. 2013;**52**: 075102

[4] Liu M, Wu Q. Research progress and prospect of water blocking agent. Guangzhou. Chemistry & Industry. 2013;**41**:32-42

[5] Nasr-EI-Din HA, Lynn JD, AI-Dossary KA. Formation damage caused by a water blockage chemical: Prevention through operator supported test programs. SPE. 2002;**33**:297-325

[6] Zhang WB, Li CF, Shi XW. Application of alcoholic acid acidification in Qinghai oil field. Oilfield Chem. 2009;**26**:24-28

[7] Bai FL. Water phase trapping damage and the study of relieving measures in gas reservoir. Petrochemical Industry Application. 2010;**29**:14-17

[8] Bang V, Pope G, Sharma M, Baran JR Jr. Development of a successful chemical treatment for gas wells with water and condensate blocking damage. In: SPE Annual Technical Conference and Exhibition, New Orleans, Louisiana, 4–7 October 2009. Paper SPE 124977. 2009

[9] Bang VSS, Yuan CW, Pope GA, Baran JR Jr, Skildum J. Improving productivity of hydraulically fractured gas condensate wells by chemical treatment. In: OSSFhore Technology Conference, Houston, Texas, 5–8 May 2008. Paper OTC 19599. 2008

[10] Bang V, Pope G, Mukul M, Baran JR Jr, Ahmadi M. A new solution to restore productivity of gas Wells with condensate and water blocks. In: SPE Annual Technical Conference and Exhibition, Denver, Colorado, 21–24 September 2008. SPE 116711. 2008

[11] Li YY, Jiang GC, Xuan Y. Synthesis and performance evaluation of drilling fluid anti-blockage agent of low porosity and low permeability reservoir. Drilling Fluid and Completion Fluid. 2014;**31**: 9-12

[12] Li XQ, Wang YC, Wang YQ. Experimental study on the effect of mixed surfactants on water blockage. Chemistry & Bioengineering. 2013: 85-89

[13] Liu Y, Guo L, Bi K, Ren W, Liu L, Li YG, et al. Synthesis of cationic gemini surfactant and its application in water blocking. Fine and Specialty Chemicals. 2011;**19**:8-10

[14] Liu XF, Kang YL, Luo PY, You LJ, Tang Y, Kong L. Wettability modification by fluoride and its application in aqueous phase trapping damage removal in tight sands to ne reservoirs. Journal of Petroleum Science and Engineering. 2015;**133**:201-207

[15] Zhang XQ. Researching on the use of new type surfactant to weaken water blocking and jamin effect. Master thesis. Qingdao, China: Ocean Univ. China; 2013

[16] Yin D, Luo P, Zhang J. Synthesis of Oligomeric silicone surfactant and its interfacial properties. Applied Sciences. 2019;**9**(3):497

[17] Jiang GC, Wei YJ, Zhang M, Luo SJ, Yang Z, Pang JT. Review of evaluation methods to gas wettability. Science Technology & Engineering. 2012;**13**: 5562-5565

[18] Owens DK, Wendt RC. Estimation of the surface free energy of polymers. Journal of Applied Polymer Science. 1969;**13**:1741-1747

[19] Dong B, Lan L, Cheng ZH, Xia HY, Niu J. Study on capillary self-absorption characteristics of dense gas reservoir rocks. Drilling and Production Technology. 2012;**35**:34-37

[20] Lai NJ, Ye ZB, Liu XJ, Yang JJ, Zhang JF. Study on water blocking damage in low permeability tight sandstone gas reservoir. Natural Gas Industry. 2005;**25**:125-127

[21] Stand. Adm. China. Determination Method of Relative Permeability of Two-Phase Fluids in Rocks (GB/T 28912–2012). Beijing, China: China Stand. Press; 2012

[22] Guo TS, Li TW, Li HB. Laboratory study of water locking damage under imbibition in volcanic gas reservoirs. Science Technology & Review. 2011;**29**: 62-66

[23] Bennion DB, Thomas FB, Ma T. Formation damage processes reducing productivity of low permeability gas reservoirs. In: SPE Rocky Mountain Regional/Low Permeability Reservoirs Symposium and Exhibition, Denver, Colorado, 12–15 March 2000. Paper SPE 60325. 2000

[24] Bennion DB, Bietz RF, Thomas FB. Reductions in the productivity of oil and gas reservoirs due to aqueous phase trapping. Petroleum Society of Canada. 1994;**33**:45-54

[25] Wu JF, Wu DQ. Surface ionization and complexation of mineral-water interface. Advances in Earth Science. 2000;**1**:90-96

[26] Cao Y, Li HL. Adsorption behavior of polymeric surfactants at solid-liquid interface. Acta Physico-Chimica Sinica. 1999;**10**:895-899

[27] Koretsky CM, Sverjensky DA, Sahai N. A model of surface site types on oxide and silicate minerals based on crystal chemistry. Implications for site types and densities, multi-site adsorption, surface infrared spectroscopy, and dissolution kinetics. American Journal of Science. 1998;**298**: 349-438

[28] Castro EAS, Gargano R, Martins JBL. Theoretical study of benzene interaction on kaolinite. Computer-Aided Materials Design. 2012;**112**:2828-2831

[29] Taghvaei E, Moosavi A, Nouri-Borujerdi A. Superhydrophobic surfaces with a dual-layer micro-and nanoparticle coating for drag reduction. Energy. 2017;**12**:1-10

Section 4

The Coating Techniques of Thin Film

Chapter 11

Electrospinning Technique as a Powerful Tool for the Design of Superhydrophobic Surfaces

Pedro J. Rivero, Adrian Vicente and Rafael J. Rodriguez

Abstract

The development of surface engineering techniques to tune-up the composition, structure, and function of materials surfaces is a permanent challenge for the scientific community. In this chapter, the electrospinning process is proposed as a versatile technique for the development of highly hydrophobic or even superhydrophobic surfaces. Electrospinning makes possible the fabrication of nanostructured ultrathin fibers, denoted as electrospun nanofibers (ENFs), from a wide range of polymeric materials that can be deposited on any type of surface with arbitrary geometry. In addition, by tuning the deposition parameters (mostly applied voltage, flow rate, and distance between collector/needle) in combination with the chemical structure of the polymeric precursor (functional groups with hydrophobic behavior) and its resultant viscosity, it is possible to obtain nanofibers with highly porous surface. As a result, functionalized surfaces with water-repellent behavior can be implemented in a wide variety of industrial applications such as in corrosion resistance, high efficient water-oil separation, surgical meshes in biomedical applications, or even in energy systems for long-term efficiency of dye-sensitized solar cells, among others.

Keywords: electrospinning, superhydrophobicity, wettability properties, polymeric precursors, industrial applications

1. Introduction

The measurement of the contact angle (CA) value is one of the most important parameters used for the determination and quantification of the wettability of solid surfaces. This CA is used to describe the behavior of a liquid droplet on a solid surface in air and is measured as the angle between the tangent at three phase points and the solid surface [1]. Accordingly, a surface is considered hydrophilic when the resultant solid surface shows a water contact angle (WCA) less than 90°, whereas a solid surface is considered hydrophobic when the WCA is higher than 90°. Nowadays, due to the development of the nanotechnology, bioinspired surfaces with special wettability properties are continuously emerging in the scientific research areas. Some representative examples are the design of novel surfaces with superhydrophilic (WCA < 10°) [2] or superhydrophobic (WCA > 150°) [3] behavior measured by water as well as surfaces with superoleophilic (CA < 10°) [4] or even superoleophobic (CA > 150°) [5] behavior measured by using oil droplets.

By convention, a superhydrophobic surface exhibits an extraordinary water contact angle value that is greater than 150° with a low sliding angle (typically less than 10°). The effect of the surface microstructure on the resultant water repellency can be explained by two distinct models depending on the degree of surface roughness such as the Wenzel model [6] and the Cassie-Baxter model [7]. According to the Wenzel model, the liquid is in contact with the entire exposed surface of the solid because the large interfacial energy at the water-solid interface induces the penetration of water into the surface cavities. However, in the Cassie-Baxter model, the liquid does not penetrate the hollows or cavities of the corrugated surface and the water droplets mostly contact air pockets that are formed between water and a rough solid surface. Consequently, in the Cassie-Baxter model, the superhydrophobic shows a lower sliding angle in comparison with the Wenzel state [8]. Till, the Cassie-Baxter state is preferred because of very small hysteresis and excellent rolling behavior even at till angles of a few degrees. In addition, it is known that some plants (i.e., lotus left), animal fur, or insect wings found in the nature can show this superhydrophobic behavior. According to this, in order to simulate this biological surface, the design of synthetic superhydrophobic surfaces is a continuous challenge in the scientific community [9].

The research is focused on the design of surfaces with a low surface energy combined with a hierarchical surface roughness on at least two different length scales (i.e., micrometric and nanometric morphology) [10]. Accordingly, multiple deposition techniques have been implemented for this specific purpose such as layer-by-layer assembly [11], sol-gel process [12], electrochemical deposition [13], chemical vapor deposition [14], lithography [15], physical vapor deposition [16], and chemical etching [17], among others. However, an interesting deposition technique is the electrospinning process because it is possible to induce the dual effect of low surface energy and the desired roughness with multiscale surface morphology, respectively [18]. In the electrospinning process, an electrostatic force is used to obtain electrically charged polymeric jet, which overcomes the surface tension of the polymeric solution. As a result, elongated fibers are accelerated from capillary tip and are then deposited onto collector with the corresponding evaporation of the solvent, thereby making possible the fabrication of fibers with a good control over their corresponding morphological, optical, and wetting properties [19]. In this sense, the fabrication of ultrathin or nanofibers can be obtained as a strict control of the several parameters such as applied voltage, flow rate, and viscosity of the polymeric precursor or distance to collector, among others [20, 21]. The surface modification to control the wettability of electrospun mats is possible due to the presence of fibers with micrometric and sub-micrometric diameter, thereby providing hierarchical surface with superhydrophobic behavior because of the small size of the resultant electrospun mats [22]. Finally, the number of scientific works based on the combination of electrospun fibers and superhydrophobic surfaces published in indexed journals has gradually increased. Potential applications can be found in areas as diverse as removal of oil from water, separation membranes, corrosion protection in metallic surfaces, or even in biomedical applications.

To sum up, this chapter is divided into the following subsections such as operational parameters in the electrospinning process, design of superhydrophobic surfaces composed of electrospun mats, and a summary table of the main applications derived from this work with their corresponding conclusions.

2. Operational parameters in electrospinning process

Electrospinning is a very versatile technique that can be implemented in a wide variety of polymeric precursors from biodegradable [23–24], copolymer [25–26], natural [27], or even synthetic nature [28–29]. The fundamentals of this deposition technique are based on the use of electrostatic forces with the aim to obtain polymeric electrospun fibers with the desired morphology (submicron or nanometric scale) as a function of the experimental parameters [30, 31]. The basic features of this deposition process are shown in **Figure 1** where a characteristic "Taylor cone" is formed during the projection of the fibers [32]. Under the action of the electric field, the droplets formed in the tip of needle are gradually elongated forming a characteristic conic shape. In addition, when the polymeric precursor has traveled through the air, the solvent is gradually evaporated during the flight of the fibers, and as a result, the fibers are finally deposited onto the corresponding collector [33, 34].

Figure 2 shows the three main key factors that have to be controlled to obtain the electrospun fibers with the desired morphology. The first factor is related to the nature of the polymeric precursor, which is associated with its molecular weight, viscosity, molar concentration, surface tension, electrical conductivity, and solvent nature [35]. The second factor is inherent to the operation of the electrospinning setup such as applied high voltage, the flow rate, and tip-to-collector distance [36]. And the third factor is derived by the external environmental conditions such as the relative humidity and temperature [37].

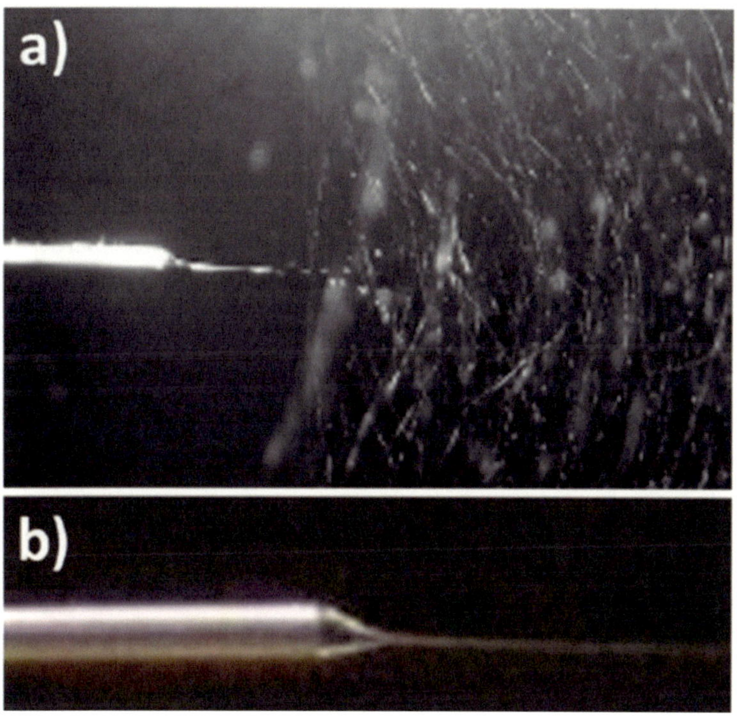

Figure 1.
(a) The aspect of the fibers being electrospun from the needle that contains the polymeric precursor solution (poly acrylic acid, PAA). (b) Detail of the "Taylor cone" formed at the tip of the needle as a function of the operational parameters. Reprinted with permission of Rivero et al. [32].

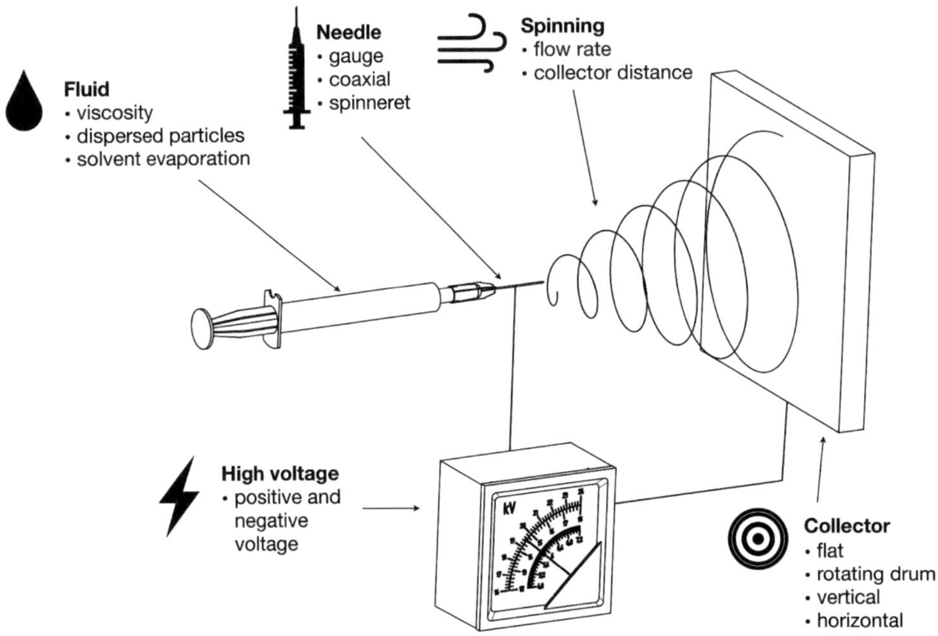

Figure 2.
Schematic representation of a electrospinning setup for the fabrication of electrospun fibers as a function of variable parameters such as nature of the fluid, nature of the solvent, type of needle, high voltage applied, flow rate, collector distance, and type of collector, respectively. Reprinted with permission of Rivero et al. [38].

3. Design of superhydrophobic surfaces obtained by electrospun fibers

By convention, a superhydrophobic surface exhibits an extraordinarily water contact angle value that is greater than 150°C with a low sliding angle (typically less than 10°). According to this, many plants (i.e., lotus left), animal fur, or insect wings found in the nature show this superhydrophobic behavior. In this sense, in order to simulate this biological surface, the design of synthetic superhydrophobic surfaces is a permanent challenge for the scientific community. The research is focused on the design of surfaces with a low surface energy combined with a hierarchical surface roughness on at least two different length scales (i.e., micrometric and nanometric morphology) [10]. And this dual effect can be perfectly controlled by the electrospinning technique because it is possible to generate continuous ultrathin fibers with micrometric and sub-micrometric diameter, thereby providing the hierarchical surface with superhydrophobic behavior because of the small size of the resultant electrospun fibers.

An interesting approach is shown in [39] because the intrinsic properties of polyvinylidene fluoride (PVDF) can be enhanced by fabricating electrospun nanocomposite PVDF fibrous mats of predetermined morphology. In this work, the effect of five different factors such as polymeric concentration, nanoparticles loading, volumetric ratio of the solvents, flow rate, and spinning distance have been analyzed on the morphology and wettability by using a screening Design of Experiments (DoE) statistical methodology. The results clearly indicate that among all the factors examined, the PVDF concentration has been found to show the most significant effect on both the morphology and wettability.

Other representative example is presented by Rawal [40] where it is theoretically demonstrated by a simple analytical mode that it is possible to obtain superhydrophobic behavior of an electrospun nonwoven mat by controlling important design parameters such as porosity and fiber orientation distribution. According to this, simply by enhancing the porosity of an electrospun fibers can show a superhydrophobic behavior. This effect of the resultant porosity is evaluated in [41] where the first example of successful production of self-standing fibrous membrane of polymers with intrinsic microporosity (PIM-2) by electrospinning technique is shown, showing a great interest as a potential membrane material for adsorption and separation applications. The fibers are composed of a mixture of 5,5′,6,6′-tetrahydroxy-3,3,3′,3′-tetramethyl-1,1′-spirobisindane (TTSBI) and decafluorobiphenyl (DFBP). The resultant coatings have shown bead-free and uniform fibers with an average diameter of 5.5 ± 1.5 μm, showing a superhydrophobic with a water contact angle value of 155 ± 6°. According to this, numerous articles can be found based on the fabrication of superhydrophobic electrospun coatings for oil-water separation. Wang et al. [42] described an electrospinning process to obtain superhydrophobic thermoplastic polyurethane nanofiber mat after decorating with modified nanosilicas. The water contact angle increases with the nanosilica concentration which it is ascribed to the increase in roughness of the TPU films as well as the low surface energy of the nanosilicas, being of great interest for separating mixtures of oil and water due to both superhydrophobicity and superoleophilicity. A similar work can be found in [43] with ultrathin electrospun fibrous PVDF membranes with both superhydrophobic and superoleophilic properties. The PVDF membranes are governed by the surface morphology and diameter of the PVDF fibers, which can be controlled by the PVDF concentration in the electrospinning solution. As the diameter of the PVDF fiber is increased, the surface roughness of the PVDF membrane is also increased. The ultrathin electrospun fibrous PVDF membranes carried out the highest water contact angle of 153° and an oil contact angle of 0°, being an interesting approach in high-efficiency liquid separation membranes for separating emulsified water-in-oil solutions. Other different types of polymeric precursor such as polymethyl methacrylate (PMMA) can be also used for the fabrication of superhydrophobic-superoleophilic fibrous membranes by electrospinning [44], thereby showing a high water contact angle up to 153.9° and nearly 0 oil contact angle. This super wettability property is associated with the hierarchical macro- and nanostructure on the surface of PMMA surface fibers, which is adjusted by weight ratio of the employed solvents. Other interesting work is presented in [45] for the preparation of a potential adsorbent with superhydrophobic (>150° in water) and superoleophilic (0° in oil) wetting properties for selective removal of crude oil from water. In this work, expanded polystyrene (EPS) was electrospun to produce beaded fibers in which zeolite was introduced to the polymeric matrix with the aim to impart rough to non-beaded fiber. Due to the use of EPS, a novel work is presented in [46] whereby the first time recycled expanded polystyrene foam with various proportions of titanium dioxide nanoparticles (TiO_2 NPs) and aluminum microparticles (Al μPs) have been successfully spun into superhydrophobic nanocomposite fibers using the electrospinning technique, showing a contact angle value of 152°.

Other interesting approach can be found in [47] where a novel method is proposed to fabricate hollow and surface porous PS fibrous membranes for the removal of oil from water. In this work, the spinning solutions were prepared by using camphene and tetraethoxysilane (TEOS) as pore-forming agents, and as a result, hollow PS fibers with 100–400 nm pores on the surface have been obtained by electrospinning and freeze-drying, being this type of membrane a great alternative and promising tool for oil of spill cleanups. In other work [48], the enhanced mechanical properties of superhydrophobic microfibrous polystyrene mats via polyamide 6 nanofibers (PA6) are evaluated. It has been corroborated that fibrous mats formed with the number ratios of jets 2/2 (PS/PA6) have shown a water contact angle of 150° with three times increased tensile strength compared with only pure fibrous

PS mat, respectively. Other important study is presented in [49] where electrospun nanofibrous mats (ENMs) have been fabricated from blends of a host hydrophilic polymer, polysulfone (PSF), and small quantities of fluorinated polyurethane additive (FPA), respectively. The ENMs have been tested in desalination by membrane distillation, thereby showing competitive permeate fluxes with stable low permeate electrical conductivities.

Other aspect to be remarked is that the electrospinning technique can be also implemented by using copolymers. A clear example can be found in [50] where segmented polyimide-siloxane copolymers have been prepared using 4,4′-oxydianiline (ODA) and 3,3′,4,4′-benzophenone tetracarboxylic dianhydride (BTDA) as hard segments and aminopropyl terminated polydimethyl-siloxane (APPS) and BTDA as soft segments. The water contact angle values of the electrospun mats have increased with increasing the siloxane content, and scanning electron microscopy (SEM) images reveal the bead formation known as pop-corn model, this finding being used specifically for self-cleaning materials.

Hang et al. [51] demonstrated the high versatility of the electrospinning because the coaxial electrospinning is presented by using Teflon AF as sheath material and polycaprolactone (PCL) as core material. The resultant electrospun fibers exhibit both superhydrophobic and oleophobic behaviors, and these coaxial fibers also preserve the core material properties as demonstrated with mechanical tensile tests. The use of coaxial electrospinning is also used in [52] where the design and fabrication of a composite electrospun membrane composed of polylactide:poly(vinyl pyrrolidone)/polylactide:poly(ethylene glycol) (PLA:PVP/PLA:PEG) core shell fibers loaded with bioactive agents, as functionally integrated wound dressing for efficient burn treatments. In addition, the electrospinning can be easily combined with other deposition techniques as it can be appreciated in [53]. In this work, superhydrophobic fabrics are produced by electrospinning and chemical vapor deposition. According to this, first electrospun PCL fibers are deposited and second, a thin layer of hydrophobic polymerized perfluoroalkylethyl methacrylate (PPFEMA) is deposited on the electrospun fibers, thereby yielding a superhydrophobicity behavior with a contact angle value of 175° and a sliding angle less than 2.5°. A similar approach based on the use of two different deposition techniques is shown in [54] where the preparation of hierarchically structured PCL superhydrophobic membranes are obtained via combination of electrospinning and electrospraying techniques. Dizge et al. [55] fabricated an electrospun cellulose nanofiber membrane and then, two different deposition techniques such as sol-gel and CVD are implemented onto this electrospun membrane, thereby showing both superhydrophobicity and oleophobicity (also known as omniophobic), as demonstrated by its wetting resistance to water, ethanol, surfactant, and mineral oil. An application of this omniophobic membrane is in direct contact membrane distillation (DCMD) to separate water from saline feed solutions containing low surface tension substances. Deka et al. [56] presented the high flux and non-wettability of electrospun nanofiber membranes fabricated by electrospraying of aerogel/polydimethylsiloxane (PDMS)/polyvinylidene fluoride (PVDF) over electrospinning polyvinylidene fluoride-co-hexafluoropropylene (PVDF-HFP) membrane. The use of aerogel and PDMS is preferred as materials with low surface energy could be a safer alternative to perfluorinated compounds for environmental applications. In addition, the experimental results demonstrate non-wetting membrane distillation performance over continuous 7 days operation of saline water (3.5% of NaCl) and high anti-wetting with harsh saline water 0.5 mM sodium dodecyl sulfate (SDS) and synthetic algal organic matter.

Other clear example of the implementation of these superhydrophobic electrospun fibers is for the development of anticorrosion surfaces where polymeric precursors with an intrinsic hydrophobic behavior are used such as polystyrene (PS), poly(vinyl chloride) (PVC), and polyvinylidene fluoride (PVDF), among others. Cui et al. [57] reported a simple and controllable electrospinning technology to fabricate polymeric nanofibers composed of polyvinylidene fluoride (PVDF)/stearic acid (SA) that are successfully deposited onto metallic substrates (aluminum sheets). The resultant electrospun nanofibers show a clearly superhydrophobic behavior with water contact angle value of 155 ± 2°C, thereby exhibiting excellent long-time corrosion resistance. In other works, metal oxide nanoparticles such as zinc oxide (ZnO) or alumina (Al_2O_3) are used, which are perfectly embedded in the electrospun fibers and act as efficient corrosion inhibitors. In addition, the presence of this type of nanoparticles between the interstices of the electrospun fibers can increase the surface roughness as well as the air entrapment, and as a result, an increase in the water repellent behavior is obtained. According to this, Radwan et al. [58] reported the addition of ZnO nanoparticles to the polymeric precursor of PVDF with the aim to improve the corrosion resistance, maintaining the same superhydrophobic behavior (155 ± 2°). In addition, by using the same corrosion inhibitor (ZnO), Iribarren et al. [59] presented multifunctional protective PVC-ZnO nanocomposite coatings deposited on aluminum alloys by electrospinning. In the first step, an exhaustive study about the evolution of the resultant fiber diameter as a function of the applied voltage and the flow rate is evaluated. **Figure 3** shows that the fiber diameter of PVC electrospun fibers is reduced when the applied voltage is increased from 8 up to 14 kV (see **Figure 3a**), whereas the resultant fiber diameter is increased when the flow rate is gradually increased (see **Figure 3b**) from 0.6 up to 1.2 mL/h. Once it has been evaluated, the operational parameters in the corresponding fiber diameter, the electrospun PVC-ZnO nanofibers were deposited by using a voltage of 14 kV and a flow rate of 0.6 mL/h because it has been demonstrated that an increase in the water repellency behavior is observed when the diameter among bead-free fibers is reduced. And this effect has been corroborated because the resultant surface wettability of the electrospun coating presents similar water contact angle values in the range of 145–155°C.

Rivero et al. [60] presented a comparative study of multifunctional coatings based on electrospun fibers with incorporated ZnO nanoparticles by using two different polymeric precursors such as PVC and PS, respectively. In order to characterize the morphology of the resultant electrospun fibers, atomic force microscopy (AFM) was performed, as it can be appreciated in **Figure 4**, showing an important difference in the fiber diameter. In this sense, 2D AFM images with their corresponding profiles (three evaluation lines) clearly reveal that a bigger size in diameter fiber is obtained for PS samples (**Figure 4a** and **b**) in comparison with PVC

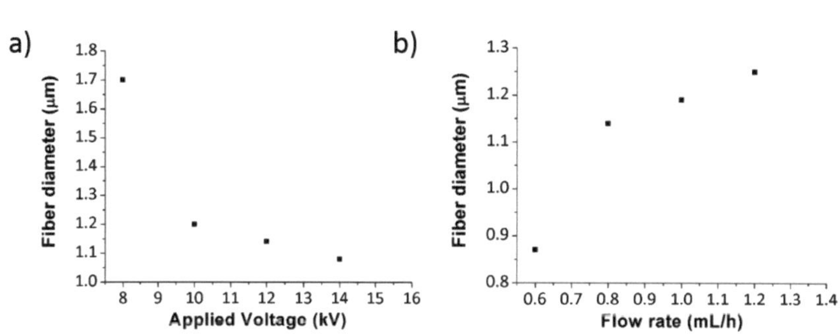

Figure 3.
(a) Evolution of the resultant fiber diameter as a function of the applied voltage with a specific fixed flow rate of 0.8 mL/h. (b) Evolution of the fiber diameter as a function of the flow rate with a specific fixed applied voltage of 12 kV. Reprinted with permission of Iribarren et al. [59].

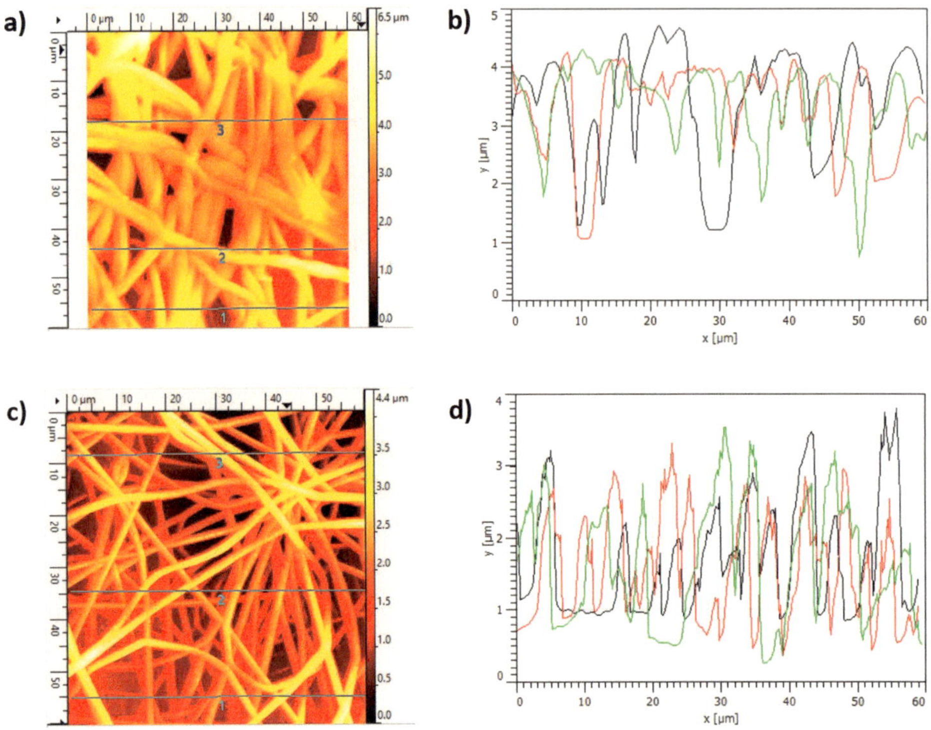

Figure 4.
2D-AFM images for a dimension of 60 × 60 μm^2 of the PS electrospun fibers (a, b) and PVC electrospun fibers (c, d) with their corresponding profiles. Reprinted with permission of Rivero et al. [60].

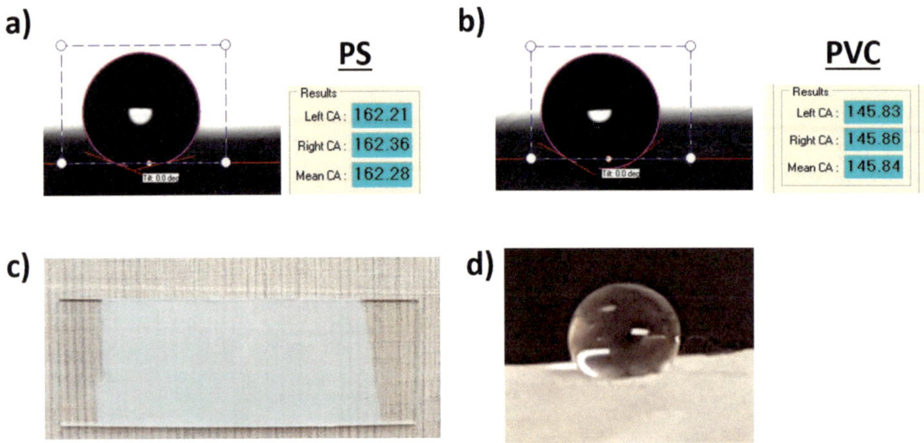

Figure 5.
Optical image of the water contact angle (WCA) value for PS (a) and PVC (b) samples; (c) aspect of the electrospun fibers with a characteristic white coloration deposited onto a standard microscope glass slide; (d) picture of the water drop onto the electrospun fibers. Reprinted with permission of Rivero et al. [60].

samples (**Figure 4c** and **d**). In addition, the arithmetic average roughness Ra of the PS coating was found to be 505.7 nm, which is higher than that in PVC coating with a value of 427.5 nm. This result is in concordance with the high value of WCA because an increase in the hydrophobicity is clearly obtained by the increment in the resultant surface roughness, and due to this, the WCA values for PS samples are higher than PVC samples (see **Figure 5**).

These two previous polymeric precursors of PS and PVC have been also studied in [61] where the superhydrophobic behavior as a function of the fiber diameter, the presence of TiO_2 nanoparticles, and the effect of heat treatment in the nanocomposite fibers are evaluated. The main purpose is for energy systems, as it can be used in the design of dye-sensitized solar cells (DSSCs) for anti-icing and self-cleaning materials with the aim to show long-term efficiency of the cells. The results show that TiO_2 nanocomposite fibers have higher contact angle values, which is attributed to the nanoscale gaps/bumps/voids formed on the fiber in the presence of TiO_2 nanoparticles, thereby reaching superhydrophobicity on the nanocomposite electrospun fiber films. Azimirad et al. [62] presented a dual layer of dip-coated TiO_2 film (top layer) and electrospun polystyrene (bottom layer) deposited onto stainless steel with the aim to show both superhydrophobicity and corrosion resistance, thereby showing great important and potential applications, especially in marine industries. Other interesting approach is presented in [63]. In this case, the optimum conditions to produce electrospun polystyrene (PS) and aluminum oxide (Al_2O_3) nanocomposite coating with the highest roughness and superhydrophobic properties are 25 kV of applied voltage and 1.5 mL/h of flow rate at 35°C. The experimental results indicate a water contact angle value of 155 ± 1.9 and contact angle hysteresis of 2 ± 4.2°.

Finally, other interesting application field for the electrospun fibers is the design of biomaterials with application as surgical meshes, sutures, or in artificial tissue [64]. In this work, a superhydrophobic poly(L-lactic acid) (PLLA) surface is obtained by dispersing synthetic talc (ts) into PLLA fibers. This synthetic talc, characterized by the presence of long aliphatic chains in the structure, is soluble in the electrospinning solvent mixture, whose viscosity is significantly modified by small amount of talc. The evaporation during the electrospinning process

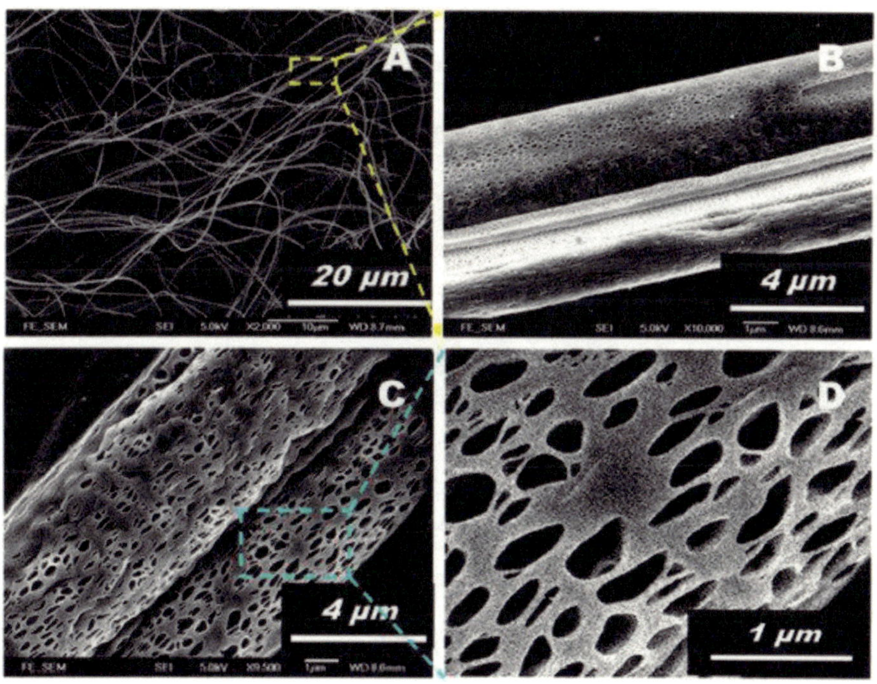

Figure 6.
(A–D) FESEM images of the prepared ES PLA/TiO_2 fibers at different magnification scales. Note that the images (B) and (D) are magnified from the locations indicated in images (A). Reprinted with permission of Wang et al. [65].

promotes the synthetic talc dispersion into the polymer matrix (PLLA), thereby obtaining a nanometric size scale distribution of the talc. In addition, among the parameters studied, the relative humidity (RH) was found significantly to affect the fiber morphology. By keeping all the electrospinning parameters constant and increasing the relative humidity (Rh), the morphology completely changes. This increment increases the homogeneity of the PLLA/ts fibers and turns out to be opposite with respect to PLLA fibers. Moreover, the influence of the electrical field by decreasing the voltage shows that the homogeneity tends to decrease. The dispersion of the synthetic talc on the nanofiber produces an evident effect in the wettability, as the PLLA/ts electrospun mat is highly hydrophobic by exhibiting an increment in the WCA value from 92 up to 140°. Other aspect to remark is that the presence of talc has promoted the development of a small amount of crystallinity

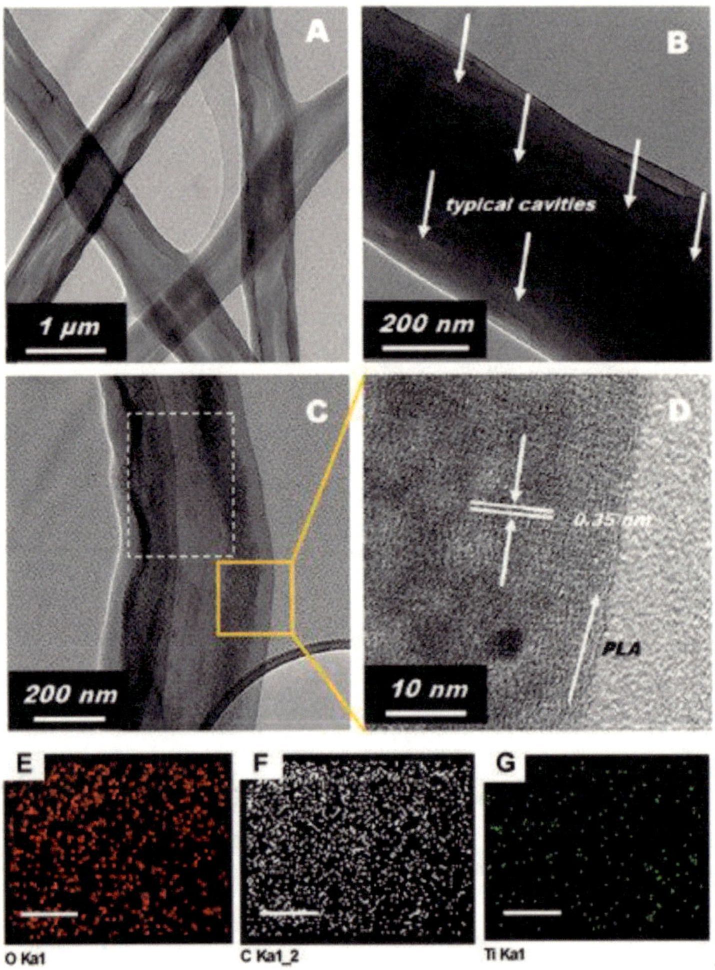

Figure 7.
TEM images of the prepared typical ES PLA/TiO_2 fibers. (A) Low-magnification TEM image of ES PLA/TiO_2 fibers. (B) Magnified TEM image of an ES PLA/TiO_2 fiber captured for the observation of typical cavities. (C) Magnified TEM image of an ES PLA/TiO_2 fiber captured for the observation of TiO_2 nanoparticles. (D) High-resolution TEM image zoomed in from the marked area in (C) for the identification of the lattice fringe of TiO_2 nanocrystallites. (E–G) The O, C, and Ti element mapping images of the location are indicated in image (C) with a dashed rectangle. The scale bars in all the mapping images (E–G) are 100 nm. Reprinted with permission of Wang et al. [65].

during the electrospinning process, thereby making possible the development of α-crystallographic form during the annealing.

Other interesting work based on the use of biodegradable polymer is presented in [65] where polylactide/TiO_2 composite fiber (PLA/TiO_2) scaffolds with both superhydrophobic and superadhesive porous surfaces have been obtained for water immobilization, antibacterial performance, and deodorization. The size of the pores of the as-fabricated PLA/TiO_2 electrospun fibers observed on the surface has shown to have a length of 200 ± 100 nm and a width of 150 ± 50 nm demonstrated by using field-emission scanning electron microscopy (FESEM) (see **Figure 6**) and transmission electron microscopy (TEM) (see **Figure 7**), respectively. As it can be appreciated in both figures, the formation of regular pores elongated along the fiber axis was observed on the fiber surface and it is associated with the rapid phase separation during the electrospinning process as induced by the voltage solvent evaporation and the fast solidification. In addition, according to the TEM image of **Figure 7**, the distribution of the TiO_2 nanoparticles over the electrospun surface fiber has been analyzed by using energy dispersive X-Ry (EDX) elemental mapping technique (**Figure 7E–G**). According to this, a uniform Ti element distribution has been observed, indicating that TiO_2 NPs have been homogeneously dispersed in the PLA electrospun fibers. In addition, a powerful adhesive force can be associated with the van der Waals forces and the accumulated negative pressure forces of the as-spun PLA/TiO_2 fibers. Other important aspect to remark is that the resultant PLA/TiO_2 composite fiber presents a high antibacterial efficiency against three different types of bacteria (*Escherichia coli, Staphylococcus aureus*, and *Candida albicans*) as well as high deodorization efficiency because the reduction of two typical pollutants such as ammonia and formaldehyde have been reduced after exposition to visible light radiation. To sum up, this multifunctional biodegradable fiber scaffold can bring promising benefits to the real world in specific biomedical or even bioengineering applications.

4. Summary table of potential industrial applications

To sum up, in order to have a better understanding of the different superhydrophobic coatings analyzed in this work, a summary of the different electrospun fibrous coatings with their corresponding employed solvents as well as their potential industrial applications can be appreciated in **Table 1**.

5. Conclusions

In this work, electrospinning is presented as a novel engineering technique for the design of superhydrophobic surfaces. In order to obtain this special wettability, it is necessary to control two crucial factors such as a low surface energy and a hierarchical surface roughness on at least two different length scales (i.e., micrometric and nanometric morphology). The electrospinning is a good candidate because it is possible to control both parameters as a function of the operational parameters such as applied voltage, polymeric precursor concentration, flow rate, and tip-to-collector distance. A good control over these parameters makes possible the fabrication of electrospun fibers with a desired morphology (mostly size, porosity, and fiber diameter).

Finally, a summary of different potential industrial applications is presented due to the design of corrosion-resistant surfaces, high-efficient water-oil separation

Electrospun fiber coating	Solvent	Research field	Ref.
TTSBI + DFBP	Tetrachloroethane (TCE)	Adsorption applications	[41]
TPU decorated with modified nanosilicas	N,N-dimethylformamide (DMF) and tetrahydrofuran (THF)	Water-oil separation	[42]
PVDF	N,N-dimethylformamide (DMF) and acetone	Water-oil separation	[43]
PMMA	N,N-dimethylacetamide (DMAc) and acetone	Water-oil separation	[44]
EPS and EPS/zeolite	Tetrahydrofuran (THF)	Water-oil separation	[45]
PS/TEOS/camphene	N,N-dimethylformamide (DMF)	Water-oil separation	[47]
PSF	N,N-dimethylacetamide (DMAc) and acetone	Desalination by membrane distillation	[49]
PAA	N,N-dimethylacetamide (DMAc) and tetrahydrofuran (THF)	Self-cleaning	[50]
Teflon AF (sheath)/PCL (core)	2,2,2-trifluoroethanol (TFE)	Microfluidics	[51]
PLA:PVP/PLA:PEG	Dichloromethane	Burn wound healing and skin regeneration	[52]
PCL	Chloroform and methanol	Fabrics	[53]
PCL	N,N-dimethylformamide (DMF) and chloroform	Water-oil separation	[54]
Cellulose acetate	N,N-dimethylacetamide (DMAc) and acetone	Water-saline solution separation	[55]
PVDF-HFP	N,N-dimethylformamide (DMF) and acetone	Desalination by membrane distillation	[56]
PVDF/SA	N,N-dimethylformamide (DMF)	Corrosion protection	[57]
PVDF-ZnO	N,N-dimethylacetamide (DMAc) and toluene	Corrosion protection	[58]
PVC-ZnO	N,N-dimethylformamide (DMF) and tetrahydrofuran (THF)	Corrosion protection	[59]
PVC-ZnO PS-ZnO	N,N-dimethylformamide (DMF) and tetrahydrofuran (THF)	Corrosion protection	[60]
PS-TiO_2 PVC-TiO_2	N,N-dimethylformamide (DMF) and N,N-dimethylacetamide (DMAc)	Energy systems (DSSC devices)	[61]
PS-TiO_2	N,N-dimethylformamide (DMF)	Corrosion protection	[62]
PS-Al_2O_3	N,N-dimethylacetamide (DMAc) and tetrahydrofuran (THF)	Corrosion protection	[63]
PLLA	Toluene and chloroform	Biomedical applications	[64]
PLA/TiO_2	Chloroform and acetone	Water immobilization, antibacterial performance, and deodorization	[65]

Table 1.
Summary of the different sensitive electrospun coatings, the metallic substrates used, and the resultant corrosion tests for the development of new and innovative protective coatings with good anticorrosion properties.

membranes, long-term efficiency of dye sensitized solar cells, or even in biomedical applications for the development of antibacterial surfaces with a high efficiency against bacteria or pollutants. To sum up, it has been demonstrated that this deposition technique can be used as a promising alternative in the real world in several disciplines of the science and technology.

Acknowledgements

This work was supported by the Ministerio de Ciencia, Innovación y Universidades-Retos (Project RTI2018-096262-B-C41-MAITAI) and by the Public University of Navarre (Project PJUPNA1929).

Author details

Pedro J. Rivero[1,2]*, Adrian Vicente[1] and Rafael J. Rodriguez[1,2]

1 Materials Engineering Laboratory, Department of Engineering, Public University of Navarre, Pamplona, Spain

2 Institute for Advanced Materials (INAMAT), Public University of Navarre, Pamplona, Spain

*Address all correspondence to: pedrojose.rivero@unavarra.es

References

[1] Zhao T, Jiang L. Contact angle measurement of natural materials. Colloids and Surfaces B: Biointerfaces. 2018;**161**:324-330

[2] Cebeci FÇ, Wu Z, Zhai L, Cohen RE, Rubner MF. Nanoporosity-driven superhydrophilicity: A means to create multifunctional antifogging coatings. Langmuir. 2006;**22**(6):2856-2862

[3] Zhang X, Shi F, Niu J, Jiang Y, Wang Z. Superhydrophobic surfaces: From structural control to functional application. Journal of Materials Chemistry. 2008;**18**(6):621-633

[4] Du C, Wang J, Chen Z, Chen D. Durable superhydrophobic and superoleophilic filter paper for oil-water separation prepared by a colloidal deposition method. Applied Surface Science. 2014;**313**:304-310

[5] Tuteja A et al. Designing superoleophobic surfaces. Science. 2007;**318**(5856):1618-1622

[6] Wenzel RN. Resistance of solid surfaces to wetting by water. Industrial and Engineering Chemistry. 1936;**28**(8):988-994

[7] Cassie ABD, Baxter S. Wettability of porous surfaces. Transactions of the Faraday Society. 1944;**40**:546-551

[8] Giacomello A, Meloni S, Chinappi M, Casciola CM. Cassie-Baxter and Wenzel states on a nanostructured surface: Phase diagram, metastabilities, and transition mechanism by atomistic free energy calculations. Langmuir. 2012;**28**(29):10764-10772

[9] Roach P, Shirtcliffe NJ, Newton MI. Progress in superhydrophobic surface development. Soft Matter. 2008;**4**(2):224

[10] Ma M et al. Decorated electrospun fibers exhibiting superhydrophobicity. Advanced Materials. 2007;**19**(2):255-259

[11] Liao K-S, Wan A, Batteas JD, Bergbreiter DE. Superhydrophobic surfaces formed using layer-by-layer self-assembly with aminated multiwall carbon nanotubes. Langmuir. 2008;**24**(8):4245-4253

[12] Rao AV, Latthe SS, Mahadik SA, Kappenstein C. Mechanically stable and corrosion resistant superhydrophobic sol-gel coatings on copper substrate. Applied Surface Science. 2011;**257**(13):5772-5776

[13] Zhao N, Shi F, Wang Z, Zhang X. Combining layer-by-layer assembly with electrodeposition of silver aggregates for fabricating superhydrophobic surfaces. Langmuir. 2005;**21**(10):4713-4716

[14] Ishizaki T, Hieda J, Saito N, Saito N, Takai O. Corrosion resistance and chemical stability of super-hydrophobic film deposited on magnesium alloy AZ31 by microwave plasma-enhanced chemical vapor deposition. Electrochimica Acta. 2010;**55**(23):7094-7101

[15] Shiu J-Y, Kuo C-W, Chen P, Mou C-Y. Fabrication of tunable superhydrophobic surfaces by nanosphere lithography. Chemistry of Materials. 2004;**16**(4):561-564

[16] Yoldi M et al. Fabrication of superhydrophobic nanostructured films by physical vapour deposition. In: Nanotechnology 2010: Advanced Materials, CNTs, Particles, Films and Composites - Technical Proceedings of the 2010 NSTI Nanotechnology Conference and Expo, NSTI-Nanotech 2010. Vol. 1. 2010. pp. 600-603

[17] Qian B, Shen Z. Fabrication of superhydrophobic surfaces by

dislocation-selective chemical etching on aluminum, copper, and zinc substrates. Langmuir. 2005;**21**(20):9007-9009

[18] Lim J-M, Yi G-R, Moon JH, Heo C-J, Yang S-M. Superhydrophobic films of electrospun fibers with multiple-scale surface morphology. Langmuir. 2007;**23**(15):7981-7989

[19] Acik G, Cansoy CE, Kamaci M. Effect of flow rate on wetting and optical properties of electrospun poly(vinyl acetate) micro-fibers. Colloid & Polymer Science. 2019;**297**(1):77-83

[20] Li D, Xia Y. Electrospinning of nanofibers: Reinventing the wheel? Advanced Materials. 2004;**16**(14):1151-1170

[21] Reneker DH, Chun I. Nanometre diameter fibres of polymer, produced by electrospinning. Nanotechnology. 1996;**7**(3):216-223

[22] Kurusu RS, Demarquette NR. Surface modification to control the water wettability of electrospun mats. International Materials Review. 2019;**64**(5):249-287

[23] Xu CY, Inai R, Kotaki M, Ramakrishna S. Aligned biodegradable nanofibrous structure: A potential scaffold for blood vessel engineering. Biomaterials. 2004;**25**(5):877-886

[24] Xu X, Yang Q, Wang Y, Yu H, Chen X, Jing X. Biodegradable electrospun poly(l-lactide) fibers containing antibacterial silver nanoparticles. European Polymer Journal. 2006;**42**(9):2081-2087

[25] Kenawy E-R et al. Release of tetracycline hydrochloride from electrospun poly(ethylene-co-vinylacetate), poly(lactic acid), and a blend. Journal of Controlled Release. 2002;**81**(1-2):57-64

[26] Thaiane da Silva T et al. Electrospun curcumin/polycaprolactone/copolymer F-108 fibers as a new therapy for wound healing. Journal of Applied Polymer Science. 2020;**137**(9)

[27] Lee KY, Jeong L, Kang YO, Lee SJ, Park WH. Electrospinning of polysaccharides for regenerative medicine. Advanced Drug Delivery Reviews. 2009;**61**(12):1020-1032

[28] Mogoşanu GD, Grumezescu AM. Natural and synthetic polymers for wounds and burns dressing. International Journal of Pharmaceutics. 2014;**463**(2):127-136

[29] Demir MM, Yilgor I, Yilgor E, Erman B. Electrospinning of polyurethane fibers. Polymer (Guildf). 2002;**43**(11):3303-3309

[30] Bhardwaj N, Kundu SC. Electrospinning: A fascinating fiber fabrication technique. Biotechnology Advances. 2010;**28**(3):325-347

[31] Rivero PJ, Garcia JA, Quintana I, Rodriguez R. Design of nanostructured functional coatings by using wet-chemistry methods. Coatings. 2018;**8**(2)

[32] Rivero PJ et al. An antibacterial submicron fiber mat with in situ synthesized silver nanoparticles. Journal of Applied Polymer Science. 2012;**126**(4):1228-1235

[33] Frenot A, Chronakis IS. Polymer nanofibers assembled by electrospinning. Current Opinion in Colloid & Interface Science. 2003;**8**(1):64-75

[34] Doshi J, Reneker DH. Electrospinning process and applications of electrospun fibers. Journal of Electrostatics. 1995;**35**(2-3):151-160

[35] Pillay V et al. A review of the effect of processing variables on the

fabrication of electrospun nanofibers for drug delivery applications. Journal of Nanomaterials. 2013;**2013**

[36] Wang X, Hsiao BS. Electrospun nanofiber membranes. Current Opinion in Chemical Engineering. 2016;**12**:62-81

[37] Deitzel JM, Kleinmeyer J, Harris D, Beck Tan NC. The effect of processing variables on the morphology of electrospun nanofibers and textiles. Polymer (Guildf). 2001;**42**(1):261-272

[38] Rivero PJ, Redin DM, Rodríguez RJ. Electrospinning: A powerful tool to improve the corrosion resistance of metallic surfaces using nanofibrous coatings. Metals (Basel). 2020;**10**(3)

[39] Al Haddad ZA, Svinterikos E, Zuburtikudis I. Designing electrospun nanocomposite poly(vinylidene fluoride) mats with tunable wettability. Colloids and Surfaces A: Physicochemical and Engineering Aspects. 2017;**523**:81-90

[40] Rawal A. Tunable superhydrophobic electrospun nonwoven mat. Materials Letters. 2011;**65**(10):1457-1459

[41] Satilmis B, Uyar T. Development of superhydrophobic electrospun fibrous membrane of polymers of intrinsic microporosity (PIM-2). European Polymer Journal. 2019;**112**:87-94

[42] Wang L, Yang S, Wang J, Wang C, Chen L. Fabrication of superhydrophobic TPU film for oil-water separation based on electrospinning route. Materials Letters. 2011;**65**(5):869-872

[43] Zhou Z, Wu X-F. Electrospinning superhydrophobic-superoleophilic fibrous PVDF membranes for high-efficiency water-oil separation. Materials Letters. 2015;**160**:423-427

[44] Liu Z, Zhao J-H, Liu P, He J-H. Tunable surface morphology of electrospun PMMA fiber using binary solvent. Applied Surface Science. 2016;**364**:516-521

[45] Alayande SO, Dare EO, Msagati TAM, Akinlabi AK, Aiyedun PO. Superhydrophobic and superoleophillic surface of porous beaded electrospun polystrene and polysytrene-zeolite fiber for crude oil-water separation. Physics and Chemistry of the Earth. 2016;**92**:7-13

[46] Uddin MN, Desai FJ, Asmatulu E. Biomimetic electrospun nanocomposite fibers from recycled polystyrene foams exhibiting superhydrophobicity. Energy, Ecology and Environment. 2020;**5**(1)

[47] Tang Y, Liu Z, Zhao K. Fabrication of hollow and porous polystyrene fibrous membranes by electrospinning combined with freeze-drying for oil removal from water. Journal of Applied Polymer Science. 2019;**136**(13)

[48] Li X, Ding B, Lin J, Yu J, Sun G. Enhanced mechanical properties of superhydrophobic microfibrous polystyrene mats via polyamide 6 nanofibers. Journal of Physical Chemistry C. 2009;**113**(47):20452-20457

[49] Khayet M, García-Payo C, Matsuura T. Super hydrophobic nanofibers electrospun by surface segregating fluorinated amphiphilic additive for membrane distillation. Journal of Membrane Science. 2019;**588**

[50] Oktay B, Toker RD, Kayaman-Apohan N. Superhydrophobic behavior of polyimide-siloxane mats produced by electrospinning. Polymer Bulletin. 2015;**72**(11):2831-2842

[51] Han D, Steckl AJ. Superhydrophobic and oleophobic fibers by coaxial electrospinning. Langmuir. 2009;**25**(16):9454-9462

[52] Li W et al. Superhydrophobic hierarchical fiber/bead composite membranes for efficient treatment of burns. Acta Biomaterialia. 2019;**92**:60-70

[53] Ma M, Mao Y, Gupta M, Gleason KK, Rutledge GC. Superhydrophobic fabrics produced by electrospinning and chemical vapor deposition. Macromolecules. 2005;**38**(23):9742-9748

[54] Zhang G, Wang P, Zhang X, Xiang C, Li L. Preparation of hierarchically structured PCL superhydrophobic membrane via alternate electrospinning/electrospraying techniques. Journal of Polymer Science Part B: Polymer Physics. 2019;**57**(8):421-430

[55] Dizge N, Shaulsky E, Karanikola V. Electrospun cellulose nanofibers for superhydrophobic and oleophobic membranes. Journal of Membrane Science. 2019;**590**

[56] Deka BJ, Lee E-J, Guo J, Kharraz J, An AK. Electrospun Nanofiber membranes incorporating PDMS-aerogel Superhydrophobic coating with enhanced flux and improved antiwettability in membrane distillation. Environmental Science & Technology. 2019;**53**(9):4948-4958

[57] Cui M, Xu C, Shen Y, Tian H, Feng H, Li J. Electrospinning superhydrophobic nanofibrous poly(vinylidene fluoride)/stearic acid coatings with excellent corrosion resistance. Thin Solid Films. 2018;**657**:88-94

[58] Radwan AB, Mohamed AMA, Abdullah AM, Al-Maadeed MA. Corrosion protection of electrospun PVDF-ZnO superhydrophobic coating. Surface and Coatings Technology. 2016;**289**:136-143

[59] Iribarren A et al. Multifunctional protective PVC-ZnO nanocomposite coatings deposited on aluminum alloys by electrospinning. Coatings. 2019;**9**(4)

[60] Rivero PJ, Iribarren A, Larumbe S, Palacio JF, Rodríguez R. A comparative study of multifunctional coatings based on electrospun fibers with incorporated ZnO nanoparticles. Coatings. 2019;**9**(6)

[61] Asmatulu R, Ceylan M, Nuraje N. Study of superhydrophobic electrospun nanocomposite fibers for energy systems. Langmuir. 2011;**27**(2):504-507

[62] Azimirad R, Safa S. Electrospun polystyrene fibres on TiO_2 nanostructured film to enhance the hydrophobicity and corrosion resistance of stainless steel substrates. Pramana-Journal of Physics. 2016;**86**(3):653-660

[63] Radwan AB, Abdullah AM, Mohamed AMA, Al-Maadeed MA. New electrospun polystyrene/Al_2O_3 nanocomposite superhydrophobic coatings: Synthesis, characterization, and application. Coatings. 2018;**8**(2)

[64] Monticelli O, Bocchini S, Gardella L, Cavallo D, Cebe P, Germelli G. Impact of synthetic talc on PLLA electrospun fibers. European Polymer Journal. 2013;**49**(9):2572-2583

[65] Wang X, Chen D, Zhang M, Hu H. Biodegradable polylactide/TiO_2 composite fiber scaffolds with superhydrophobic and superadhesive porous surfaces for water immobilization, antibacterial performance, and deodorization. Polymers (Basel). 2019;**11**(11)

Chapter 12

Smart Coatings with Carbon Nanoparticles

Xoan Xosé Fernández Sánchez-Romate, Alberto Jiménez Suárez and Silvia González Prolongo

Abstract

Smart coatings based on polymer matrix doped with carbon nanoparticles, such as carbon nanotubes or graphene, are being widely studied. The addition of carbon nanofillers into organic coatings usually enhances their performance, increasing their barrier properties, corrosion resistance, hardness, and wear strength. Moreover, the developed composites provide a new generation of protective organic coatings, being able to intelligently respond to damage or external stimuli. Carbon nanoparticles induce new functionalities to polymer coatings, most of them related to the higher electrical conductivity of nanocomposite due to the formation of percolation network. These coatings can be used as strain sensors and gauges, based on the variation of their electrical resistance (structural health monitoring, SHM). In addition, they act as self-heaters by the application of electrical voltage associated to resistive heating by Joule effect. This opens new potential applications, particularly deicing and defogging coatings. Superhydrophobic and self-cleaning coatings are inspired from lotus effect, designing micro- and nanoscaled hierarchical surfaces. Coatings with self-healable polymer matrix are able to repair surface damages. Other relevant smart capabilities of these new coatings are flame retardant, lubricating, stimuli-chromism, and antibacterial activity, among others.

Keywords: polymer nanocomposite, structural health monitoring, self-heating, self-healing, anti- and deicing

1. Introduction

Smart coatings are special covering materials which are able to sense and respond to an external stimulus. They are made with programmable materials, which respond to changes in light, chemical, thermal, or other stimuli. This brings them new performances, typically self-healing, self-cleaning, self-sensors, etc. due to their piezoelectric, thermoelectric, piezoresistive, and chemical properties (**Figure 1**). Most of the current smart coatings are based on nanoreinforced polymers. The incorporation of functional organic and inorganic nanofillers usually improves the thermal and mechanical properties of polymers, providing them new functionalities. As it is well-known, one of the main advantages to add nanofillers is their high specific area, which reduces significantly the nanofiller content and enhances the load transfer from the matrix, when the interface is suitable.

In this work, we focused on the addition of carbon nanoparticles, mainly graphene (G), graphene nanoplatelets (GNP), and carbon nanotubes (CNT). They

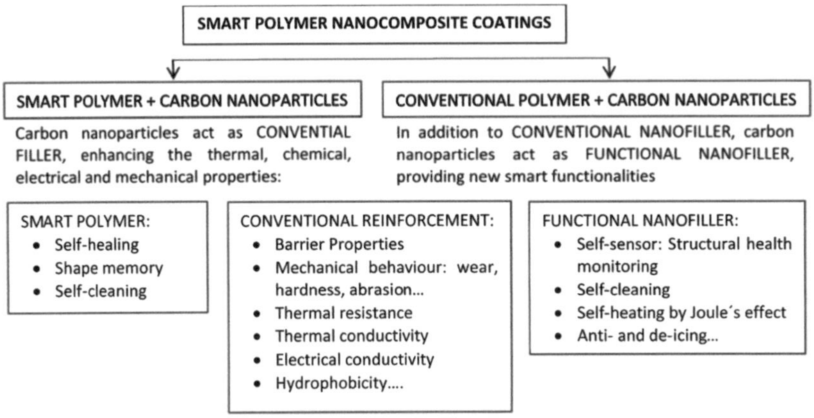

Figure 1.
Summary of smart coatings with carbon nanoparticles.

have extraordinary electrical and thermal conductivity and a unique combination of mechanical properties with great stiffness and high toughness [1–6]. They are composed of carbon, exhibiting low toxicity and environmental friendliness. For all these reasons, they are considered as multifunctional fillers of polymer matrix. In fact, polymer nanocomposites reinforced with carbon nanoparticles usually present enhanced mechanical, electrical, and thermal properties together with new performance as smart materials.

They can act as strain sensors due to their piezoresistive behavior, varying the electrical resistance of composite induced by the deformation of the electrical network formed by graphitic nanofillers. On the other hand, the nanofillers can be used as actuators, for example, as self-heater due to Joule's heating or as chemical absorbents. In this case, the matrix is a neat stimulus-responsive polymer, while the carbon nanofillers provide the stimuli to induce the polymer response.

2. Synthesis, processing, and main properties of polymer coatings

2.1 Synthesis and processing

The synthesis and processing of nanocomposite polymer coatings usually have at least two separated stages: the dispersion of nanofillers into the monomers, prepolymers, or polymers and the coating manufacture.

There are numerous different techniques to disperse the carbon nanofiller into monomers and polymers. In fact, there are numerous articles and reviews published. For this reason, in this chapter, they are only mentioned. It is well-known that the improvement of properties, in special mechanical ones, on nanocomposites is strongly dependent on the dispersion quality together with the polymer-nanofiller interface, which relies on the chemical and physical interaction between functionalized nanofillers and polymer matrix. Good dispersion of nanoparticles is critical to achieve high-performance nanocomposite. The most common **processing techniques of nanocomposites** can be organized on three ways: direct mixing, in situ polymerization in the presence of nanoparticles, and solution mixing. One more processing way is the in situ synthesis of particles, which is usually based on in situ sol–gel process inside polymers, but it is only used for inorganic nanofillers.

Carbon nanofillers must be dispersed on polymer or prepolymers depending on the polymer nature. The dispersion of nanofillers on thermoplastic polymers often carries out during the polymer manufacturing process, as extrusion or calandering. However, the nanofiller dispersion on rubber and thermosetting polymers is usually carried out in a previous step of curing process into monomers or prepolymers. In this last case, different dispersion techniques can be also applied, based on the application of mechanical forces or an electric or magnetic field.

As it was mentioned above, the second step consists on manufacturing the own coating, applying different common **processing techniques of coatings**. Cold spray process is commonly used for processing polymer nanocomposite coatings, avoiding the thermal deterioration of substrate. Dispersion, emulsion, and latex in situ polymerizations are other applied manufacturing processes.

2.2 Properties of nanocomposite polymer coatings

Graphitic nanofillers are often used to improve the **mechanical properties** of polymer coatings. The poor tribological performance of polymer coatings can be improved by adequately addition of graphitic nanofillers into the matrix because graphite is a solid lubricant. Polymer coating containing graphene can present excellent tribological properties, with low friction coefficient and reduced wear rate [7]. The increment of graphene content gradually decreases both friction coefficient and wear rate of composite coating. Under high temperature, graphene-reinforced thermosetting coatings show better friction reduction and wear resistance than neat coating. The values of these properties are enhanced by the increase of graphene content. Meanwhile, the friction coefficient and the wear rate of the graphene/composite coatings do not show a clear tendency with the increase of temperature. This behavior could be explained by the formation of a transfer film on the surface, which suppresses the huge heat and contact pressure [7]. CNT/polymer coatings can induce anti-friction, wear-proof, and self-lubrication performance [8], reducing the friction and improving the wear resistance. However, numerous factors affect their tribological behavior, such as the composition and properties of sliding pairs, such as their surface roughness and main mechanical properties (hardness, stiffness, and fracture toughness) and the sliding parameters, such as load, speed, temperature, and lubrication state, among others. This behavior is explained by the different involved mechanisms: bridge crack of CNT and lock the propagation of cracks, lubricant effect by dislodgement of individual graphene layers, strengthening of reinforced polymer matrix and dissipation of heat, and reducing the temperature induced wear [8]. It is worthy to note that there is an optimum carbon nanoparticle content to achieve the best tribological properties. However, this value depends on many factors such as aspect ratio of nanofiller, the dispersion degree and orientation of nanofillers, and the interactions with polymer matrix at interfaces.

The incorporation of carbon nanoparticles into polymer composites also increases their hardness. Increasing nanofiller content leads to improvement of hardness; however, the slope of the curve is reduced as the amount of graphitic nanofiller increases, which is attributed to agglomerations in the composite coating [9].

One of the most important applications of polymers reinforced with graphitic nanofillers is as **anticorrosive coatings**. The anticorrosive coatings can be classified in accordance to the protection mechanism against corrosion [10]: barrier protection, cathodic protection, anodic passivation, electrolytic inhibition, and active corrosion inhibition.

Graphitic nanoparticles enhances the **barrier properties** of polymer coating due to the "torturous path effect" and "nano-barrier wall effect," which strongly depends on exfoliation, dispersion, and orientation degree of nanofiller, their aspect ratio, the polymer-nanofiller interface, and the crystallinity of thermoplastic polymer or cross-linking degree of thermosetting resins. The presence of nanofillers constructs tortuous paths, decreasing the diffusion coefficient. The orientation of graphene and their high surface area forms a zigzag diffusion pathway hindering the diffusion of corrosion species. In addition, their excellent electrical conductivity prevents the electrons form the cathodic site by providing an alternative path [11]. The functionalization of graphitic nanofillers with polar groups [12] enhances the ionic resistance of coating by the creation of negative charge on the graphitic nanofillers surface when exposed to alkaline and neutral environment, preventing the diffusion of chlorine and hydroxyl anions.

In the last years, the development of **superhydrophobic** surfaces is being extensively researched. Superhydrophobic coatings have a wide range of applications in textile, automotive parts, construction, agriculture, optical, and maritime industry. It is well established that three factors are needed to create a superhydrophobic surface: low surface energy, microscaled roughness, and nano- and microscaled hierarchical surfaces [13]. The hierarchical multiscale surface can be achieved in coatings with high CNT content, forming CNT agglomerates at the microscale and CNT themselves at nanoscale. Polyvinylidine fluoride (PVDF) are commonly used as hydrophobic polymer coating. Water contact angle increases from 105° for neat PVDF to 170° with very high content of CNT [14]. Similar enhancement of hydrophobicity is reached by the addition of graphene nanoplatelets [15]. The superhydrophobicity is usually requested together with other functionalities, such as self-cleaning, anti-icing, and deicing, which will be addressed further on.

The incorporation of graphene into polymers allows increasing the **thermal stability** of polymer composite [16]. Higher specific area and aspect ratio of carbon nanofillers induce higher stability [17]. This enhancement is also accompanied with an important increase of the low thermal conductivity of polymers. Very high enhancements have been reported, from 0.2 W/mK for neat epoxy coating to 20 W/mK for composite with 30% graphene [16]. Graphene is more effective nanofiller than CNT in order to enhance the thermal conductivity of polymer composites. The thermal conductivity of graphene is attributed to phonons and electrons [18], justifying their excellent thermal conductivity (2000 W/mK). For example, copper is considered a good thermal conductor, whose conductivity is 400 W/mK. Here, the contribution of phonons is limited to 1–2% of total. The addition of nonfunctionalized graphene can induce an enhancement of 50% of thermal conductivity. However, when graphene is modified with functional groups, which is able to form covalent and non-covalent bonds with the polymer matrix, the increment can reach 100%. Moreover, a higher 300% of thermal conductivity increment is reached when graphene is modified with a titanium-coupling agent [19]. The researchers [18, 19] explained it by the reduction of interfacial thermal resistance between graphene and polymer due to the surface modification. For the same reason, higher dispersion degree of graphene into polymer matrix also enhances the thermal conductivity of composite.

It is well-known that the **electrical conductivity** is also increased by the addition of graphitic nanofillers. Higher specific area and higher aspect ratio of nanofillers decrease the nanoparticles which lead to reach an important enhancement of electrical conductivity, from 10^{-8} to 10^{10} S/m for the isolating polymers to 0.01–10 S/m for nanocomposites. Here, the functionalization of nanofillers usually implies a low enhancement of electrical conductivity of composites. This is associated to the reduction of the electrical conductivity of neat nanofillers due to the partial breakage of some C-C structure during the functionalization and the surrounding isolating polymer layer formed over the functionalized nanofillers, which hinders the direct contacts between electrical nanoparticles.

Nanoreinforced coatings are being studied to improve the **flame retardancy** (FR) of flammable substrates [20]. The addition of graphene usually reduces the total heat release (THR) because they reduce the release of deleterious gas during thermal decomposition, arising from the radical trapping and layered hindering effect. CNT also enhances the FR behavior due to the strengthening of carbonized layers [21]. Also, they act as excellent physical barrier, reducing significantly the peak heat release rate (PHRR).

3. Smart nanocomposite coatings: Self-sensing

Among other functionalities, the concept of structural health monitoring (SHM) is of great interest in polymer coatings. It is based in an online inspection of the damage extent. In this context, a proper SHM technique must accomplish the following four levels, established by Rytter et al. [22]: (1) detection, (2) localization, (3) quantification of the damage, and (4) the estimation of remaining life, also known as *prognosis* (**Figure 2**).

Nowadays, there are a lot of different SHM techniques such as lamb waves, fiber optics, and acoustic emission, among others. However, they usually involve complex mathematical and statistical tools and do not often give an overall information of the health of the structure [23, 24]. Therefore, the development of SHM techniques is now gaining a great deal of attention.

3.1 Fundamentals of SHM with carbon nanoparticles

As commented before, carbon nanoparticles present unique mechanical and, especially, electrical properties in comparison to other materials [25]. Therefore, their addition into an insulating media promotes the creation of electrical networks. This fact induces an enhancement of the electrical conductivity of several orders of magnitude, becoming the polymer coating electrically conductive [26, 27].

Here, it is important, firstly, to define the concept of **percolation threshold**. It is the critical volume fraction of nanoparticles in which an efficient electrical network is formed, allowing the current flow. It depends on several factors mainly related to the geometry of the nanofiller (including their 0D, 1D, or 2D nature) as well as their dispersion state, that is, their distribution inside the nanocomposite. The determination of the percolation threshold is a crucial factor that determines the minimum content of nanofiller that is needed for electrical applications.

Furthermore, the influence of the different parameters of the nanoparticle network in the percolation threshold has been widely studied in the last years. Li et al. [28] proposed a simple analytical model correlating the geometry, aspect ratio, and

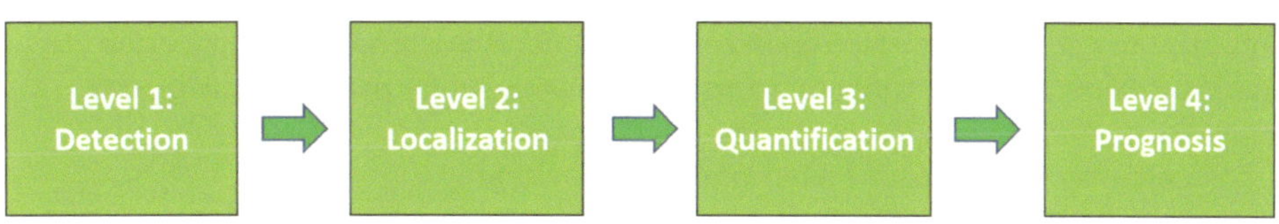

Figure 2.
Schematics of the four SHM levels established by Rytter et al.

dispersion state with the value of the percolation threshold. They concluded that the lower the aspect ratio and the higher the degree of agglomeration of nanoparticles, the higher the percolation threshold. In this context, carbon black (CB) reinforced polymers show very high values of percolation threshold [29] due to their low aspect ratio Bauhofer et al. [30] did an extensive review of percolation threshold in carbon nanotube (CNT)-based polymer nanocomposites by analyzing the effect of nanofiller geometry and dispersion technique. It was observed that the most aggressive dispersion procedures, such as ultrasonication, although leading to the most homogeneous distribution of nanoparticles, lead to a very significant breakage of the CNTs. This prevalent reduction of the aspect ratio leads to increasing values of percolation threshold.

However, the effect of dispersion procedure varies depending on the nanofiller. In this case, ultrasonication has proved to be a good dispersion technique for graphene nanoplatelet (GNP)-based nanocomposites. GNPs are formed by several layers of graphene, and it is widely used as reinforcement in polymer nanocomposites because of the lower cost. Here, ultrasonication promotes the exfoliation of the graphene layers [31]. Therefore, these exfoliating mechanisms induce a reduction of the percolation threshold due to an increase of the aspect ratio of the nanofillers. The combination of an ultrasonication stage with three roll milling can promote the creation of GNP nanocomposites with very low percolation threshold due to the combination of exfoliation stretching effects.

The concept of SHM with nanoparticles, therefore, is based in the monitoring of the changes of the electrical network when subjected to strain or damage. However, for a better understanding of these SHM techniques, it is important to know which are the main conducting mechanisms in the electrical network. Here, three different mechanisms can be identified: the intrinsic conductivity of the nanofiller, the contact between adjacent nanoparticles, and the tunneling transport that takes places between two neighboring particles that are not in intimate contact. Among them, the tunneling transport plays a dominant role in the electrical network of the nanocomposite [32]. It is explained because the associated tunneling resistance is several orders of magnitude higher than the intrinsic and contact resistance. Therefore, the variations of the electrical network when subjected to strain or damage will be ruled by the variation of the tunneling distance between nanoparticles.

In this regard, Simmons [33] established a linear-exponential correlation between the tunneling resistance and the interparticle distance, also known as tunneling distance. It means that the higher the separation between neighboring nanoparticles, the higher the electrical resistance is. More specifically, when subjected to strain, there is a variation of the electrical resistance that is correlated to an increase of the tunneling distance between adjacent nanoparticles:

$$R_{tunnel} = \frac{h^2 t}{Ae^2\sqrt{2m\varphi}} exp\left(\frac{4\pi t}{h}\sqrt{2m\Phi}\right) \tag{1}$$

where R_{tunnel} is the tunneling resistance; t is the tunneling distance; h is the Planck's constant; m and e are the electron mass and charge; A is the area in which electrical transport takes places or tunneling area; and φ is the height barrier of the matrix.

3.2 Sensitivity of polymer-based nanocomposites

In this context, the concept of percolation threshold that has been discussed before plays a key role. In fact, the linear-exponential dependence means that the higher the tunneling distance at the initial situation, that is, when no strain is applied, the higher will be the electrical resistance variation associated to the variation in the tunneling distance when applying strain, as can be observed in **Figure 3**.

Here, it is necessary to define the concept of gauge factor (GF), as the variation of the normalized resistance $\Delta R/R_0$ divided by the applied strain, ε:

$$GF = \frac{\Delta R/R_0}{\varepsilon} \quad (2)$$

Therefore, in order to achieve the highest GFs, it would be necessary to work with volume fractions of nanofiller near the percolation threshold, as the distance between adjacent nanoparticles will be the highest possible to form an efficient electrical network and, thus, the variation of the tunneling distance will be the highest. This has been observed in both GNP and CNT nanocomposites, where the contents near the percolation threshold achieved the highest GFs [34, 35].

However, in this sense, there are significant differences among the different nanoparticles. For example, GNP-based nanocomposites have shown a more accused exponential behavior of the electrical resistance with applied strain [35] than CNT-based ones [36]. It means that the values of GF at low strain levels are much lower than at high strain levels.

This accused exponential behavior of GNP nanocomposites can be explained accordingly to the different interactions that take place inside the electrical network. In fact, the tunneling area of these 2D particles is, generally, much higher than in the case of CNTs, and it leads to the fact that the value of the interparticle distance can be much higher for an efficient tunneling transport. Therefore, as explained before, the higher the tunneling distance, the more accused the exponential correlation between the electrical resistance and the applied strain will be.

Moreover, there is also a correlation between the exponential behavior and the sensitivity of the nanocomposite, and, thus, GNP-based ones show higher GF values than CNT nanocomposites (from 12–15 to 2–4 at low strain levels, respectively, for nanocomposites manufactured following similar techniques) [35, 36].

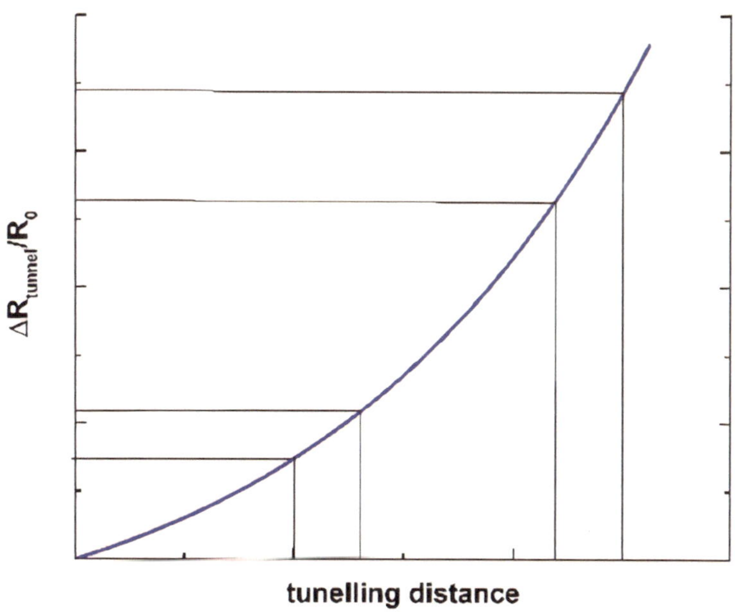

Figure 3.
Variation of the tunneling resistance as a function of the tunneling distance between nanofillers.

3.3 SHM in nanocomposite coatings

The enormous potential of the nanoparticles for SHM applications has been widely exploited in the development of surface sensors and smart coatings. More specifically, their use as substitutes of strain gages is gaining attention nowadays.

Basically, a strain gauge is a device for indicating the strain level of a structure at the point of attachment. To date, the most used are based in conventional metallic foils in which the strain is obtained from the electrical resistance variation due to the deformation of the metallic foil when subjected to this strain level. GF values of conventional strain gauges are around 2 and usually show a very linear dependence of electrical resistance change with applied strain.

The research in strain-sensing devices with carbon-based nanocomposites is mainly focused in the development of highly stretchable sensors. In this regard, graphene, carbon nanotubes, and other carbon nanoparticles, such as carbon black, have been widely explored, among others. The addition of these nanoparticles to polymers with high strain capabilities such as fluoroelastomers [37], thermoplastic polyurethane [38], or vulcanized silicone [39] has demonstrated excellent sensing capabilities. More specifically, they present enormous potential for human motion sensing or wearable electronics [37] as the GF at high strain levels (>20%) can be in the range of 400–4000 depending on the content and morphology of the carbon nanofiller. Here, a highly accused exponential behavior is observed at higher strain levels due to the prevalence of tunneling mechanisms in the carbon nanoparticle network. In addition, they can be also used as pressure sensors, with excellent sensing capabilities in comparison to others [40] as they are able to detect very small pressure changes due to the strain induced fail that they promote.

Here, it can be stated that the 2D nature of the electrical network in a nanocomposite coating promotes an increase of the percolation threshold when compared to a bulk nanocomposites, where a 3D uniform disposition of the nanofillers is supposed to be [41]. Moreover, the cross-sectional area of the coatings is obviously much lower than 3D nanocomposites, so the electrical resistance is

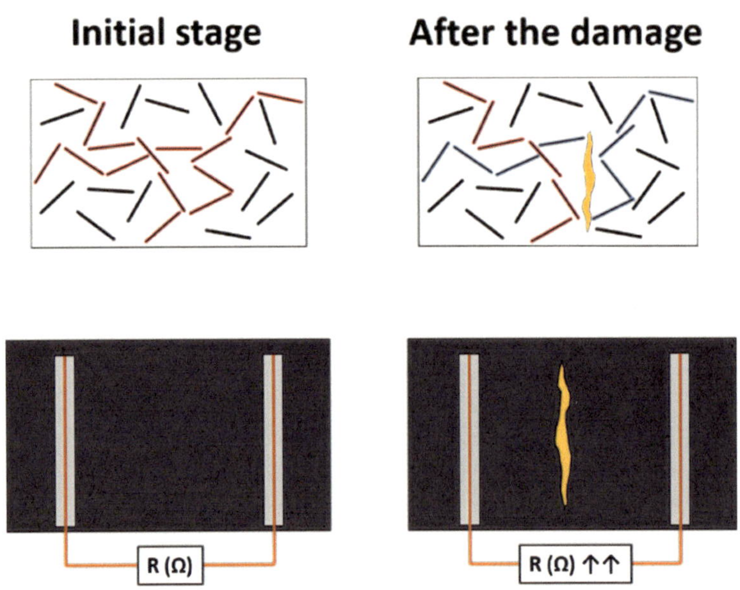

Figure 4.
Schematics of the effect of a crack in the electrical network.

much higher. Therefore, the amount of nanofiller necessary for SHM purposes is much higher in polymer coatings.

Furthermore, apart from flexible devices such as strain gauges, wearable electronics, or human motion sensors, their SHM capabilities in other polymer coating based on thermosetting polymers have been widely demonstrated. More specifically, GNP-based coatings have proved to be very sensitive at low strain levels to both tensile and compressive loading, as well as a good repeatability under cyclic loads [42]. In addition, CNT-based ones have also demonstrated good sensing properties with a high linearity [43].

Although the interest as strain sensing devices is very significant, their crack sensing capabilities can be even of more interest. Here, the crack detection is based in a sudden breakage of the electrical pathways due to the presence of the crack itself. It will be reflected in a sharp increase of the electrical resistivity of the coating, and, thus, the electrical resistance during the measurement will increase as well, as shown in the schematics of **Figure 4**.

In this regard, electrical impedance tomography is gaining a great deal of attention as an SHM technique. It is based in a mapping of the electrical conductivity of a structure based in the electrical resistance measurements on its surface. Therefore, by using this technique, it will be possible to detect, locate, and even quantify

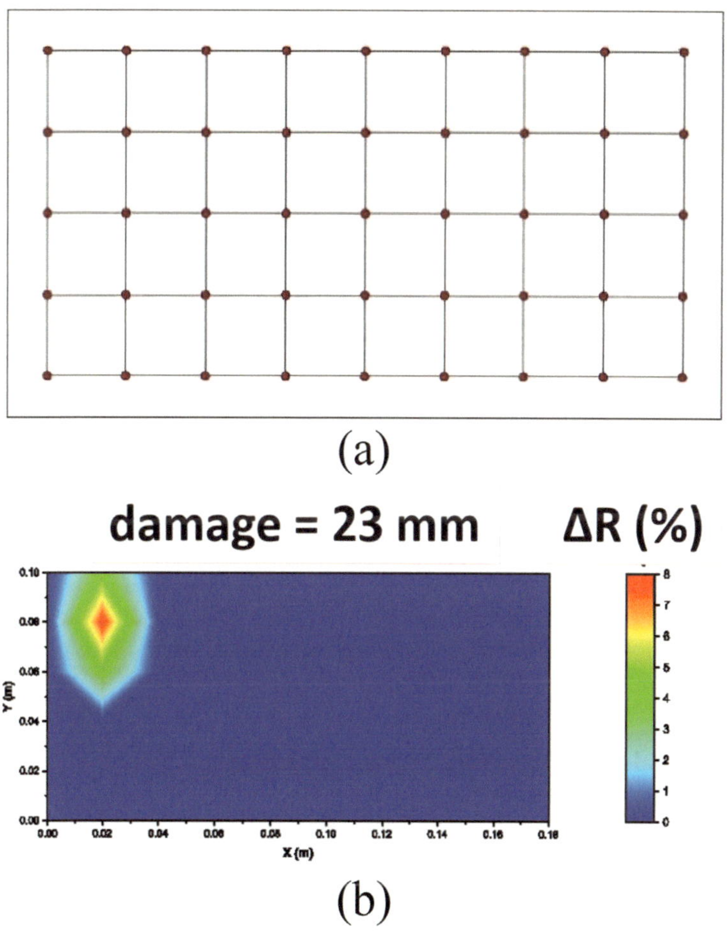

Figure 5.
(a) Schematics of electrode disposition and (b) mapping of variation of the electrical resistance correlated to an induced damage [43].

superficial defects by analyzing the changes of its surface resistivity. In this context, their effectiveness has been widely demonstrated in polymer coatings for the detection of superficial cracks [43], where the electrical resistance measurement between adjacent channels can easily detect an artificial damage (**Figure 5**), as well as in sensing skins for spatial pressure mapping, where the strain induced by the applied pressure is monitored [44]. Here, the main issue is correlated to the positioning of the electrodes and the data processing, which usually involves the use of complex mathematical tools. However, the results for SHM applications are very promising and give a new functionality to nanoreinforced polymer coatings.

4. Smart nanocomposite coatings: Self-heating

Surface heating is a challenge for several applications, and it is usually carried out by different approaches such as external heating source (portable equipment) or heating elements positioned on the surface (i.e., electrical resistances) which, in fact, modifies the surface quality of the parts where they are positioned. Heat is required in organic coatings for different purposes:

- Deicing systems: ice accretion to surfaces when subjected to cold and humid environments is something very common that requires the use of deicing alternatives, and, among them, heat of the surface to create a liquid film can be of great interest.

- Self-curing coatings: the use of coatings with curing temperatures above 23°C is often limited because of the need of external heating sources that makes the production more difficult. Nevertheless, the glass transition temperatures or hardness are usually higher for higher cross-linked coating, thus making more interesting the use of higher curing/post-curing temperatures.

- Heat activated self-healing mechanisms: as previously mentioned, the main source for self-healing activation is UV radiation and temperature. Nevertheless, the requirement of a heating source limits the application of these promising coatings to structures with easy access to be heated.

These applications will be further explored after an explanation of basic aspects of self-heating coatings based on the introduction of conductive nanoparticles inside polymer matrices, such as CNTs and GNPs. Nevertheless, these types of coatings are not limited to these applications, and they can find a potential field to be implemented in any product that need to be heated such as heating seats for commercial vehicles [45], floor heating, heating textiles, etc. where temperatures required are usually below 100°C [46].

4.1 Fundamentals of self-heating by Joule effect

The addition of carbon nanoparticles inside a polymer matrix above the percolation threshold, which has been previously explained, allows getting an electrically conductive material. The electrical current that flows through the material will generate heat according to Joule's law (Eq. (1)), which is commonly known as Joule effect in materials:

$$Q = i^2 R t \tag{3}$$

where Q is the heat generated, i is the current flow, R is the electrical resistance, and t is the time the current is applied.

The first thing that can be analyzed from Eq. (1) is that higher current intensity would lead to higher heat generated and, consequently, higher contents of carbon nanoparticles will be desired for this purpose in order to increase the temperature reached or to reduce the voltage required. Although all common carbon nanostructures can be used for this purpose (carbon black, carbon nanotubes, graphene nanoplatelets, or even graphite flakes), the importance of reaching high intensity values usually gives the best results for CNT-filled materials [47]. In fact, very high CNT amounts can be found in the literature in order to increase the electrical conductivity and, consequently, the current flowing at lower voltages applied. This is the case of the study based on ABS as matrix where CNT was added up to 15 wt.% in order to allow reaching temperatures over 200°C when voltages of only 12 V were applied [48] or the research carried out by Chu et al. where similar results in terms of temperature and voltage at contents of 7.5 wt.% of CNT in PDMS were found [49]. The interest in the use of low voltages is based on the use of batteries commonly installed in cars and trucks, among others.

Apart from the heat generated, there is an important fact regarding these percolated electrically conductive networks, which is the homogeneous distribution of heat through the coating. Two important effects must be taken into account for this aspect: (1) thermal conductivity of polymers which is particularly low, thus making heat transfer through the coating more difficult, and (2) homogeneous presence of the carbon nanoparticles through the polymer matrix, which is not always easily reached.

Both CNTs and GNPs show extremely good thermal conductivity individually. Nevertheless, in spite of their similar intrinsic thermal conductivity, the morphology of GNPs makes them more interesting for this purpose, even when compared to SWCNT [50]. Even at the same content of both types of nanoparticles, Zakaria et al. found that the thermal conductivity was higher for GNP nanocomposites than the MWCNT ones. In fact, although for electrical properties, higher contents of GNPs are usually required to meet similar properties to the ones found for MWCNT, in that case, at only 3 wt.% of nanoreinforcement, GNP nanocomposites showed an increase of 126.4% in thermal conductivity while 3 wt.% of MWCNT only increased this property by 60.2% [51]. In fact, experimental values of thermal conductivity are usually lower than those predicted theoretically, and it has been attributed mainly to waviness, dispersion, alignment, interfacial resistance, and contact resistance [52].

Proper exfoliation of GNPs causes an important increment on thermal conductivity related to an increase of the aspect ratio. Chu et al. [53] proposed a model to calculate the thermal conductivity of nanocomposites based on randomly oriented nanoparticles which takes into account geometrical aspects of the nanoparticles (aspect ratio) as well as differences in the intrinsic thermal conductivity of the nanofiller in each direction. In the case of GNPs, these aspects will be strongly related to their exfoliation and dispersion in the polymer matrix. On the other hand, the waviness of the nanoreinforcement may reduce the effective aspect ratio of the nanofillers which lead to propose few layers GNPs as an optimal solution instead of individual monolayers that tend to roll up easier during dispersion stage.

These self-heating coatings do not require extremely high thermal conductivities, but they should be high enough to ensure good heat transfer through the whole surface for the purposes mentioned above.

The formation of aggregates is very common in this type of materials, and this may cause that at very low carbon nanoparticle loadings, some resin areas are free of nanoreinforcement, which leads to nonuniform heating of the samples. In that

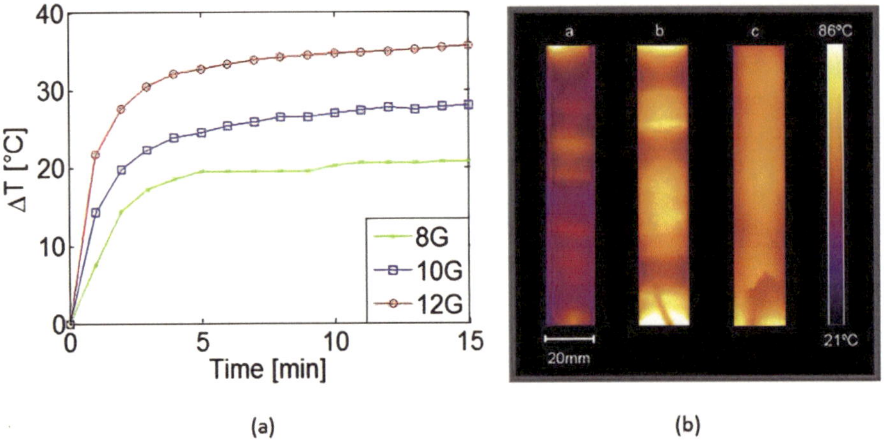

Figure 6.
Joule effect heating of epoxy nanocomposites containing (a) 8 wt.%, (b) 10 wt.%, and (c) 12 wt.% of GNPs [54].

cases, the thermal conductivity of the sample is even more important, as heat will not be homogeneously generated, thus making more important its thermal conduction. Prolongo et al. showed this effect when comparing MWCNT loaded with GNP ones, and they found that differences between maximum and minimum temperatures were much higher in those specimens based on MWCNT [47]. When adding GNPs, also this effect was found as lower contents of GNPs lead to areas with lower nanoreinforcement concentration, thus leading to higher temperature differences between different areas in the same sample, while the samples containing higher GNP contents (12 wt.%) showed more uniform heating (**Figure 6**).

4.2 Self-heating as deicing system

Icing on structure surface can seriously affect the function of the system, and, even, it may cause its damage and consequently the need for replacement which leads to economic, environmental, and security issues. Wind turbines or aircraft surfaces are examples in which ice accretion has a detrimental effect on operation conditions by modifying the aerodynamic profile, structural weight, etc. [55, 56]. Most strategies currently used are based on two different approaches that affect the coatings used:

- Passive methods which do not require external energy source. The main advantage of these methods is their lack of energy consumption to operate; nevertheless, their effectiveness is usually lower, so they are commonly used in combination with active methods to reduce the power needed [57]. Hydrophobic coatings are one of the passive methods most widely accepted which can be achieved by different approaches such as the addition of nanoparticles or tuning of surface textures in multiple length scales based, among others, in biomimetic techniques [58, 59]. This aspect has been discussed in Section 2.2.

- Active methods which require external energy source. These methods are usually more effective to avoid icing problem and can be used in combination with passive methods. Among these active methods, heating systems are the most reliable ones in spite of the power consumption although mechanical ice breakage by means of inflatable rubber boots can be found in small airplanes. When heating the surface, several approaches can be used (infrared heating and warm air conduction, among others), but electrical resistance heating seems to be one of the most promising ones. To avoid the use of additional membranes or layers, multifunctional coatings with heating capability by thermoresistive methods at the same time they protect the underlying structural material are very interesting.

The use of carbon nanostructures has been studied for this purpose trying to create an active method based on Joule effect heating at the same time that hydrophobicity is increased by the addition of these nanoreinforcements. With this purpose, several research efforts have been already done reaching very promising results in terms of temperature, homogeneity, heat rate, and power consumption. By the use of GNPs, Redondo et al. achieved 35°C of temperature increments at 800 V with a heat rate of 13.6°C/min and a power consumption of less than 3 W. This temperature increment should be enough to produce ice melting on the coating and, consequently, separation of the ice accreted to the surface, even at severe ambient temperatures below −20°C [54]. When using CNTs, the electrical conductivity of the materials is usually higher than GNPs, thus increasing the value of the intensity at a constant voltage which is useful for heat generation according to Eq. (1). CNT/PVA films with very high CNT concentration have been produced in order to get very low electrical resistance and, consequently, higher electroresistive heating. The same procedure, using high MWCNT loadings (10 wt.% in poly-1,3,4-oxadiazole), allowed to reach temperatures above 100°C by the application of only 40 V in these nanoreinforced films. Prolongo et al. compared the temperature reached by the use of CNTs and GNPs as nanoreinforcements, and with contents of 0.5 wt.% of CNTs, the temperature reached almost 100°C with only 90 V applied, while 300 V were required to go over 65°C when 8% GNP were added [47]. Nevertheless, the authors claimed that temperature was more homogeneous when GNPs were used as nanoreinforcement.

Finally, the authors are currently exploring the possibility of using the electrical network of CNTs to sense the temperature and, consequently, activate the voltage application when required because of the weather conditions measured by the coating itself. Coatings based on CNT/PDMS were manufactured and showed an effect-denominated negative temperature coefficient (NTC). This effect on electrical response was nonlinear with temperature changes, being the sensitivity more than six times higher in the range −5 to 5°C than at room temperature, which makes them potential candidates for temperature measurement for smart coatings being able to detect temperature and activate the voltage required accordingly [60].

4.3 Self-curing coatings

Based on the same basics of previous applications, self-heating nanoreinforced polymers have been developed in order to allow curing by electrical voltage application. This issue has been already explored to take advantage of the uniform heat as far as the heat is generated from the material itself and, also, of the absence of power loses associated to heat transmission from the heating element to the material itself, as it happens when ovens are used [61]. Mas et al. proved the efficiency of this curing method by the addition of MWCNT to an epoxy matrix, and they found uniform thermal properties in the resin cured by Joule heating. In fact, they were able to control the real curing temperature by the coupling of thermocouples to the voltage source with a PID controller. The input of the thermocouples was used by the PID to adjust the power supply in order to keep the curing temperature constant during the process. One of the main advantages they found was the high heating rate as the heat emerges from the material itself.

This fact was also proved by Jang et al. in the research conducted to prove that curing by Joule heating of thin films of PDMS reinforced with high loadings of

CNT (7 wt.%) could lead to even better performance than the equivalent oven-cured samples. They measured the mechanical properties of both materials, and they found that stiffness of samples cured by Joule effect heating was slightly higher, which was associated to faster and more uniform heating of the whole material volume. So they conclude that this curing technique could allow obtaining materials at shorter curing times, more homogenous, and with higher cross-linked structures [62].

Other carbon nanostructures different from CNTs have been also studied to cure thermoset materials out of the oven, such as the previously study from Mas et al. established [61]. Xia et al. proposed the use of GNP to cure epoxy matrices by adding contents over 8.5 wt.% as they found this value to be the percolation threshold for the GNP morphology and dispersion technique used. They found curing degrees similar to those found for oven-cured samples but with much faster heating rates and more homogenous curing, similar to the studies previously mentioned. Nevertheless, they found preferential orientation of the GNPs which was associated to the presence of an electric field during curing stage as it was not observed in oven-cured samples. This fact is very important as they found improved electrical and mechanical performance in this direction, but this anisotropic behavior must be taken into account when designing elements with these materials [63].

So, the application of curing by Joule effect heating has been proved as possible in electrically conductive networks based on carbon nanoparticles inside polymer matrices with particular interest due to faster curing cycles, more uniform curing degree, and easy application in large structures/surfaces.

4.4 Self-healing coatings

Any thermally activated mechanism could be beneficiated from the previously explained self-heating by thermoresistive heating. Thermo-reversible Diels-Alder reactions are one of these examples, in which the Diels-Alders and retro-Diels-Alder reactions are favored at different temperatures allowing the restoration of the covalent bonds, thus repairing the cracks generated. This method was used by Willocq et al. to produce MWCNT nanoreinforced polymer matrices with self-healing capabilities by Joule effect heating at low voltages of 25 V which were enough to reach the retro-Diels-Alder reaction temperature in the vicinity of the macroscopic damage (crack) due to local higher heating around the crack. This local temperature increment around the crack was strongly dependent on the position of the electrodes with respect to the direction of the crack [64].

Huang et al. used GNPs dispersed in a thermoplastic polyurethane (TPU) matrix in order to create a percolated network and reach 98% self-healing efficiency by the application of 220 V to the material. The use of GNPs in this case allowed improving alternative self-healing approaches such as heating by IR radiation absorption which is also enhanced by the presence of GNPs in the matrix, thus allowing different alternatives to improve the thermal self-healing process of the TPU [65].

Also, this Joule effect-based heating can be used to activate self-healing in thermosetting/thermoplastic blends which self-healing capabilities. This mechanism is activated by a temperature increase above the melting temperature of the thermoplastic material used and high enough to allow its proper flow through the cracks created. Several materials have been studied for this purpose, and, among them, polycaprolactone (PCL) has been widely studied due to its low melting temperature which can allow self-healing mechanism to take place at temperatures around 100°C to allow its proper flow [66]. Zhang et al. proved the efficiency of the addition of CNT to EVA/PCL composites with shape memory purposes, which allowed heating the blends up to 100°C which would be enough to activate self-healing mechanisms in blends containing PCL as healing phase [67]. All these recent studies reveal that the use of self-heating can help to develop self-healing polymer-based materials with the main advantages to avoid the use of external heat sources, use electrical voltage that could be remotely activated, allow heat to emerge from the material itself avoiding heating other materials parts, and reduce losses due to heat transfer.

5. Smart nanocomposite coatings: Self-cleaning

Self-cleaning is a surface property consisting in keeping the surface clean under severe environmental conditions [68]. It is inspired from lotus effect with a specific micro- and nano-hierarchical surface morphology and low surface energy, providing superhydrophobicity. A surface can be considered as superhydrophobic when the water contact angle is higher than 150° (WCA > 150°) and the low sliding angle is lower than 10° (SA < 10°). Another opposite approach for getting self-cleaning surfaces is the incorporation of photocatalytic fillers affording hydrophilic surfaces (WCA < 5–10°) able to keep free of organic contaminants and moisture. The self-cleaning surfaces can be developed with nanodoped polymer coatings and with neat graphitic coating, such as it is shown in **Figure 7**.

One of the most studied self-cleaning materials is based on titanium oxide (TiO_2) and zinc oxide (ZnO) due to their superhydrophobicity and photocatalytic decomposition ability of organic pollutants. These coatings are usually manufactured by sol-gel process. Modifying these nanofillers with graphene or its derivatives can enhance their visible light response. Hybrid TiO_2/graphene nanofillers can exhibit strong electronic overlap and high interfacial binding energy; thus, photoexcited carriers can transfer from TiO_2 to graphene, and its band gap is reduced, improving the visible light photoresponse [69, 70]. Nevertheless, graphene enhances the photocatalytic efficiency of ZnO due to that graphene accepts the electron from ZnO nanoparticles, preventing the recombination of photo-generated electron hole in the semiconductor. These nanocoatings show superhydrophobicity when they are irradiated with visible light. An interesting alternative approach is the impregnation of cotton fibers to manufacture industrial self-cleaning textiles [71]. The treated fabrics exhibit an increase anti-bacteriological behavior and high biocompatibility.

Another possibility to develop superhydrophobic self-cleaning coatings is the use of hydrophobic polymer matrix, such as polysiloxanes or fluoro-polymers,

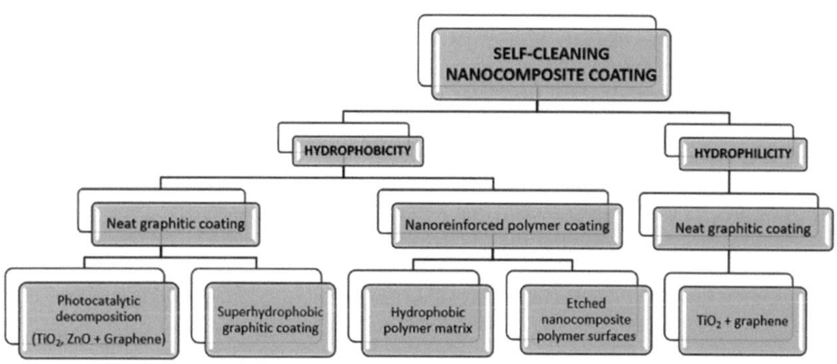

Figure 7.
Scheme of the main approaches of self-cleaning coatings with carbon nanoparticles.

doped with graphitic nanofillers, giving high electrical conductivity [72]. These coatings present high water contact angle and low sliding angle. In addition, they exhibit high wear resistance and excellent anticorrosion performance [73]. The presence of polymer matrix enhances the adherence with substrate and increases the chemical resistance. When the matrix is not hydrophobic polymer, the nanocomposite coating can be chemically etched (i.e., fatty acids such as stearic acid, palmitic acid, and oleic acid) in order to decrease the surface energy. This approach allows using very different matrices (i.e., acrylic resin, epoxy, etc.), looking for other advantages, such as superior adhesion, mechanical properties, low cost, etc., making it suitable for many industrial fields. Together with self-cleaning behavior, other related properties are commonly developed in smart coating based on carbon nanoparticles, such as antifouling, antimicrobial, anticorrosion, anti-soiling, etc. Graphetic nanostructures hold antimicrobial and antibiofilm activities, although the involved mechanisms are not completely understood and accepted [74]. For these applications, other nanoparticles of gold, silver, copper, TiO_2, ZnO, Fe_2O_3, and CuS, are anchored to the graphene nanosheets to enhance the bactericidal efficiency. Now, current efforts consist on developing membranes and films with polymer matrix, such as chitosan, poly(vinyl alcohol), polyamides, polyethyleneimine, etc., for applications in disinfection, gas separation, and water desalination.

6. Conclusions

The development of polymer coatings with carbon nanoparticles is now gaining a lot of interest. They present enhanced electrical, thermal, and mechanical properties that can be used to confer new functionalities, turning into smart materials, able to interact with the environment, responding appropriately to external stimuli. These new functionalities are possible thanks to the creation of an effective electrical, thermal, or mechanical network inside the polymer matrix, making it possible for their use for structural health monitoring applications, resistive heating by Joule's effect among others such us self-cleaning and self-healing.

Acknowledgements

This work was supported by the Young Researchers R&D Project (SMART-MULTICOAT M2183) financed by the Comunidad de Madrid government and Rey Juan Carlos University.

Author details

Xoan Xosé Fernández Sánchez-Romate*, Alberto Jiménez Suárez and Silvia González Prolongo
Materials Science and Engineering Area, University Rey Juan Carlos, Madrid, Spain

*Address all correspondence to: xoan.fernandez.sanchezromate@urjc.es

References

[1] Pham VP, Jang H, Whang D, Choi J. Direct growth of graphene on rigid and flexible substrates: Progress, applications, and challenges. Chemical Society Reviews. 2017;**46**(20): 6276-6300

[2] Pham VP, Nguyen MT, Park JW, Kwak SS, Nguyen DHT, Mun MK, et al. Chlorine-trapped CVD bilayer graphene for resistive pressure sensor with high detection limit and high sensitivity. 2D Materials. 2017;**4**(2):025049

[3] Pham VP, Kim KN, Jeon MH, Kim KS, Yeom GY. Cyclic chlorine trap-doping for transparent, conductive, thermally stable and damage-free graphene. Nanoscale. 2014;**6**(24): 15301-15308

[4] Pham PV. A library of doped-graphene images via transmission electron microscopy. C—Journal of Carbon Research. 2018;**4**(2):34

[5] Pham PV. Hexagon flower quantum dot-like Cu pattern formation during low-pressure chemical vapor deposited graphene growth on a liquid Cu/W substrate. ACS Omega. 2018;**3**(7): 8036-8041

[6] Pham VP, Jo YW, Oh JS, Kim SM, Park JW, Kim SH, et al. Effect of plasma-nitric acid treatment on the electrical conductivity of flexible transparent conductive films. Japanese Journal of Applied Physics. 2013;**52**(7R): 075102

[7] Zhang Y, Zhang D, Wei X, Zhong S, Wang J. Enhanced tribological properties of polymer composite coating containing graphene at room and elevated temperatures. Coatings. 2018; **8**(3):91

[8] Zhang W, Ma G, Wu C. Anti-friction, wear-proof and self-lubrication application of carbon nanotubes. Reviews on Advanced Materials Science. 2014;**36**:75-88

[9] Ayatollahi MR, Doagou-Rad S, Shadlou S. Nano-/microscale investigation of tribological and mechanical properties of epoxy/MWNT nanocomposites. Macromolecular Materials and Engineering. 2012;**297**(7):689-701

[10] Othman NH, Ismail MC, Mustapha M, Sallih N, Kee KE, Jaal RA. Graphene-based polymer nanocomposites as barrier coatings for corrosion protection. Progress in Organic Coatings. 2019;**135**:82-99

[11] Liu D, Zhao W, Liu S, Cen Q, Xue Q. Comparative tribological and corrosion resistance properties of epoxy composite coatings reinforced with functionalized fullerene C60 and graphene. Surface and Coatings Technology. 2016;**286**:354-364

[12] Ramezanzadeh B, Niroumandrad S, Ahmadi A, Mahdavian M, Moghadam MM. Enhancement of barrier and corrosion protection performance of an epoxy coating through wet transfer of amino functionalized graphene oxide. Corrosion Science. 2016;**103**:283-304

[13] Roach P, Shirtcliffe NJ, Newton MI. Progress in superhydrophobic surface development. Soft Matter. 2008;**4**(2): 224-240

[14] Prasad G, Anand Prabu A. Effect of stirring on hydrophobicity of PVDF/ CNT nanocomposite coatings. Advanced Materials Research. 2014;**938**:199-203

[15] Jin J, Wang X, Song M. Graphene-based nanostructured hybrid materials for conductive and superhydrophobic functional coatings. Journal of Nanoscience and Nanotechnology. 2011; **11**(9):7715-7722

[16] Bustero I, Gaztelumendi I, Obieta I, Mendizabal MA, Zurutuza A, Ortega A, et al. Free-standing graphene films embedded in epoxy resin with enhanced thermal properties. Advanced Composites and Hybrid Materials. 2020;**3**:31-40

[17] Norkhairunnisa M, Azizan A, Mariatti M, Ismail H, Sim L. Thermal stability and electrical behavior of polydimethylsiloxane nanocomposites with carbon nanotubes and carbon black fillers. Journal of Composite Materials. 2012;**46**(8):903-910

[18] Li A, Zhang C, Zhang Y. Thermal conductivity of graphene-polymer composites: Mechanisms, properties, and applications. Polymers. 2017;**9**(9):437

[19] Tian L, Wang Y, Li Z, Mei H, Shang Y. The thermal conductivity-dependent drag reduction mechanism of water droplets controlled by graphene/silicone rubber composites. Experimental Thermal and Fluid Science. 2017;**85**:363-369

[20] Guo Y, Bao C, Song L, Qian X, Yuan B, Hu Y. Radiation cured epoxy acrylate composites based on graphene, graphite oxide and functionalized graphite oxide with enhanced properties. Journal of Nanoscience and Nanotechnology. 2012;**12**(3):1776-1791

[21] Qiu J, Zhang S, Wang G, Gong Y. Surface modification and application of multi-walled carbon nanotubes in fire-retardant coatings. New Carbon Materials. 2009;**4**:344-348

[22] Rytter A. Vibrational Based Inspection of Civil Engineering Structures. [PhD Thesis]. University of Aalborg; 1993

[23] Basri R, Chiu W. Numerical analysis on the interaction of guided lamb waves with a local elastic stiffness reduction in quasi-isotropic composite plate structures. Composite Structures. 2004;**66**(1):87-99

[24] Di Sante R. Fibre optic sensors for structural health monitoring of aircraft composite structures: Recent advances and applications. Sensors. 2015;**15**(8): 18666-18713

[25] De Volder MFL, Tawfick SH, Baughman RH, Hart AJ. Carbon nanotubes: Present and future commercial applications. Science. 2013; **339**(6119):535-539

[26] Hamidinejad M, Zhao B, Zandieh A, Moghimian N, Filleter T, Park CB. Enhanced electrical and electromagnetic interference shielding properties of polymer-graphene nanoplatelet composites fabricated via supercritical-fluid treatment and physical foaming. ACS Applied Materials & Interfaces. 2018;**10**(36):30752-30761

[27] Chen J, Han J, Xu D. Thermal and electrical properties of the epoxy nanocomposites reinforced with purified carbon nanotubes. Materials Letters. 2019;**246**:20-23

[28] Li J, Ma PC, Chow WS, To CK, Tang BZ, Kim J. Correlations between percolation threshold, dispersion state, and aspect ratio of carbon nanotubes. Advanced Functional Materials. 2007; **17**(16):3207-3215

[29] Zhao J, Dai K, Liu C, Zheng G, Wang B, Liu C, et al. A comparison between strain sensing behaviors of carbon black/polypropylene and carbon nanotubes/polypropylene electrically conductive composites. Composites Part A: Applied Science and Manufacturing. 2013;**48**:129-136

[30] Bauhofer W, Kovacs JZ. A review and analysis of electrical percolation in carbon nanotube polymer composites. Composites Science and Technology. 2009;**69**(10):1486-1498

[31] Moriche R, Prolongo SG, Sánchez M, Jiménez-Suárez A, Sayagués MJ, Ureña A. Morphological

changes on graphene nanoplatelets induced during dispersion into an epoxy resin by different methods. Composites Part B: Engineering. 2015;**72**:199-205

[32] Li C, Thostenson ET, Chou T. Dominant role of tunneling resistance in the electrical conductivity of carbon nanotube-based composites. Applied Physics Letters. 2007;**91**(22):223114

[33] Simmons JG. Generalized formula for the electric tunnel effect between similar electrodes separated by a thin insulating film. Journal of Applied Physics. 1963;**34**(6):1793-1803

[34] Kuronuma Y, Takeda T, Shindo Y, Narita F, Wei Z. Electrical resistance-based strain sensing in carbon nanotube/polymer composites under tension: Analytical modeling and experiments. Composites Science and Technology. 2012;**72**(14):1678-1682

[35] Moriche R, Sanchez M, Jimenez-Suarez A, Prolongo SG, Urena A. Strain monitoring mechanisms of sensors based on the addition of graphene nanoplatelets into an epoxy matrix. Composites Science and Technology. 2016;**123**:65-70

[36] Sánchez-Romate XF, Artigas J, Jiménez-Suárez A, Sánchez M, Güemes A, Ureña A. Critical parameters of carbon nanotube reinforced composites for structural health monitoring applications: Empirical results versus theoretical predictions. Composites Science and Technology. 2019;**171**:44-53

[37] Shajari S, Mahmoodi M, Rajabian M, Karan K, Sundararaj U, Sudak LJ. Highly sensitive and stretchable carbon nanotube/fluoroelastomer nanocomposite with a double-percolated network for wearable electronics. Advanced Electronic Materials. 2020;**6**(2):1901067

[38] Li Y, Wang S, Xiao Z, Yang Y, Deng B, Yin B, et al. Flexible TPU strain sensors with tunable sensitivity and stretchability by coupling AgNWs with rGO. Journal of Materials Chemistry C. 2020;**8**(12):4040-4048

[39] Yang H, Gong LH, Zheng Z, Yao XF. Highly stretchable and sensitive conductive rubber composites with tunable piezoresistivity for motion detection and flexible electrodes. Carbon. 2020;**158**:893-903

[40] Wagner S, Yim C, McEvoy N, Kataria S, Yokaribas V, Kuc A, et al. Highly sensitive electromechanical piezoresistive pressure sensors based on large-area layered $PtSe_2$ films. Nano Letters. 2018;**18**(6):3738-3745

[41] Lee BM, Loh KJ. A 2D percolation-based model for characterizing the piezoresistivity of carbon nanotube-based films. Journal of Materials Science. 2015;**50**(7):2973-2983

[42] Sánchez-Romate XXF, Moriche R, Jiménez-Suárez A, Sánchez M, Prolongo S, Ureña A. Sensitive response of GNP/epoxy coatings as strain sensors: Analysis of tensile-compressive and reversible cyclic behavior. Smart Materials and Structures. 2020;**29**: 065012

[43] Sánchez-Romate XF, Moriche R, Jiménez-Suárez A, Sánchez M, Prolongo SG, Güemes A, et al. Highly sensitive strain gauges with carbon nanotubes: From bulk nanocomposites to multifunctional coatings for damage sensing. Applied Surface Science. 2017; **424**:213-221

[44] Dai H, Thostenson ET. Large-area carbon nanotube-based flexible composites for ultra-wide range pressure sensing and spatial pressure mapping. ACS Applied Materials & Interfaces. 2019;**11**(51):48370-48380

[45] Park HK, Kim SM, Lee JS, Park J, Hong Y, Hong CH, et al. Flexible plane heater: Graphite and carbon nanotube

hybrid nanocomposite. Synthetic Metals. 2015;**203**:127-134

[46] Jeong YG, An J. UV-cured epoxy/graphene nanocomposite films: Preparation, structure and electric heating performance. Polymer International. 2014;**63**(11):1895-1901

[47] Prolongo S, Moriche R, Del Rosario G, Jiménez-Suárez A, Prolongo M, Ureña A. Joule effect self-heating of epoxy composites reinforced with graphitic nanofillers. Journal of Polymer Research. 2016;**23**(9):189

[48] Dorigato A, Moretti V, Dul S, Unterberger S, Pegoretti A. Electrically conductive nanocomposites for fused deposition modelling. Synthetic Metals. 2017;**226**:7-14

[49] Chu K, Kim D, Sohn Y, Lee S, Moon C, Park S. Electrical and thermal properties of carbon-nanotube composite for flexible electric heating-unit applications. IEEE Electron Device Letters. 2013;**34**(5):668-670

[50] Yu A, Ramesh P, Itkis ME, Bekyarova E, Haddon RC. Graphite nanoplatelet-epoxy composite thermal interface materials. The Journal of Physical Chemistry C. 2007;**111**(21):7565-7569

[51] Zakaria MR, Kudus MHA, Akil HM, Thirmizir MZM. Comparative study of graphene nanoparticle and multiwall carbon nanotube filled epoxy nanocomposites based on mechanical, thermal and dielectric properties. Composites Part B: Engineering. 2017;**119**:57-66

[52] Han Z, Fina A. Thermal conductivity of carbon nanotubes and their polymer nanocomposites: A review. Progress in Polymer Science. 2011;**36**(7):914-944

[53] Chu K, Jia C, Li W. Effective thermal conductivity of graphene-based composites. Applied Physics Letters. 2012;**101**(12):121916

[54] Redondo O, Prolongo S, Campo M, Sbarufatti C, Giglio M. Anti-icing and de-icing coatings based Joule's heating of graphene nanoplatelets. Composites Science and Technology. 2018;**164**:65-73

[55] Bragg M, Gregorek G, Lee J. Airfoil aerodynamics in icing conditions. Journal of Aircraft. 1986;**23**(1):76-81

[56] Dalili N, Edrisy A, Carriveau R. A review of surface engineering issues critical to wind turbine performance. Renewable and Sustainable Energy Reviews. 2009;**13**(2):428-438

[57] Parent O, Ilinca A. Anti-icing and de-icing techniques for wind turbines: Critical review. Cold Regions Science and Technology. 2011;**65**(1):88-96

[58] Cao L, Jones AK, Sikka VK, Wu J, Gao D. Anti-icing superhydrophobic coatings. Langmuir. 2009;**25**(21):12444-12448

[59] Peng C, Xing S, Yuan Z, Xiao J, Wang C, Zeng J. Preparation and anti-icing of superhydrophobic PVDF coating on a wind turbine blade. Applied Surface Science. 2012;**259**:764-768

[60] Jang S, Park Y. Carbon nanotube-reinforced smart composites for sensing freezing temperature and deicing by self-heating. Nanomaterials and Nanotechnology. 2018;**8**:18479804 18776473

[61] Mas B, Fernández-Blázquez JP, Duval J, Bunyan H, Vilatela JJ. Thermoset curing through joule heating of nanocarbons for composite manufacture, repair and soldering. Carbon. 2013;**63**:523-529

[62] Jang S, Kim D, Park Y. Accelerated curing and enhanced material properties of conductive polymer nanocomposites

by joule heating. Materials. 2018; **11**(9):1775

[63] Xia T, Zeng D, Li Z, Young RJ, Vallés C, Kinloch IA. Electrically conductive GNP/epoxy composites for out-of-autoclave thermoset curing through joule heating. Composites Science and Technology. 2018;**164**: 304-312

[64] Willocq B, Bose R, Khelifa F, Garcia S, Dubois P, Raquez J. Healing by the joule effect of electrically conductive poly (ester-urethane)/carbon nanotube nanocomposites. Journal of Materials Chemistry A. 2016;**4**(11):4089-4097

[65] Huang L, Yi N, Wu Y, Zhang Y, Zhang Q, Huang Y, et al. Multichannel and repeatable self-healing of mechanical enhanced graphene-thermoplastic polyurethane composites. Advanced Materials. 2013;**25**(15): 2224-2228

[66] Karger-Kocsis J. Self-healing properties of epoxy resins with poly (ε-caprolactone) healing agent. Polymer Bulletin. 2016;**73**(11):3081-3093

[67] Zhang Z, Wang W, Yang J, Zhang N, Huang T, Wang Y. Excellent electroactive shape memory performance of EVA/PCL/CNT blend composites with selectively localized CNTs. The Journal of Physical Chemistry C. 2016;**120**(40): 22793-22802

[68] Dalawai SP, Aly MAS, Latthe SS, Xing R, Sutar RS, Nagappan S, et al. Recent advances in durability of superhydrophobic self-cleaning technology: A critical review. Progress in Organic Coatings. 2020;**138**:105381

[69] Kavitha M, Rolland L, Johnson L, John H, Jayaraj M. Visible light responsive superhydrophilic TiO_2/reduced graphene oxide coating by vacuum-assisted filtration and transfer method for self-cleaning application. Materials Science in Semiconductor Processing. 2020;**113**:105011

[70] Gillespie PN, Martsinovich N. Electronic structure and charge transfer in the TiO_2 rutile (110)/graphene composite using hybrid DFT calculations. The Journal of Physical Chemistry C. 2017;**121**(8):4158-4171

[71] Stan MS, Badea MA, Pircalabioru GG, Chifiriuc MC, Diamandescu L, Dumitrescu I, et al. Designing cotton fibers impregnated with photocatalytic graphene oxide/Fe, N-doped TiO_2 particles as prospective industrial self-cleaning and biocompatible textiles. Materials Science and Engineering: C. 2019;**94**:318-332

[72] Wei D, Grande L, Chundi V, White R, Bower C, Andrew P, et al. Graphene from electrochemical exfoliation and its direct applications in enhanced energy storage devices. Chemical Communications. 2012;**48**(9): 1239-1241

[73] Lv C, Wang H, Liu Z, Wang C, Zhang W, Li M, et al. Fabrication of durable fluorine-free polyphenylene sulfide/silicone resin composite superhydrophobic coating enhanced by carbon nanotubes/graphene fillers. Progress in Organic Coatings. 2019;**134**: 1-10

[74] Szunerits S, Boukherroub R. Antibacterial activity of graphene-based materials. Journal of Materials Chemistry B. 2016;**4**(43):6892-6912

Chapter 13

Advances in Dropwise Condensation: Dancing Droplets

Rongfu Wen and Xuehu Ma

Abstract

Vapor condensation is a ubiquitous phase change phenomenon in nature, as well as widely exploited in various industrial applications such as power generation, water treatment and harvesting, heating and cooling, environmental control, and thermal management of electronics. Condensation performance is highly dependent on the interfacial transport and its enhancement promises considerable savings in energy and resources. Recent advances in micro/nano-fabrication and surface chemistry modification techniques have not only enabled exciting interfacial phenomenon and condensation enhancement but also furthered the fundamental understanding of interfacial wetting and transport. In this chapter, we present an overview of dropwise condensation heat transfer with a focus on improving droplet behaviors through surface design and modification. We briefly summarize the basics of interfacial wetting and droplet dynamics in condensation process, discuss the underlying mechanisms of droplet manipulation for condensation enhancement, and introduce some emerging works to illustrate the power of surface modification. Finally, we conclude this chapter by providing the perspectives for future surface design in the field of condensation enhancement.

Keywords: condensation, micro/nanostructures, droplet, wetting, nucleation, heat transfer enhancement

1. Introduction

Condensation heat transfer has been at the forefront of both fundamental and engineering research due to its significance in many conventional and emerging industrial applications. For example, vapor condensation has been widely exploited for thermal management of high-power systems to maintain adequate performance and system reliability, such as advanced lasers, light-emitting diodes, radars, microprocessors, electrical machines, and power inverters [1, 2]. Meanwhile, the thermal efficiency of steam cycles, responsible for a major fraction of electricity production, is highly dependent on the heat transfer performance of vapor condensation [3]. Furthermore, condensation performance strongly influences the energy and infrastructure costs of water treatment and desalination technologies, which is becoming increasingly important due to water scarcity and world population growth [4]. Recent advances in condensation processes on the micro/nanostructured surfaces have also enabled many emerging applications in water and energy systems, such as atmospheric water harvesting [5], solar steam generation [6, 7], humidity control of building environment [8], droplet-jumping-induced gas absorption [9], jumping-droplet electronics cooling [10, 11], and jumping-droplet electrostatic energy harvesting [12]. More natural phenomena and emerging applications in the field can be found in **Figure 1**.

Vapor preferentially condenses on a solid substrate rather than directly homogeneous nucleation in the vapor phase due to the smaller energy barrier [13]. Once vapor condensation occurs on a solid substrate, the wetting of liquid condensate, determined by surface topography and chemical compositions, plays a key role in heat transfer performance as it is involved in the whole cycle of nucleation, growth, and departure of a liquid phase [14]. Dropwise condensation on a hydrophobic surface, where the gravity-driven droplet roll-off frequently refresh the surface, has an order of magnitude higher heat transfer efficiency than that of filmwise condensation on a hydrophilic surface, where a continuously thickening liquid film covers on the condensing surface [15]. Dropwise condensation is an intrinsically multi-scale energy transfer process, involving the initial formation of droplets at a length scale at a few nanometers, then growth and coalescence, and final droplet departure/shedding of at the millimeter scale, as shown in **Figure 2**. Besides, each sub-process of condensation has different preferred wettability for accelerating the whole cycle of condensed droplets, for example, easier initial nucleation on a hydrophilic substrate with low energy barrier, and faster surface refreshing on a hydrophobic surface with low surface adhesion [16–19]. Thus, the requirements on the surface topography and chemical compositions are dynamically varying from the initial nucleation to final droplet departure.

Despite that superhydrophobic surfaces with micro/nanostructures can further promote surface refreshing with spectacular water repellency, for example, self-propelled droplet jumping, the presence of vapor layer within the micro/nanostructures that is beneficial for reduce solid-liquid adhesion, brings in an additional thermal resistance to hinder droplet growth [20, 21]. Even worse is that the nucleation of nanoscale droplets within micro/nanostructures can cause unwanted pinned states of condensed droplets, which can lead to flooding phenomenon and ultimately heat transfer degradation of superhydrophobic surfaces [22–26]. To

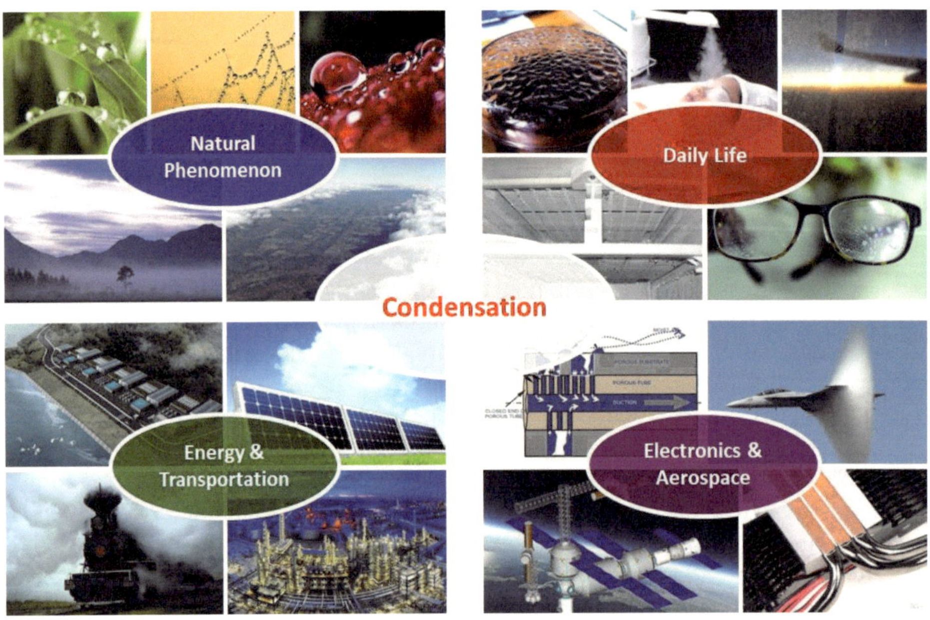

Figure 1.
Condensation phenomenon in nature and daily lives, and its applications in various conventional and emerging industrial systems. The high efficiency of vapor condensation heat transfer is critical in water and energy fields.

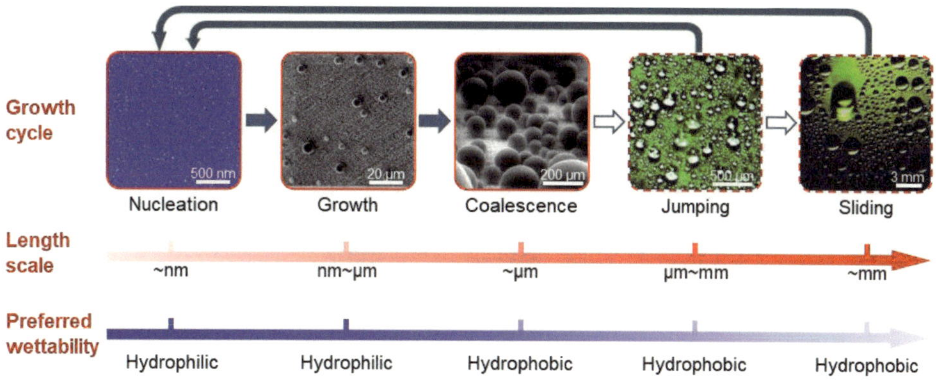

Figure 2.
Multiscale characteristics of dropwise condensation and differential preference on wettability for various subprocesses. Droplet cycle from nucleation to departure has the characteristic length scales from a few nanometers to several millimeters and the preferred wettability from hydrophilicity to hydrophobicity.

prevent the nucleation-induced flooding, some emerging techniques, for example, hybrid surfaces with wettability contrast and slippery liquid-infused porous (SLIP) surfaces, have been proposed to manipulate droplet behaviors in condensation process. For example, the different wettability on a hybrid surface can spatially control initial nucleation, addressing the random nucleation on a uniform surface [18, 19, 27]. On a SLIP surface, the lubricating fluid is immiscible with liquid condensate while preferentially wets the micro/nanostructures on the substrate, creating a lubricating fluid layer between the substrate and condensed droplets to reduce surface adhesion for high droplet mobility [5, 28–30].

Given the importance of surface structures and wettability on condensation processes, intensive efforts have been devoted to understanding the physics of dropwise condensation and to developing various micro/nanostructures and functional coatings to control droplet behaviors. In this chapter, we present an overview of the advance in dropwise condensation with a focus on improving droplet dynamics by surface modification. We briefly summarize the surface fabrication and modification, introduce droplet nucleation and size distribution in dropwise condensation, discuss the underlying mechanisms of droplet manipulation using micro/nanostructures, and introduce some typical works to illustrate the power of surface modification. We also discuss several emerging strategies to enhance condensation that could break the limit of conventional dropwise condensation. Finally, we conclude this chapter by providing the perspectives for future surface design in the field of condensation enhancement.

2. Basics of dropwise condensation

Wetting behavior of liquid condensate on the substrate is critical in condensation process. For a smooth surface without roughness, a droplet forms an intrinsic contact angle, defined as the angle between the solid-liquid and liquid-vapor interfaces within the liquid (**Figure 3**). The intrinsic contact angle θ is a force balance at the tri-phase contact line, which can be described by the Young equation,

$$\sigma_{\mathrm{lv}} \cos\theta = \sigma_{\mathrm{sv}} - \sigma_{\mathrm{sl}} \quad (1)$$

where σ_{lv}, σ_{sv}, and σ_{sl} are the liquid-vapor, solid-vapor, and solid-liquid interfacial tensions, respectively. On a real surface with roughness, there is a contact angle

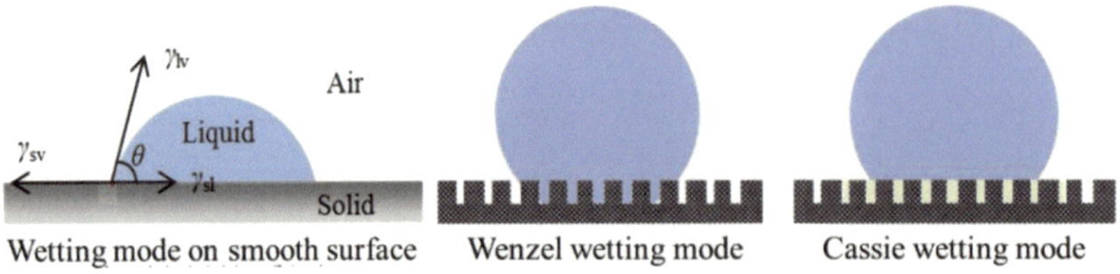

Figure 3.
Wetting states of a droplet on the smooth and structured surfaces. Intrinsic contact angle of a droplet on a smooth surface and apparent contact angle of a droplet in Wenzel and Cassie states on a structured surface.

hysteresis for an incipient droplet motion $\Delta\theta$, defined as the difference between advancing contact angle θ_{adv} and receding contact angle θ_{rec}. When a water contact angle on a surface is typically smaller or larger than 90°, the surface is defined as hydrophilic or hydrophobic, respectively. Further introducing the roughness, for example, micro/nanostructures, can increase surface hydrophobicity to superhydrophobicity, defined with a contact angle larger than 150° and a contact angle hysteresis smaller than 5° [31].

During condensation process, the wetting of condensed droplets on a rough surface can be differentiated into the highly pinned Wenzel state with large adhesion and suspended Cassie state with high mobility. Both Wenzel and Cassie states can be understood as a global energy minimization of a droplet. On such a surface with roughness r defined as the ratio of the total surface area to projected area, the apparent contact angle of a droplet in Wenzel state can be expressed as,

$$\cos\theta_W = r\cos\theta \tag{2}$$

For a droplet in Cassie state on a rough surface, the apparent contact angle is defined by,

$$\cos\theta_C = \varphi(\cos\theta + 1) - 1 \tag{3}$$

Nucleation is the first step in dropwise condensation to create a new solid-liquid interface, followed by droplet growth and shedding [16]. The critical nucleation radius r_{min} is determined by both liquid properties and surface subcooling and it can be given by the classical nucleation equation [32],

$$r_{min} = \frac{2\sigma_{lv}T}{\rho_l h_{fg}\Delta T} \tag{4}$$

where ρ_l and h_{fg} are the liquid density and latent heat of the vapor-to-liquid phase transition, respectively. The energy barrier ΔG_e for droplet formation on a substrate should be overcome to activate the nucleation process [19],

$$\Delta G_e = \frac{\pi\sigma_{lv}r_{min}^2(2 - 3\cos\theta + \cos^3\theta)}{3} \tag{5}$$

Compared to a hydrophobic surface with a larger contact angle θ, vapor nucleation occurs more easily on a hydrophilic surface with a smaller intrinsic contact angle. The intrinsic wettability of a surface also has a strong effect on the nucleation rate J via the inverse exponential dependence on ΔG_e [19],

$$J = J_0\exp\left(-\frac{\Delta G_e}{kT}\right) = J_0\exp\left(-\frac{\pi\sigma_{lv}r_{min}^2(2 - 3\cos\theta + \cos^3\theta)}{kT}\right) \tag{6}$$

where J_0 is a kinetic constant. Once a droplet nucleates on the surface, it grows by direct vapor-to-liquid condensation on the droplet surface. During initial growth without coalescence, droplets grow with an expected radius as a function of time as,

$$R = t^{\alpha} \tag{7}$$

where α is the power-law exponent, which ranges from 0 to 1 depending on the surface property, surface subcooling, and vapor conditions. If non-condensable gas (NCG) is present in the vapor, a mass transfer boundary layer will be established near the solid surface, resulting in vapor molecules diffusion through the boundary layer to the droplet surface. The mass transfer flux can be expressed as,

$$w = D_{12}\frac{(p_r - p_s)}{\delta_0 RT} \tag{8}$$

where D_{12} is the diffusion constant of vapor molecules in the gas. p_r and p_s are the vapor pressure and saturation pressure, respectively. δ_0 is the thickness of the boundary layer. Here, the boundary layer thickness can be related to the free-stream velocity U, the kinematic viscosity μ, and the Schmidt number, $Sc = \mu/D_{12}$, by

$$\delta_0 = \frac{(x\mu/U)^{1/2}}{CSc^{1/3}} \tag{9}$$

When a droplet grows large enough to contact with adjacent droplets, they merge and speed up droplet growth, stabilizing the surface coverage to a constant value, known as the self-similarity in dropwise condensation [14, 15, 17, 33, 34]. On a vertical or inclined hydrophobic surface, condensed droplets are generally removed by the gravity-driven shedding. Droplet departure radius r_{max} can thus be estimated by the force balance between the surface tension and gravity [35],

$$r_{max} = \left(\frac{6c(\cos\theta_{rec} - \cos\theta_{adv})\sin\theta}{\pi(2 - 3\cos\theta + \cos^3\theta)}\frac{\sigma_{lv}}{\rho_l g}\right)^{1/2} \tag{10}$$

Once the droplets begin to departure from the surface driven by the gravity, other droplets in the path of droplet departure can be effectively swept by coalescence, refreshing the condensing surface. New small droplets then re-nucleate, grow, and coalescence on the fresh surface, which is responsible for the high-efficiency heat transfer performance of dropwise condensation [36, 37]. Compared with the time scale of initial nucleation, the duration of droplet growth, coalescence, and departure usually dominates the whole cycle [1, 37]. Increasing apparent contact angle θ and decreasing contact angle hysteresis $(\theta_{adv} - \theta_{rec})$ can effectively reduce droplet departure size for accelerating surface refreshing.

Understanding the cycle of condensed droplets on a solid surface, a classical model was proposed by Le Fevre and Rose to predict dropwise condensation heat transfer where the heat transfer through individual droplet is calculated first and the average heat flux is then obtained by integrating over all the distributed droplets on the condensing surface [38].

$$q = \int_{r_{min}}^{r_{max}} q_d(r)N(r)\mathrm{d}r \tag{11}$$

where q_d is the heat flux through an individual droplet and N is the droplet size distribution on the surface. Subsequently, corresponding modifications of dropwise condensation heat transfer model to include more accurate expressions for the heat transfer through an individual droplet [35, 39–48] and droplet size distribution [14, 33, 39, 43, 46, 49–51] are developed, for example, the conduction resistance of the liquid droplet, the thermal resistance of a hydrophobic coating, mass transfer on the liquid–vapor interface, and the effect of interface curvature.

3. Surface fabrication for dropwise condensation

In most thermal systems, liquid condensate typically forms a liquid film on the heat transfer surface because of the high surface energy of common industrial components such as clean metals. To achieve dropwise condensation for high-efficient heat transfer, various hydrophobic coatings such as long-chain fatty acid, polymer materials, rare-earth oxide ceramics, and self-assembled monolayers [14, 37, 52–58], are usually applied to increase hydrophobicity for high water repellency. Monolayer coatings, typically a few nanometers in thickness, for example, long-chain fluorocarbons and fatty acids, can increase water repellency of the surface with a negligible additional thermal resistance (**Figure 4a**). However, they are generally not durable due to the low chemical stability and low bonding strength with the substrate [14, 52]. With a similar hydrophobicity, polytetrafluoroethylene (PTFE) coatings have been attempted to increase surface repellency while maintaining low thermal resistance (**Figure 4b**) [59]. Thicker polymer coatings have been shown to maintain robust water repellency during vapor condensation (**Figure 4c**). However, they typically have a large thermal resistance that can even negate the heat transfer enhancement achieved by dropwise condensation [54]. In addition, the initiated chemical vapor deposition (iCVD) and plasma-enhanced CVD techniques have been used to grow ultrathin conformal polymer coatings to achieve hydrophobicity (**Figure 4d**). However, further study is necessary to evaluate the durability of these ultrathin coatings for condensation heat transfer enhancement [55, 58]. Recently, ultrathin graphene (**Figure 4e**) with low thermal

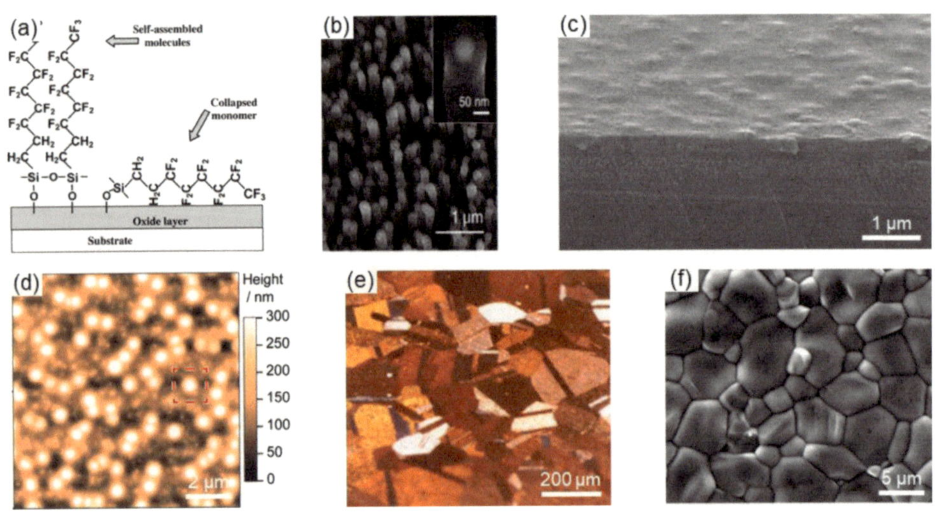

Figure 4.
Hydrophobic coatings for achieving dropwise condensation. (a) Self-assembled monolayer coating [52]. (b) Polytetrafluoroethylene (PTFE) coating on carbon nanotubes [53]. (c) Fluorocarbon film from perfluorocyclobutane precursors [54]. (d) Thin film of poly-(1H,1H,2H,2H-perfluorodecyl acrylate)-co-divinyl benzene p(PFDA-co-DVB) grafted to a substrate by iCVD [55]. (e) Graphene coatings on copper substrate [56]. (f) Rare-earth oxide ceramics [57].

resistance and good chemical stability has been used to obtain stable dropwise condensation, without obvious heat transfer degradation in a 2-week measurement [56]. Another typical material is the rare-earth oxide (REO), which can be potentially used as a hydrophobic material at scale due to the development of ceramic processing techniques (**Figure 4f**) [57]. Note that the wettability of REO is reported to be intrinsically hydrophilic and the hydrophobicity of REO is due to the adsorbed hydrocarbon species [60–62]. Despite that many new functional coatings mentioned above have shown some potential to achieve dropwise condensation, a cost-effective, low-thermal resistance, and robust hydrophobic coating to promote sustainable dropwise condensation has proven to be exceedingly challenging, resulting in ubiquitous filmwise condensation in real industrial applications.

To further improve droplet mobility, various micro/nanostructured surfaces are developed using advanced fabrication technologies. Micro/nanostructured silicon surfaces are fabricated using both wet and dry etching methods [63]. Typically, silicon nanowires synthesized by wet etching methods are vertically aligned [64]. Due to the surface tension of water during the drying process of nanowire synthesis, a large number of micro-defects are naturally formed where the closely aligned silicon nanowires cannot individually stand but form clusters [65]. Compared with the wet etching methods, finer structure geometries can be fabricated by dry etching where the etching rate can be controlled more precisely. In addition to the nanowires with uniform diameters, conical silicon nanowires were also fabricated to promote the formation of high-mobility droplets with the auxiliary Laplace pressure difference (**Figure 5a**) [25]. To meet the need of multiple length scales in manipulating droplet growth, hierarchical silicon nanowires with both microscale and nanoscale features have been fabricated by coupling micro-patterns and nanostructures [66]. **Figure 5b** shows a hierarchical surface with parallel micro-grooves that are formed by patterning silicon nanowire arrays with different lengths [67, 68]. **Figure 5c** shows another hierarchical surface, consisting of micro-pyramids covered by silicon nanowires [69]. Compared to the silicon, metal materials, for example, aluminum, stainless steel, and copper, have better physical and thermal properties, to be exploited for improving heat transfer such as high thermal conductivity, stability, and machinability. Among various surface fabrication

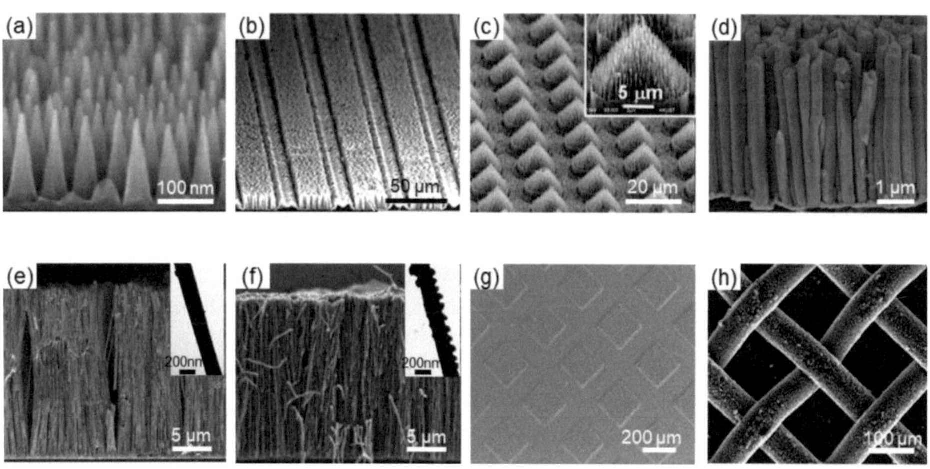

Figure 5.
Micro/nanostructured surfaces for condensation enhancement. (a) Conical silicon nanowire arrays [25]. (b) Microgroove silicon nanowires [67]. (c) Micro-pyramids covered by silicon nanowires [69]. (d) Gold nanowires [70]. (e) Closely spaced copper nanowire arrays [65]. (f) 3D copper nanowire networks [71]. (g) Hierarchical surface with micro-patterned copper nanowire arrays [21]. (h) Hierarchical copper mesh-covered structure [72].

methods for metal micro/nanostructures, the electrochemical deposition (or electroplating) method with the assistance of templates is considered to be a convenient and versatile approach [21, 65, 70, 71, 73]. **Figure 5d** shows the gold nanowires fabricated to promote droplet coalescence without direct coalescence [70]. Using a two-step template-assisted electrochemical deposition method, closely spaced copper nanowires were fabricated to promote the formation of mobile droplets in suspended wetting states (**Figure 5e**) [65]. To prevent the formation of microdefects between agglomerated nanowire arrays, the 3D copper nanowire networks consisting of interconnected nanowires with nanoscale bumps were fabricated using 3D porous anodic alumina oxide templates (**Figure 5f**). These nanobumps on the nanowire walls serve as anchors to fix adjacent nanowires [71]. **Figure 5g** shows a hierarchical surface with micro-patterned copper nanowire arrays [21]. In addition, the commercial metal materials, such as copper mesh, foam, and particles, have been applied to develop low-cost superhydrophobic surfaces and to provide a solution for large-scale deployment of enhanced heat transfer surfaces in a diverse range of technologies (**Figure 5h**) [72]. Despite that diverse micro/nanostructured surfaces have been fabricated to meet the multiple length scale need in manipulating droplet growth as mentioned above, there remains a need for advanced surface features to optimize droplet behaviors for future vapor condensation enhancement.

4. Nucleation and droplet size distribution

According to the classic nucleation theory, a substrate with low contact angles can decrease energy barrier and increase nucleation rate. The geometrical structure can also enable spatial distribution of nucleation sites on a micro/nanostructured surface. More interestingly, the structure feature plays an important role in determining the wetting mode of a nucleated droplet and subsequent droplet growth and dynamics. For example, a droplet in Cassie state has low adhesion and high mobility, which is favorable for dropwise condensation. Various theoretical and semiempirical models have been proposed to describe the relationship between surface structure and nucleation, for example, interfacial free energy analysis and cluster theory [16, 17, 19, 74]. To obtain more detailed information on the droplet nucleation in micro/nanostructures, many numerical modeling methods have been developed such as the Gibbs free energy analyses, finite element method, lattice Boltzmann method, lattice density functional method, and molecular dynamics (MD) simulations [13, 18, 75–80].

Nucleation of water droplet on the surfaces with different wettability and structural features has been widely investigated by MD simulations. The nucleation process and wetting state of nucleated droplets were obtained by manipulating the interaction potentials between surface atoms and vapor molecules [13]. The near-constant contact angles can be established for nanoscale nucleated droplets on a surface, decreasing as the solid-liquid interaction intensities linearly. On an uniform high-energy surface (**Figure 6a**), the interactions between surface atoms and water molecules are strong to capture water molecules rapidly once they get close to the surface. A large number of water molecules are absorbed on the surface and the nucleation occurs almost instantaneously. On a hybrid surface with different surface energies (**Figure 6b**), where high-energy sites distributed on a low-surface energy surface, the nucleation preferably initiates on the high-energy sites. The clusters are trapped on the high-energy sites instead of migrating around observed on an uniform surface (**Figure 6a**). The growth of nucleated droplets on high-energy particles can be divided into three stages: the formation of a wet-spot,

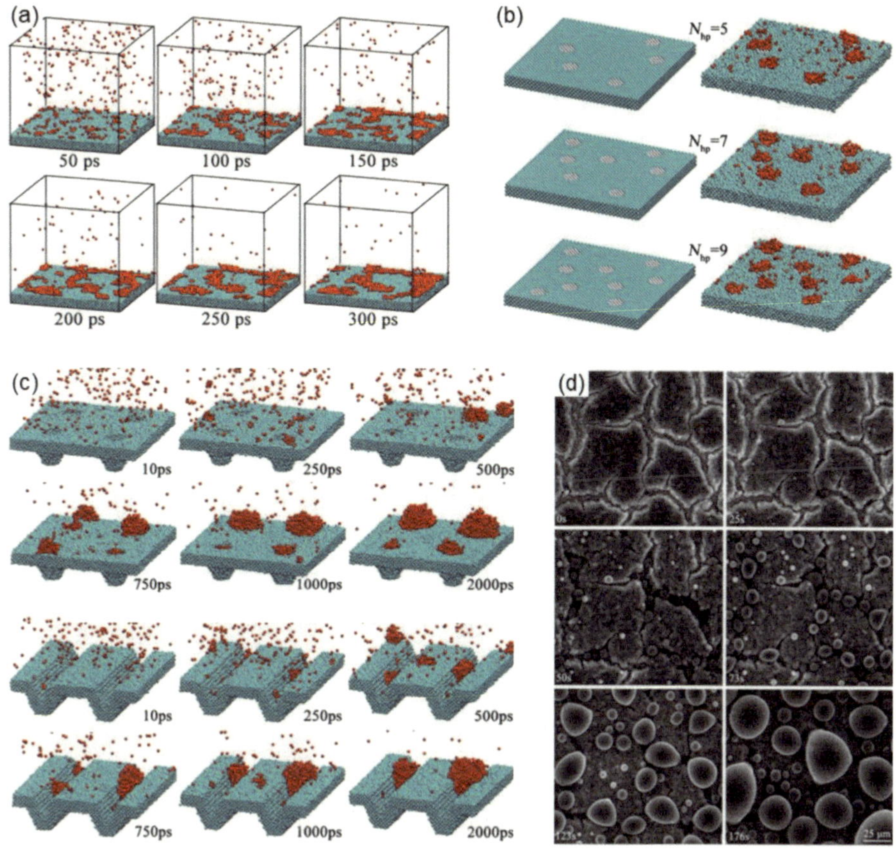

Figure 6.
Effect of surface energy and structure feature on the nucleation. (a) Random nucleation occurring on a uniform smooth surface. (b) Initial nucleation occurs selectively in the high-energy sites on the hybrid surfaces with wetting contrast [13]. (c) Effect of structure features (micropores and microgrooves) on the initial nucleation. (d) ESME images showing droplets prefer to nucleate in the micro-defects on a nanostructured surface.

increase of contact angle with near-constant contact line, and finally growth with constant contact angle. The droplet growth rate can be improved by increasing the size of high-energy sites. Moreover, the microstructures (**Figure 6c**), for example, micro-pores and microgrooves, can promote droplet nucleation due to the lower energy barrier caused by the larger solid-liquid contact area when compared with a smooth surface. The preferred nucleation in the microstructures was also observed on the nanostructured surfaces with micro-defects (**Figure 6d**).

After droplets nucleate on a solid surface, they can grow through two mechanisms: direct growth without coalescence for the small droplets by vapor condensing on the liquid-vapor interface, and coalescence-dominated growth for the large droplets by merging with neighboring droplets. As a result, condensed droplets with a wide range of sizes exist on a surface, forming a unique self-similarity. The transient characteristic of droplet size distribution has been widely studied to understand the heat transfer mechanism of dropwise condensation [14, 17, 32, 34, 51, 81, 82]. In the initial droplet growth stage after nucleation, all the droplet sizes are uniform and droplet size distribution has a peak value, showing the lognormal distribution (**Figure 7a**). With droplets growing up, the average droplet size increases. Then, the coalescences among droplets lead to a wider span of droplet size and decline of the right peak. With the increasing space among the first generation droplets, new droplets nucleate and grow at the exposed substrate, forming a bimodal distribution. With further coalescence and nucleation, the

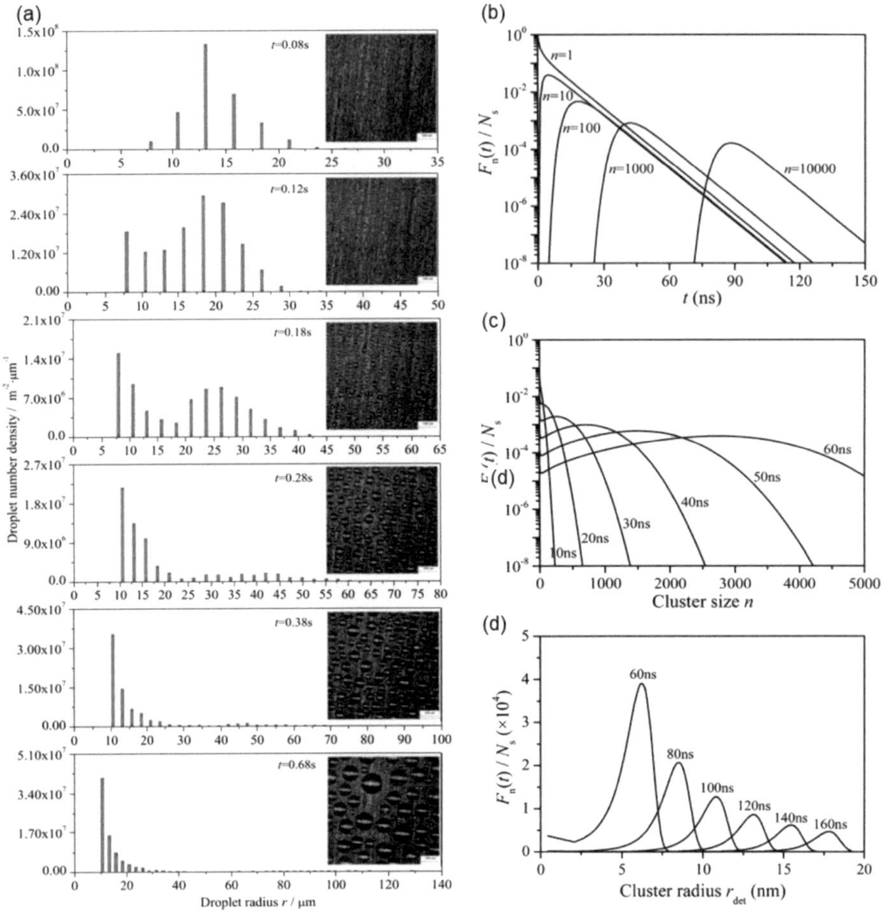

Figure 7.
Evolution of droplet size distribution on a plain hydrophobic surface. (a) Experimental results of the transient droplet size distribution in condensation [14]. (b–d) Evolution of cluster size distribution by MD simulation: (b) cluster size distribution as a function of time, (c) cluster size distribution as a function of time by MD simulation and (d) evolution of cluster radius distribution [17].

number of larger droplets decreases while the number of smaller droplets increases, and droplet size range becomes wider. Finally, the right peak vanishes and the distribution becomes unimodal. Due to the self-similarity feature of droplet size distribution, a representative large enough surface area contains various growth stages, resulting in a steady-state droplet size distribution [14, 82].

To further understand the evolution of droplet size distribution, a transient cluster size distribution model was introduced to investigate the kinetics of the initial condensation stage by MD simulation method [17]. It is well known that the growth/decay of clusters is significantly affected by the cluster size and contact angles of the condensing surface. With the increase of cluster sizes, more surface area of the cluster is exposed to the vapor, as well as the increased attachment/detachment frequencies. When the contact angle decreased to a certain value, the attachment frequency becomes larger than detachment frequency, resulting in the continuous growth of the clusters and subsequent nucleation. The results of the evolution of cluster/droplet size distribution indicate that the transient cluster size distribution translates from a monotonic decreasing distribution to a unimodal distribution with time (**Figure 7b,c**). The cluster radius at the peak of cluster/droplet size distribution curve shifts to the larger cluster sizes with time. However, the peak value decreases as the size distribution curve is close to a lognormal distribution with time, which is distinctly different from the homogeneous equilibrium distribution (**Figure 7d**).

5. Condensation on superhydrophobic surfaces

Surface modification plays a crucial role in manipulating droplet wetting and dynamics. When micro/nanostructures are covered with hydrophobic coatings, the resulted superhydrophobic surfaces can promote the formation of highly mobile droplets in the Cassie states. More interestingly, it has been demonstrated that on such micro/nanostructured superhydrophobic surfaces small microdroplets can undergo coalescence-induced droplet jumping due to the release of excess surface energy (**Figure 8a**), which is independent of gravity [75, 76, 83, 84]. Jumping

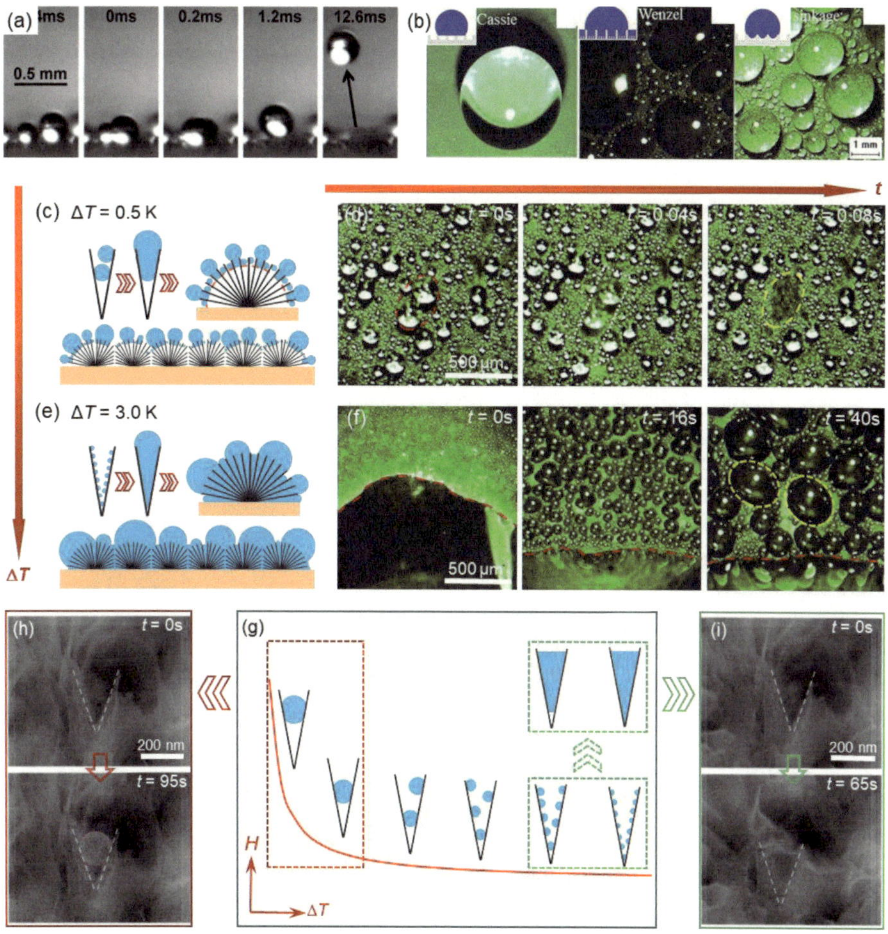

Figure 8.
Wetting and dynamics of the droplet on the superhydrophobic surfaces. (a) Droplet jumping due to the merge of two small droplets [83]. (b) Various wetting modes of condensed droplets on a nanostructured surface [85]. (c–i) Wetting transition of condensed droplets with the increase of surface subcooling: (c) schematic illustrating suspended droplets under a small subcooling, (d) time-lapse images of jumping droplets, (e) schematic illustrating immersed droplets under a large subcooling, (f) time-lapse images of flooding phenomenon, (g) schematic illustrating droplet nucleation in the nanostructure as a function of subcooling. ESEM images of a suspended (h) and immersed (i) droplet in nanostructures at a small and a large surface subcooling, respectively [24].

droplets have been shown to enhance dropwise condensation through accelerating surface refreshing and reducing time-averaged droplet thermal resistance on the surface [22, 66, 86, 87]. Different from the wetting states of the injected droplets on a superhydrophobic surface, such as the millimeter-scale raindrops that are much larger than the micro/nanoscale structure features of lotus leaves, vapor condensation starts with the formation of nanoscale droplets, which can result in loss of non-wettability and flooding phenomenon on micro/nanostructured surfaces. Various droplet wetting modes (**Figure 8b**), for example, Wenzel state, Cassie state, and partially wetting state, have been observed on the superhydrophobic surface [85, 88]. Moreover, the wetting transition of condensed droplets can cause a large degradation of heat transfer performance [22, 89]. A considerable amount of work has focused on understanding the effect of surface structure and wettability on wetting transition with a variety of explanations including surface free energy models [88, 90, 91], Laplace pressure instabilities [92–94], and thermodynamic models [95, 96].

Figure 8c–i shows that the condensation mode transitions from jumping droplet condensation to flooding condensation on a nanostructured superhydrophobic surface as the surface subcooling increases from 0.3 to 3 K [24]. To illustrate the mechanism of wetting transition, a spatial confinement effect on the droplet nucleation was proposed to explain the wetting states of condensed droplets under different surface subcooling. Condensation begins with nucleation with the formation of nanoscale condensed droplets, with the diameter of several nanometers at large subcooling to hundreds of nanometers at small subcooling. Besides, the nucleation site density is inversely proportional to the critical droplet nucleation radius, $N_s \sim 0.037/r_{min}$ [97]. Under a small subcooling, sparse large nucleated droplets on the nanostructures tend to form as the suspended Cassie state (**Figure 8c**), promoting the subsequent self-droplet jumping and effective surface refreshing (**Figure 8d**). As the surface subcooling increases, the critical droplet radius rapidly decreases, leading to the preferred droplet nucleation at the bottom of nanostructures (**Figure 8e**). With the further increase of surface subcooling, large pinned droplets can form on the nanostructures (**Figure 8f**). This is due to the coalescence between a large number of small droplets nucleated within the structures, filling the nanostructures with liquid condensate. To validate the proposed effect of spatial confinement on droplet formation on superhydrophobic surfaces, **Figure 8g,h** shows the experimental observation of the wetting states of nucleated droplets in the nanoscale gaps at different surface subcooling. Small surface subcooling promotes the formation of suspended droplets (**Figure 8h**) due to a larger critical nucleation size (about 166 nm), while large surface subcooling results in the immersed droplet (**Figure 8i**) due to a smaller critical nucleation size. Thus, the wetting states of condensed droplets on the micro/nanostructured surfaces can be controlled by manipulating initial nucleation through adjusting the structure feature and length scale of surface structures.

To delay the occurrence of flooding for achieving stable jumping droplets, various superhydrophobic surfaces with closely spaced nanostructures (**Figure 9a**) have been recently proposed to spatially control nucleation [65, 98, 99]. Minimizing the spacings using high-aspect ratio nanowires can promote to obtain a vapor density difference between the inside and outside of the nanoscale spacing. By this strategy of controlling vapor density, the closely spaced nanowires have been demonstrated to mitigate droplet nucleation in the spacing with a larger energy barrier compared with the top surface of nanowires. The formation of mobile droplets in suspended or partially wetting states is favored for jumping droplet condensation. Compared to the droplet sliding on a plain hydrophobic surface driven by gravity (**Figure 9b**), droplets on the superhydrophobic nanowired surface can be

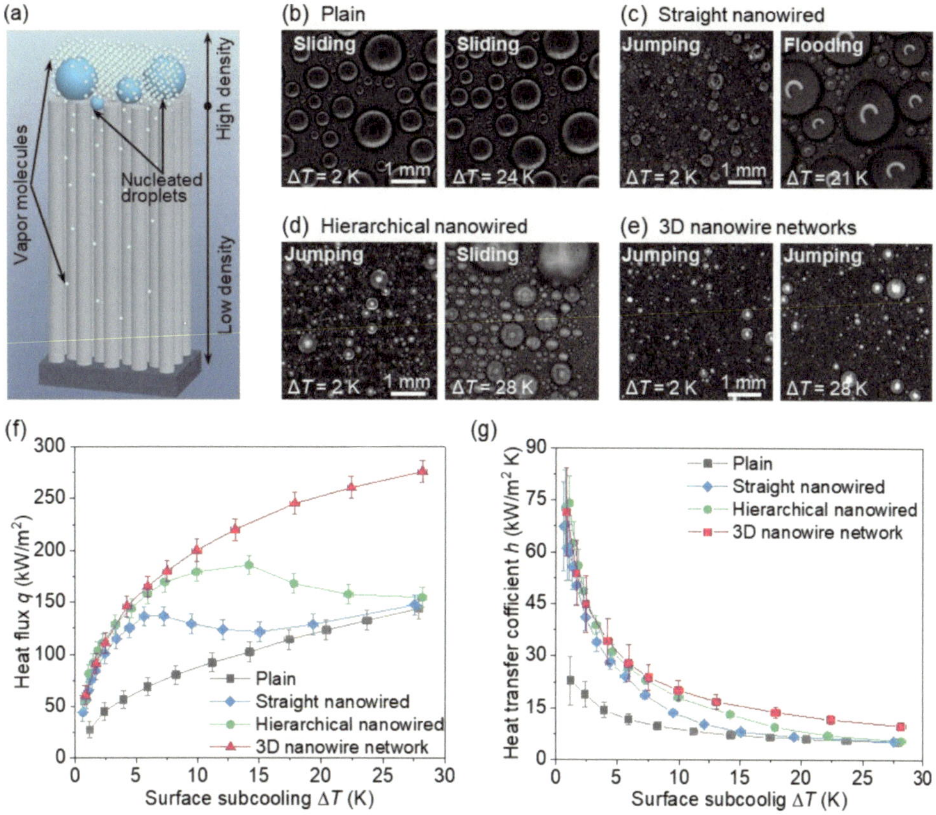

Figure 9.
Superhydrophobic copper nanowires for enhancing condensation heat transfer. (a) Schematic illustrating the spatial control of nucleation for mobile droplets on the top of closely spaced nanowires [65]. (b–e) Droplet behaviors on the plain hydrophobic, straight nanowired, and hierarchical nanowired surfaces, and 3D nanowire networks [71]. (f–g) Heat flux and heat transfer coefficient as a function of surface subcooling. Stable jumping droplet condensation enables the highest heat flux and heat transfer coefficient on the surface with 3D nanowire networks [71].

efficiently removed by efficient jumping (**Figure 9c**) at a small surface subcooling of 2 K. The enhanced condensation heat flux and heat transfer coefficient on the straight nanowired surfaces as a function of surface subcooling have been experimentally demonstrated compared with conventional dropwise condensation on the plain hydrophobic surface (**Figure 9f,g**). However, condensation heat flux reaches its maximum as surface subcooling increases to 6 K and then slightly reduces as surface subcooling is increased to 15 K. Further increase of surface subcooling can increase heat flux monotonically with a similar trend as dropwise condensation on the plain hydrophobic surface. Subsequently, a hierarchical superhydrophobic surface with micro-patterned copper nanowire arrays was developed to improve droplet dynamics [21]. Micro-valleys, consisting of short nanowire arrays between long nanowire arrays, were fabricated to enable spontaneous droplet movement during droplet growing process. For the droplets formed in the micro-valleys, they can rapidly grow on the short nanowire arrays to a critical size where the droplet dimension is comparable to micro-valleys. Further growth of these droplets can push droplet out from the micro-valleys with a Laplace pressure difference between the bottom and top of the droplet, which accelerates droplet re-nucleation and growth.

Although the range of surface subcooling for jumping droplet condensation has been extended by the spatial control of nucleation using the closely arranged nanowires, droplet wetting transition still occurs on both the uniform and hierarchical nanowired surfaces (**Figure 9c,d**) with the increase of the surface subcooling, resulting in the heat transfer degradation [21, 65]. This is because the active nucleation size decreases with the increase of surface subcooling while the straight nanowires tend to agglomerate and form a large number of micro-defects on the surface in fabrication processes [65, 73]. For condensation on such structured surfaces, vapor prefers to nucleate first in the micro-defects with a smaller energy barrier, resulting in the formation of large pinned droplets and flooding phenomenon [67, 68]. To avoid the agglomeration of straight nanowires, a 3D copper nanowire network fabricated has recently been demonstrated to eliminate the micro-defects between nanowires [71]. Benefiting from the interconnections between the nanowires, the surface morphology of the 3D nanowire network appears to be homogeneous without micro-defects. Compared to the droplet wetting transition on the uniform and hierarchical nanowired surfaces (**Figure 9c,d**), sustainable droplet jumping phenomena is obtained on the superhydrophobic surface with 3D nanowire networks even at a large surface subcooling of 28 K (**Figure 9e**). Benefiting from the formation of suspended and partially wetting droplets, a significant enhancement in heat transfer is realized on the 3D nanowire networks throughout the wide range of surface subcooling experimented (**Figure 9f,g**).

6. Potential strategies for condensation enhancement

In addition to the condensation enhancement on the superhydrophobic surfaces as mentioned above, several other potential strategies and surfaces have recently been proposed to enhance condensation by manipulating droplet behaviors [5, 27, 28, 30, 69, 72, 100–105]. For example, a hybrid structured surface with wettability contrast was proposed to spatially control nucleation, where droplets preferentially form on the hydrophilic region [27]. By confining the hydrophilic regions on the top of the micropillars that are surrounded by superhydrophobic nanowires, such a hybrid surface can achieve both efficient droplet nucleation and jumping removal. Similar design of hybrid micro/nanostructured surfaces with wetting contrast can be found in the literature [69, 106, 107]. The slippery liquid-infused porous surface has been widely studied to promote droplet mobility [5, 28, 108–112]. A hydrophilic directional slippery rough surface was recently fabricated to improve droplet nucleation and removal [5]. Such surface consists of nanotextured directional microgrooves in which the nanotextures alone are infused with hydrophilic liquid lubricant. The surface has hydrophilic surface chemistry, a slippery interface, directional structures, and large surface areas for droplet nucleation and motion. Further experimentation under atmospheric pressure and pure vapor are needed to elucidate the condensation enhancement. Recently, another strategy was presented to enable thin film condensation on a porous superhydrophilic surface with a hydrophobic coating as a nucleation deterrent and energy barrier for the liquid film to propagate above the surface [101, 113, 114]. Further experimental demonstration of heat transfer enhancement of such thin film condensation is needed. It is noted that most of the surfaces mentioned above require precision fabrication, which is difficult to scale up cost-effectively to meet the large area applications.

Some emerging strategies and surface design provide the potential to achieve condensation enhancement by using macrotextures or low-cost commercial materials [72, 100, 103, 104, 115–117]. The plain hybrid surfaces, consisting of plain hydrophilic and hydrophobic regions, were proposed to achieve dropwise-filmwise condensation for reducing the droplet departure size [23, 100, 115, 116, 118–123].

On such a hybrid surface (**Figure 10a**), the growing droplets in the hydrophobic regions are removed when they grow large enough to contact with the liquid film in the hydrophilic regions, resulting in a smaller removal size than that of gravity-driven departure. Another attempt in coupling droplets with a liquid film for enhanced condensation was recently reported on a hierarchical mesh-covered surface [72]. The typical structural feature of the hierarchical mesh-covered surface is composed of an interweaving microchannel network between a superhydrophobic woven mesh layer and a flat substrate, and a large number of micropores among mesh wires (**Figure 10b**). Vapor permeates freely through the mesh layer and condenses on the substrate to form a thin liquid condensate film that is confined in the interconnected channel network. When the condensate film grows out of the micropores due to the increased condensation heat flux, the surrounding liquid condensate can be drained out by the liquid film flow and eventually leaves the superhydrophobic hierarchical mesh-covered surface in the form of gravity-driven falling droplets. The thin liquid film in the interweaving microchannel network not only provides efficient low-thermal resistance condensation interfaces but also continuously transports liquid condensate to be drained out from the substrate.

When the NCG presences in water vapor, condensation performance is highly dependent on the initial nucleation and mass transfer of vapor molecules in the diffusion layer near the condensing surface [103, 124, 125]. A hydrophilic copper surface with interval fluorocarbon-coated hydrophobic bumps to enable falling droplet-enhanced filmwise condensation (**Figure 10c**). Due to the decreased nucleation energy barrier on the hydrophilic regions, water vapor can rapidly nucleate, grow, and form a thin liquid film on the vertical surface. To prevent continuous thickness growth of condensate liquid film along the vertical tube, interval hydrophobic bumps are designed to remove liquid film periodically, which is enabled by the reduction of surface adhesion for condensate liquid on the hydrophobic bumps. More importantly, the condensate liquid departing from the hydrophobic bumps in the form of falling droplets can strongly disturb the NCG diffusion boundary layer. High-performance condensation heat transfer in the presence of NCG was

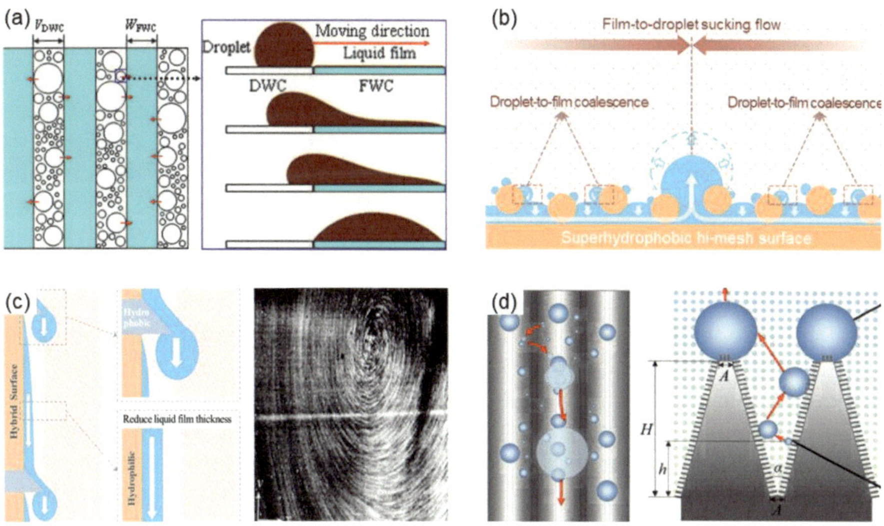

Figure 10.
Emerging strategies and surface design for enhancing condensation heat transfer. (a) Hybrid surface with hydrophobic and hydrophilic stripes to enable dropwise-filmwise condensation [100]. (b) Sucking flow condensation on a hierarchical mesh structured surface [72]. (c) A hydrophilic copper tube surface with interval fluorocarbon-coated hydrophobic bumps [103]. (d) A superhydrophobic surface covered with macrogroove arrays [104].

experimentally demonstrated on the vertical hydrophilic copper tube with hydrophobic fluorocarbon-coated bumps, which is better than both the conventional filmwise and dropwise condensation while avoiding the durability issues of ultrathin hydrophobic coatings. Recently, a structured surface with macrogroove arrays was proposed to improve droplet jumping dynamics in the presence of NCG by coupling rapid droplet growth and efficient droplet jumping relay (**Figure 10d**) [104]. The droplets formed on top of the cones and the bottom of the grooves play different roles during condensation process. Specifically, the cones can promote droplet formation and growth by breaking through the limitation of NCG layer. The droplets with higher mobility can be formed on the bottom of the grooves, resulting in series of coalescence-induced droplet jumping. Such a droplet jumping relay can enable a considerable vibration to trigger the jumping removal of droplets on top of the cones.

7. Summary

This chapter reviewed the recent advances in the fundamental understanding and performance enhancement of dropwise condensation by dancing droplets, as well as some other emerging enhancement strategies and surface design. Various micro/nanostructured surfaces, along with functional wetting coatings, have been developed with designed morphology for diverse surface features. Addressing the intrinsic requirements on multiple length scales in the nucleation, growth, merge/coalescence and departure of the dynamic droplets, unprecedented enhancement in heat transfer performance has been demonstrated for dropwise condensation processes.

An efficient condensing surface should enable both rapid droplet growth and frequent surface refreshing. Due to the excellent surface refreshing capability, jumping droplet condensation on the superhydrophobic surfaces is one of the most active research areas over the last decade on enhancing condensation heat transfer. However, as surface subcooling increases, the mobile droplets in the suspended Cassie state can transition to the highly pinned Wentzel state due to the nucleation occurring within the structures, resulting in the flooding phenomenon and performance degradation. By decreasing the structure scale to be comparable with critical nucleation size, superhydrophobic surfaces with closely spaced nanowires have been demonstrated to minimize droplet nucleation within the structures and to promote the formation of mobile droplets on the surface. On such a nanowired surface, excellent water repellency has been demonstrated to enable efficient jumping droplets without flooding phenomenon, even at a large surface subcooling. A significant enhancement in heat transfer was also achieved under a very wide range of surface subcooling experimented. In addition to the superhydrophobic nanowired surfaces, several other strategies have also been proposed to enhance condensation processes, for example, improving droplet nucleation and jumping by designing hydrophilic patterns on the superhydrophobic surfaces, accelerating droplet removal through liquid film sucking on the hybrid surfaces with hydrophilic and hydrophobic strips, promoting thin film condensation using hybrid nanostructured surfaces, enhancing droplet mobility and transport using slippery liquid-infused porous surfaces, and improving liquid condensate removal using hierarchical mesh-covered surfaces.

Acknowledgements

The authors acknowledge the continuous support from National Natural Science Foundation of China (Nos. 51836002, 51706031, and 51236002), the Fundamental

Research Funds for the Central Universities (No. DUT20RC(3)016), and thank their collaborators who made important contributions to the work reviewed in this chapter including Ronggui Yang, Zhong Lan, Yung-Cheng Lee, Benli Peng, Wei Xu, Sifang Wang.

Conflict of interest

The authors declare no conflict of interest.

Nomenclature

D_{12}	diffusion constant
ΔG_e	nucleation energy, J
h_{fg}	latent heat of vaporization, J/kg
J	nucleation rate
J_0	kinetic constant
k	Boltzmann constant
N	droplet population density, m^{-3}
p_r	vapor pressure, Pa
p_s	saturation pressure, Pa
q_d	heat transfer through individual droplet, W
q	heat flux, W/m^2
r	surface roughness
r_{min}	minimum droplet nucleation radius, m
r_{max}	droplet departure radius, m
Sc	Schmidt number
T	temperature, K
ΔT	surface subcooling, K
U	free-stream velocity, m/s
w	mass transfer flux, mol/m^2
θ	contact angle, °
θ_{adv}	advancing contact angle, °
θ_C	contact angle of Cassie droplet, °
θ_{rec}	receding contact angle, °
θ_W	contact angle of Wenzel droplet, °
$\Delta\theta$	contact angle hysteresis, °
ψ	solid fraction
σ_{lv}	liquid-vapor interfacial tension, N/m
σ_{sv}	solid-vapor interfacial tension, N/m
σ_{sl}	solid-liquid interfacial tension, N/m
μ	kinematic viscosity, N s/m^2
ρl	liquid density, kg/m^3
$\boldsymbol{\delta_0}$	thickness of boundary layer, m

Author details

Rongfu Wen* and Xuehu Ma*
School of Chemical Engineering, Dalian University of Technology, Dalian, P.R. China

*Address all correspondence to: rongfuwen@dlut.edu.cn and xuehuma@dlut.edu.cn

References

[1] Attinger D, Frankiewicz C, Betz AR, Schutzius TM, Ganguly R, Das A, et al. Surface engineering for phase change heat transfer: A review. MRS Energy & Sustainability. 2014;**1**:1-40

[2] Chen X, Ye H, Fan X, Ren T, Zhang G. A review of small heat pipes for electronics. Applied Thermal Engineering. 2016;**96**:1-17

[3] Stark AK, Klausner JF. An R&D strategy to decouple energy from water. Joule. 2017;**1**:416-420

[4] Urban JJ. Emerging scientific and engineering opportunities within the water-energy nexus. Joule. 2017;**1**(4): 665-688

[5] Dai X, Sun N, Nielsen SO, Stogin BB, Wang J, Yang S, et al. Hydrophilic directional slippery rough surfaces for water harvesting. Science Advances. 2018;**4**(3):eaaq0919

[6] Ni G, Li G, Boriskina L, Svetlana V, Li H, Yang W, et al. Steam generation under one sun enabled by a floating structure with thermal concentration. Nature Energy. 2016;**1**(9):16126

[7] Zhao F, Zhou X, Shi Y, Qian X, Alexander M, Zhao X, et al. Highly efficient solar vapour generation via hierarchically nanostructured gels. Nature Nanotechnology. 2018;**13**(6): 489-495

[8] Yeow JTW, She JPM. Carbon nanotube-enhanced capillary condensation for a capacitive humidity sensor. Nanotechnology. 2006;**17**(21): 5441-5448

[9] Hao T, Wang K, Chen Y, Ma X, Lan Z, Bai T. Multiple bounces and oscillatory movement of a microdroplet in superhydrophobic minichannels. Industrial and Engineering Chemistry Research. 2018;**57**(12):4452-4461

[10] Boreyko JB, Zhao YJ, Chen CH. Planar jumping-drop thermal diodes. Applied Physics Letters. 2011;**99**(23): 234105

[11] Oh J, Birbarah P, Foulkes T, Yin SL, Rentauskas M, Neely J, et al. Jumping-droplet electronics hot-spot cooling. Applied Physics Letters. 2017;**110**(12): 123107

[12] Miljkovic N, Preston DJ, Enright R, Wang EN. Jumping-droplet electrostatic energy harvesting. Applied Physics Letters. 2014;**105**(1):013111

[13] Xu W, Lan Z, Peng BL, Wen RF, Ma XH. Effect of surface free energies on the heterogeneous nucleation of water droplet: A molecular dynamics simulation approach. The Journal of Chemical Physics. 2015;**142**(5):054701

[14] Wen RF, Lan Z, Peng BL, Xu W, Ma XH. Droplet dynamics and heat transfer for dropwise condensation at lower and ultra-lower pressure. Applied Thermal Engineering. 2015;**88**: 265-273

[15] Rose JW. Dropwise condensation theory and experiment: A review. Proceedings of the Institution of Mechanical Engineers, Part A. 2002;**216** (A2):115-128

[16] Lan Z, Wen RF, Wang AL, Ma XH. A droplet model in steam condensation with noncondensable gas. International Journal of Thermal Sciences. 2013;**68**:1-7

[17] Xu W, Lan Z, Peng BL, Wen RF, Ma XH. Evolution of transient cluster/ droplet size distribution in a heterogeneous nucleation process. RSC Advances. 2014;**4**(60):31692-31699

[18] Xu W, Lan Z, Peng BL, Wen RF, Ma XH. Heterogeneous nucleation capability of conical microstructures for

water droplets. RSC Advances. 2015; **5**(2):812-818

[19] Varanasi KK, Hsu M, Bhate N, Yang WS, Deng T. Spatial control in the heterogeneous nucleation of water. Applied Physics Letters. 2009;**95**(9): 094101

[20] Miljkovic N, Enright R, Wang EN. Effect of droplet morphology on growth dynamics and heat transfer during condensation on superhydrophobic nanostructured surfaces. ACS Nano. 2012;**6**(2):1776-1785

[21] Wen RF, Xu SS, Zhao DL, Lee YC, Ma XH, Yang RG. Hierarchical superhydrophobic surfaces with micropatterned nanowire arrays for high-efficiency jumping droplet condensation. ACS Applied Materials & Interfaces. 2017;**9**(51):44911-44921

[22] Miljkovic N, Enright R, Nam Y, Lopez K, Dou N, Sack J, et al. Jumping-droplet-enhanced condensation on scalable superhydrophobic nanostructured surfaces. Nano Letters. 2013;**13**(1):179-187

[23] Narhe RD, Beysens DA. Nucleation and growth on a superhydrophobic grooved surface. Physical Review Letters. 2004;**93**(7):076103

[24] Wen RF, Lan Z, Peng BL, Xu W, Yang RG, Ma XH. Wetting transition of condensed droplets on nanostructured superhydrophobic surfaces: Coordination of surface properties and condensing conditions. ACS Applied Materials & Interfaces. 2017;**9**(15): 13770-13777

[25] Mouterde T, Lehoucq G, Xavier S, Checco A, Black CT, Rahman A, et al. Antifogging abilities of model nanotextures. Nature Materials. 2017;**16**: 658-663

[26] Hou Y, Yu M, Shang Y, Zhou P, Song R, Xu X, et al. Suppressing ice nucleation of supercooled condensate with biphilic topography. Physical Review Letters. 2018;**120**(7):075902

[27] Hou YM, Yu M, Chen XM, Wang ZK, Yao SH. Recurrent filmwise and dropwise condensation on a beetle mimetic surface. ACS Nano. 2015;**9**(1): 71-81

[28] Park KC, Kim P, Grinthal A, He N, Fox D, Weaver JC, et al. Condensation on slippery asymmetric bumps. Nature. 2016;**531**(7592):78-82

[29] Rykaczewski K, Paxson AT, Staymates M, Walker ML, Sun XD, Anand S, et al. Dropwise condensation of low surface tension fluids on omniphobic surfaces. Scientific Reports. 2014;**4**:4158

[30] Anand S, Paxson AT, Dhiman R, Smith JD, Varanasi KK. Enhanced condensation on lubricant-impregnated nanotextured surfaces. ACS Nano. 2012; **6**(11):10122-10129

[31] Lafuma A, Quere D. Superhydrophobic states. Nature Materials. 2003;**2**(7):457-460

[32] Maa JR. Drop size distribution and heat-flux of dropwise condensation. Chemical Engineering Journal. 1978; **16**(3):171-176

[33] Ma XH, Song TY, Lan Z, Bai T. Transient characteristics of initial droplet size distribution and effect of pressure on evolution of transient condensation on low thermal conductivity surface. International Journal of Thermal Sciences. 2010; **49**(9):1517-1526

[34] Xu W, Lan Z, Liu Q, Du B, Ma X. Droplet size distributions in dropwise condensation heat transfer: Consideration of droplet overlapping and multiple re-nucleation. International Journal of Heat and Mass Transfer. 2018;**127**:44-54

[35] Kim S, Kim KJ. Dropwise condensation modeling suitable for superhydrophobic surfaces. Journal of Heat Transfer. 2011;**133**(8):081502

[36] Ma XH, Rose JW, Xu DQ, Lin JF, Wang BX. Advances in dropwise condensation heat transfer: Chinese research. Chemical Engineering Journal. 2000;**78**(2–3):87-93

[37] Wen RF, Lan Z, Peng BL, Xu W, Ma XH, Cheng YQ. Droplet departure characteristic and dropwise condensation heat transfer at low steam pressure. Journal of Heat Transfer. 2016; **138**:071501

[38] Le Ferve EJ, Rose JW. A theory of heat transfer by dropwise condensation. In: Proceedings of the Third International Heat Transfer Conference. Chicago, IL: American Institute of Chemical Engineers; 1966. pp. 362-375

[39] AbuOrabi M. Modeling of heat transfer in dropwise condensation. International Journal of Heat and Mass Transfer. 1998;**41**(1):81-87

[40] Miljkovic N, Enright R, Wang EN. Modeling and optimization of superhydrophobic condensation. Journal of Heat Transfer. 2013;**135**(11): 111004

[41] Vemuri S, Kim KJ. An experimental and theoretical study on the concept of dropwise condensation. International Journal of Heat and Mass Transfer. 2006;**49**(3–4):649-657

[42] Lee S, Yoon HK, Kim KJ, Kim S, Kennedy M, Zhang BJ. A dropwise condensation model using a nano-scale, pin structured surface. International Journal of Heat and Mass Transfer. 2013;**60**:664-671

[43] Liu X, Cheng P. Dropwise condensation theory revisited part II. Droplet nucleation density and condensation heat flux. International Journal of Heat and Mass Transfer. 2015; **83**:842-849

[44] Zheng S, Eimann F, Philipp C, Fieback T, Gross U. Modeling of heat and mass transfer for dropwise condensation of moist air and the experimental validation. International Journal of Heat and Mass Transfer. 2018;**120**:879-894

[45] Niu D, Guo L, Hu HW, Tang GH. Dropwise condensation heat transfer model considering the liquid-solid interfacial thermal resistance. International Journal of Heat and Mass Transfer. 2017;**112**:333-342

[46] Shang Y, Hou Y, Yu M, Yao S. Modeling and optimization of condensation heat transfer at biphilic interface. International Journal of Heat and Mass Transfer. 2018;**122**:117-127

[47] Chavan S, Cha H, Orejon D, Nawaz K, Singla N, Yeung YF, et al. Heat transfer through a condensate droplet on hydrophobic and nanostructured superhydrophobic surfaces. Langmuir. 2016;**32**(31): 7774-7787

[48] Phadnis A, Rykaczewski K. The effect of Marangoni convection on heat transfer during dropwise condensation on hydrophobic and omniphobic surfaces. International Journal of Heat and Mass Transfer. 2017;**115**:148-158

[49] Tanaka H. A theoretical study of dropwise condensation. Journal of Heat Transfer. 1975;**97**(1):72

[50] Mei MF, Yu BM, Cai JC, Luo L. A fractal analysis of dropwise condensation heat transfer. International Journal of Heat and Mass Transfer. 2009;**52**(21–22):4823-4828

[51] Weisensee PB, Wang Y, Qian H, Daniel S, William PK, Miljkovic N. Condensate droplet size distribution on lubricant-infused surfaces. International

Journal of Heat and Mass Transfer. 2017; **109**:187-199

[52] Kulinich SA, Farzaneh M. Hydrophobic properties of surfaces coated with fluoroalkylsiloxane and alkylsiloxane monolayers. Surface Science. 2004;**573**(3):379-390

[53] KKS L, Bico J, KBK T, Chhowalla M, GAJ A, Milne WI, et al. Superhydrophobic carbon nanotube forests. Nano Letters. 2003;**3**:1701-1705

[54] Gnanappa AK, O'Murchu C, Slattery O, Peters F, Aszalós-Kiss B, Tofail SAM. Effect of annealing on hydrophobic stability of plasma deposited fluoropolymer coatings. Polymer Degradation and Stability. 2008;**93**(12):2119-2126

[55] Paxson AT, Yague JL, Gleason KK, Varanasi KK. Stable dropwise condensation for enhancing heat transfer via the initiated chemical vapor deposition (iCVD) of grafted polymer films. Advanced Materials. 2014;**26**(3): 418-423

[56] Preston DJ, Mafra DL, Miljkovic N, Kong J, Wang EN. Scalable graphene coatings for enhanced condensation heat transfer. Nano Letters. 2015;**15**(5): 2902-2909

[57] Azimi G, Dhiman R, Kwon HM, Paxson AT, Varanasi KK. Hydrophobicity of rare-earth oxide ceramics. Nature Materials. 2013;**12**(4): 315-320

[58] Khalil K, Soto D, Paxson A, Katmis AU, Gleason K, Varanasi KK. Grafted nanofilms promote dropwise condensation of low-surface-tension fluids for high-performance heat exchangers. Joule. 2019;**3**(5):1377-1388

[59] McCarthy M, Gerasopoulos K, Maroo SC, Hart AJ. Materials, fabrication, and manufacturing of micro/nanostructured surfaces for phase-change heat transfer enhancement. Nanoscale and Microscale Thermophysical Engineering. 2014; **18**(3):288-310

[60] Lundy R, Byrne C, Bogan J, Nolan K, Collins MN, Dalton E, et al. Exploring the role of adsorption and surface state on the hydrophobicity of rare earth oxides. ACS Applied Materials & Interfaces. 2017;**9**(15): 13751-13760

[61] Fu S-P, Rossero J, Chen C, Li D, Takoudis CG, Abiade JT. On the wetting behavior of ceria thin films grown by pulsed laser deposition. Applied Physics Letters. 2017;**110**(8): 081601

[62] Cha H, Wu A, Kim MK, Saigusa K, Liu A, Miljkovic N. Nanoscale-agglomerate-mediated heterogeneous nucleation. Nano Letters. 2017;**17**(12): 7544-7551

[63] Roach P, Shirtcliffe NJ, Newton MI. Progess in superhydrophobic surface development. Soft Matter. 2008;**4**(2): 224-240

[64] Lu MC, Chen R, Srinivasan V, Carey VP, Majumdar A. Critical heat flux of pool boiling on Si nanowire array-coated surfaces. International Journal of Heat and Mass Transfer. 2011; **54**(25–26):5359-5367

[65] Wen RF, Li Q, Wu JF, Wu GS, Wang W, Chen YF, et al. Hydrophobic copper nanowires for enhancing condensation heat transfer. Nano Energy. 2017;**33**:177-183

[66] Li Y, Duan C. Bubble-regulated silicon nanowire synthesis on micro-structured surfaces by metal-assisted chemical etching. Langmuir. 2015; **31**(44):12291-12299

[67] Lo CW, Wang CC, Lu MC. Scale effect on dropwise condensation on

superhydrophobic surfaces. ACS Applied Materials & Interfaces. 2014; **6**(16):14353-14359

[68] Lo CW, Wang CC, Lu MC. Spatial control of heterogeneous nucleation on the superhydrophobic nanowire array. Advanced Functional Materials. 2014; **24**(9):1211-1217

[69] Chen XM, Wu J, Ma RY, Hua M, Koratkar N, Yao SH, et al. Nanograssed micropyramidal architectures for continuous dropwise condensation. Advanced Functional Materials. 2011; **21**(24):4617-4623

[70] Anderson DM, Gupta MK, Voevodin AA, Hunter CN, Putnam SA, Tsukruk VV, et al. Using amphiphilic nanostructures to enable long-range ensemble coalescence and surface rejuvenation in dropwise condensation. ACS Nano. 2012;**6**(4):3262-3268

[71] Wen RF, Xu SS, Ma XH, Lee YC, Yang RG. Three-dimensional superhydrophobic nanowire networks for enhancing condensation heat transfer. Joule. 2018;**2**(2):269-279

[72] Wen R, Xu S, Zhao D, Yang L, Ma X, Liu W, et al. Sustaining enhanced condensation on hierarchical mesh-covered surfaces. National Science Review. 2018;**5**(6):878-887

[73] Wen RF, Li Q, Wang W, Latour B, Li CH, Li C, et al. Enhanced bubble nucleation and liquid rewetting for highly efficient boiling heat transfer on two-level hierarchical surfaces with patterned copper nanowire arrays. Nano Energy. 2017;**38**:59-65

[74] Lan Z, Chen F, Qiang W, Xue Q, Ma X. Direct observation of water clusters for surface design. Chemical Engineering Science. 2020;**217**:115475

[75] Xu W, Lan Z, Peng BL, Wen RF, Ma XH. Effect of nano structures on the nucleus wetting modes during water vapour condensation: From individual groove to nano-array surface. RSC Advances. 2016;**6**(10):7923-7932

[76] Peng BL, Wang SF, Lan Z, Xu W, Wen RF, Ma XH. Analysis of droplet jumping phenomenon with lattice Boltzmann simulation of droplet coalescence. Applied Physics Letters. 2013;**102**(15):151601

[77] Huang D, Quan X, Cheng P. An investigation on vapor condensation on nanopillar array surfaces by molecular dynamics simulation. International Journal of Heat and Mass Transfer. 2018;**98**:232-238

[78] Liu X, Cheng P, Quan X. Lattice Boltzmann simulations for self-propelled jumping of droplets after coalescence on a superhydrophobic surface. International Journal of Heat and Mass Transfer. 2014;**73**:195-200

[79] Niu D, Tang G. Molecular dynamics simulation of droplet nucleation and growth on a rough surface: Revealing the microscopic mechanism of the flooding mode. RSC Advances. 2018; **8**(43):24517-24524

[80] Gao S, Liao Q, Liu W, Liu Z. Effects of solid fraction on droplet wetting and vapor condensation: A molecular dynamic simulation study. Langmuir. 2017;**33**:12379-12388

[81] Zhang L, Xu Z, Lu Z, Du J, Wang EN. Size distribution theory for jumping-droplet condensation. Applied Physics Letters. 2019;**114**:163701

[82] Mei MF, Hu F, Han C, Cheng Y. Time-averaged droplet size distribution in steady-state dropwise condensation. International Journal of Heat and Mass Transfer. 2015;**88**:338-345

[83] Boreyko JB, Chen CH. Self-propelled dropwise condensate on superhydrophobic surfaces. Physical Review Letters. 2009;**103**(18):184501

[84] Yan X, Zhang L, Sett S, Feng L, Zhao C, Huang Z, et al. Droplet jumping: Effects of droplet size, surface structure, pinning, and liquid properties. ACS Nano. 2019;**13**(2): 1309-1323

[85] Ma XH, Wang SF, Lan Z, Peng BL, Ma HB, Cheng P. Wetting mode evolution of steam dropwise condensation on superhydrophobic surface in the presence of noncondensable gas. Journal of Heat Transfer. 2012;**134**(2):021501

[86] Preston DJ, Wang EN. Jumping droplets push the boundaries of condensation heat transfer. Joule. 2018; **2**(2):205-207

[87] Enright R, Miljkovic N, Alvarado JL, Kim K, Rose JW. Dropwise condensation on micro- and nanostructured surfaces. Nanoscale and Microscale Thermophysical Engineering. 2014;**18**(3):223-250

[88] Enright R, Miljkovic N, Al-Obeidi A, Thompson CV, Wang EN. Condensation on superhydrophobic surfaces: The role of local energy barriers and structure length scale. Langmuir. 2012;**28**(40):14424-14432

[89] Zhang P, Maeda Y, Lv F, Takata Y, Orejon D. Enhanced coalescence-induced droplet-jumping on nanostructured superhydrophobic surfaces in the absence of microstructures. ACS Applied Materials & Interfaces. 2017;**9**(40):35391-35403

[90] Xu W, Lan Z, Peng BL, Wen RF, Chen YS, Ma XH. Directional movement of droplets in grooves: Suspended or immersed? Scientific Reports. 2016;**6**:18836

[91] Rykaczewski K, Paxson AT, Anand S, Chen XM, Wang ZK, Varanasi KK. Multimode multidrop serial coalescence effects during condensation on hierarchical superhydrophobic surfaces. Langmuir. 2013;**29**(3):881-891

[92] Park KC, Choi HJ, Chang CH, Cohen RE, McKinley GH, Barbastathis G. Nanotextured silica surfaces with robust superhydrophobicity and omnidirectional broadband supertransmissivity. ACS Nano. 2012; **6**(5):3789-3799

[93] Checco A, Ocko BM, Rahman A, Black CT, Tasinkevych M, Giacomello A, et al. Collapse and reversibility of the superhydrophobic state on nanotextured surfaces. Physical Review Letters. 2014;**112**(21):216101

[94] Feng J, Qin ZQ, Yao SH. Factors affecting the spontaneous motion of condensate drops on superhydrophobic copper surfaces. Langmuir. 2012;**28**(14): 6067-6075

[95] Raj R, Enright R, Zhu YY, Adera S, Wang EN. Unified model for contact angle hysteresis on heterogeneous and superhydrophobic surfaces. Langmuir. 2012;**28**(45):15777-15788

[96] Rykaczewski K, Osborn WA, Chinn J, Walker ML, Scott JHJ, Jones W, et al. How nanorough is rough enough to make a surface superhydrophobic during water condensation? Soft Matter. 2012;**8**(33):8786-8794

[97] Rose JW. Further aspects of dropwise condensation theory. International Journal of Heat and Mass Transfer. 1976;**19**(12):1363-1370

[98] Seo D, Shim J, Moon B, Lee K, Lee J, Lee C, et al. Passive anti-flooding superhydrophobic surfaces. ACS Applied Materials & Interfaces. 2020; **12**(3):4068-4080

[99] Xing D, Wu F, Wang R, Zhu J, Gao X. Microdrop-assisted microdomain hydrophilicization of superhydrophobic surfaces for

high-efficiency nucleation and self-removal of condensate microdrops. ACS Applied Materials & Interfaces. 2019; **11**(7):7553-7558

[100] Peng BL, Ma XH, Lan Z, Xu W, Wen RF. Experimental investigation on steam condensation heat transfer enhancement with vertically patterned hydrophobic-hydrophilic hybrid surfaces. International Journal of Heat and Mass Transfer. 2015;**83**:27-38

[101] Oh J, Zhang R, Shetty PP, Krogstad JA, Braun PV, Miljkovic N. Thin film condensation on nanostructured surfaces. Advanced Functional Materials. 2018;**28**:1707000

[102] Yao Y, Aizenberg J, Park K-C. Dropwise condensation on hydrophobic bumps and dimples. Applied Physics Letters. 2018;**112**(15):151605

[103] Wen R, Zhou X, Peng B, Lan Z, Yang R, Ma X. Falling-droplet-enhanced filmwise condensation in the presence of non-condensable gas. International Journal of Heat and Mass Transfer. 2019;**140**:173-186

[104] Cheng Y, Du B, Wang K, Chen Y, Lan Z, Wang Z, et al. Macrotextures-induced jumping relay of condensate droplets. Applied Physics Letters. 2019; **114**(9):093704

[105] Bintein P-B, Lhuissier H, Mongruel A, Royon L, Beysens D. Grooves accelerate dew shedding. Physical Review Letters. 2019;**122**(9): 098005

[106] Mondal B, Mac Giolla Eain M, Xu Q, Egan VM, Punch J, Lyons AM. Design and fabrication of a hybrid superhydrophobic-hydrophilic surface that exhibits stable dropwise condensation. ACS Applied Materials & Interfaces. 2015;7(42):23575-23588

[107] Tang Z, He M, Bian R, Duan Z, Luan K, Hou J, et al. Multiple superwettable nanofiber arrays prepared by a facile dewetting strategy via controllably localizing a low-energy compound. Advanced Functional Materials. 2019;**29**(30):1900060

[108] Hao C, Li J, Liu Y, Zhou X, Liu Y, Liu R, et al. Superhydrophobic-like tunable droplet bouncing on slippery liquid interfaces. Nature Communications. 2015;**6**:7986

[109] Preston DJ, Lu Z, Song Y, Zhao Y, Wilke KL, Antao DS, et al. Heat transfer enhancement during water and hydrocarbon condensation on lubricant infused surfaces. Scientific Reports. 2018;**8**(1):540

[110] Seo D, Shim J, Lee C, Nam Y. Brushed lubricant-impregnated surfaces (BLIS) for long-lasting high condensation heat transfer. Scientific Reports. 2020;**10**(1):2959

[111] Sett S, Sokalski P, Boyina K, Li L, Rabbi KF, Auby H, et al. Stable dropwise condensation of ethanol and hexane on rationally designed ultrascalable nanostructured lubricant-infused surfaces. Nano Letters. 2019; **19**(8):5287-5296

[112] Jing X, Guo Z. Durable lubricant impregnated surfaces for water collection under extremely severe working conditions. ACS Applied Materials & Interfaces. 2019;**11**(39): 35949-35958

[113] Preston DJ, Wilke KL, Lu Z, Cruz SS, Zhao Y, Becerra LL, et al. Gravitationally driven wicking for enhanced condensation heat transfer. Langmuir. 2018;**34**(15):4658-4664

[114] Wang R, Antao DS. Capillary-enhanced filmwise condensation in porous media. Langmuir. 2018;**34**: 13855-13863

[115] Alwazzan M, Egab K, Peng B, Khan J, Li C. Condensation on hybrid-

patterned copper tubes (I): Characterization of condensation heat transfer. International Journal of Heat and Mass Transfer. 2017;**112**:991-1004

[116] Winter RL, McCarthy M. Dewetting from amphiphilic minichannel surfaces during condensation. ACS Applied Materials & Interfaces. 2020;**12**(6):7815-7825

[117] Cheng Y, Wang Z. New approach for efficient condensation heat transfer. National Science Review. 2019;**6**(2): 185-186

[118] Peng BL, Ma XH, Lan Z, Xu W, Wen RF. Analysis of condensation heat transfer enhancement with dropwise-filmwise hybrid surface: Droplet sizes effect. International Journal of Heat and Mass Transfer. 2014;**77**:785-794

[119] Wu J, Zhang L, Wang Y, Wang P. Efficient and anisotropic fog harvesting on a hybrid and directional surface. Advanced Materials Interfaces. 2017; **4**(2):1600801

[120] Yang K-S, Huang Y-Y, Liu Y-H, Wu S-K, Wang C-C. Enhanced dehumidification via hybrid hydrophilic/hydrophobic morphology having wedge gradient and drainage channels. Heat and Mass Transfer. 2019; **55**:3359-3368

[121] Gou X, Guo Z. Hybrid hydrophilic-hydrophobic CuO@TiO2-coated copper mesh for efficient water harvesting. Langmuir. 2020;**36**(1):64-73

[122] Sharma CS, Lam CWE, Milionis A, Eghlidi H, Poulikakos D. Self-sustained cascading coalescence in surface condensation. ACS Applied Materials & Interfaces. 2019;**11**(30):27435-27442

[123] Boreyko JB, Hansen RR, Murphy KR, Nath S, Retterer ST, Collier CP. Controlling condensation and frost growth with chemical micropatterns. Scientific Reports. 2016; **6**:19131

[124] Ma XH, Zhou XD, Lan Z, Li YM, Zhang Y. Condensation heat transfer enhancement in the presence of non-condensable gas using the interfacial effect of dropwise condensation. International Journal of Heat and Mass Transfer. 2008;**51**(7–8):1728-1737

[125] Zhao Y, Preston DJ, Lu Z, Zhang L, Queeney J, Wang EN. Effects of millimetric geometric features on dropwise condensation under different vapor conditions. International Journal of Heat and Mass Transfer. 2018;**119**: 931-938

Section 5

Imaging of Silicon Pillars by MWCNT Tip

Chapter 14

Measuring the Blind Holes: Three-Dimensional Imaging of through Silicon via Using High Aspect Ratio AFM Probe

Imtisal Akhtar, Malik Abdul Rehman and Yongho Seo

Abstract

Three-dimensional integration and stacking of semiconductor devices with high density, its compactness, miniaturization and vertical 3D stacking of nanoscale devices highlighted many challenging problems in the 3D parameter's such as CD (critical dimension) measurement, depth measurement of via holes, internal morphology of through silicon via (TSV), etc. Current challenge in the high-density 3D semiconductor devices is to measure the depth of through silicon via (TSV) without destructing the sample; TSVs are used in 3D stacking devices to connect the wafers stacked vertically to reduce the wiring delay, power dissipation, and of course, the form factor in the integration system. Special probes and algorithms have been designed to measure 3D parameters like wall roughness, sidewall angle, but these are only limited to deep trench-like structures and cannot be applied to structures like via holes and protrusions. To address these problems, we have proposed an algorithm based nondestructive 3D Atomic Force Microscopy (AFM). Using the high aspect ratio (5, 10, 20, 25) multiwall carbon nanotubes (MWCNTs) AFM probe, the depth of holes up to 1 micron is faithfully obtained. In addition to this, internal topography, side walls, and location of via holes are obtained faithfully. This atomic force microscopy technique enables to 3D scan the features (of any shape) present above and below the surface.

Keywords: algorithms, AFM, surface characterization, carbon nanotube, 3D AFM, sidewall, through silicon via (TSV)

1. Introduction

Scanning probe microscopy (SPM) over the former 20 years has been intensively used by many research groups due to its applicability in extensive filed of materials. In past 50 years, the most significant advancement in AFM was ranked at second place [1–8]. This is because the application of AFM not limited to the semiconductor field but also covers many wide range fields like chemical group identification [9], cell biological [10], semiconductor to study the properties of the materials at nanoscale [11]. Although, applications and techniques of SPM are diverse in nature, but they do share a common feature, i.e., probe to sense localized or confined signal, because in many cases the probe that confines the spatial accuracy.

AFM [12] is the most common form of SPM in which the probe is usually in the form of sharp, rigid and having long-life time mounted on the cantilever or tuning fork to transduce the tip-sample force. However, depending upon the application of AFM probe, it might have different shapes, length but the constant geometry over a long-time is common thing in all types of probes. As the probe scanning over the sample, the topography of the surface is constructed depending upon the mode of operation of AFM. The ensued image depicts the geometry of tip corresponding to the surface geometry.

Noteworthy improvement and advancement in the instrumentation, correction of artifacts and scanning algorithms has been accomplished. In addition to this, different type of probes has been prepared to obtain the deep trench (DT) [13] and sidewall roughness (SWR) [14], tracing vertical sidewalls [15], overhang and undercut features [16], etc. In addition to this, many useful techniques such as reflectometry [17] to measure the depth of via holes, X-Ray methods [18] and atomic force microscopy (AFM). Each method has its own pros and cons. For examples, conventional AFM is unable to image the features having high aspect ratio such a silicon pillars or deep via holes. In addition to this, reflectometry is also a useful and non-destructive technique. However, its resolution is limited to optical microscopy. Moreover, in some of the measurement instruments prior knowledge of sample surface is required.

In silicon industry, many improvements in the instrumentation has been done to measure the dimensions up to several hundred nanometer scale [19], lithography, manipulation of individual etc. In nanotechnology, AFM always is the essential contrivance especially when it comes to the height measurement of the feature. Because, measuring the deep, narrow via holes always been difficult and mainly unsolved problem. Measuring the depth of the hole is not enough and situation become worse and complicated when the sidewalls of the hole needs to be scanned. Many other groups also tried to increase the functionality and versatility of AFM either by combining the optical instrument [20] with atomic force microscopy but that is just limited to scan the trench to smaller dimension or limited to visible spectrum. In addition to this, flair shape probe [21], trident probe [22], bi-pod, capped probe [23], Osborne [22], hammerhead [24] and many other shaped probes are designed to address problem related to scanning. However, there is not one-size-fit-all answer. Seo et al. [25] has proposed general algorithm for scanning the deep via holes. However, that general algorithm has not been implemented on samples. One of the major problem associated with the algorithm was it does not stored the motion of probe when it find the sidewall wall in either direction which is necessary to keep following the boundary of the sample. Also, in case of no wall around, the tip moved one step further along the previous direction of motion. In this case, at some point, tip will lost the boundary of sample.

In this work we methodically explorer the scanning algorithm to scan the different types of features present above (protrusion) or below (holes) the surface and removed the problem associated with [24]. We have used the single MWCNT as a scanning probe attached by dielectrophoresis and FIB treatment. In addition to this, first time we have successfully implemented the non-destructive technique with high aspect ratio on real samples to obtain the three-dimension topography, depth measurement of holes, position of via holes by force-distance mapping for not only holes as well as protrusion.

2. Experimental method

There are mainly two aspects of fabrication of quartz tuning fork sensor followed by ball capped nanotube: nanotube attachment on electrochemically etched tungsten probe and modification using FIB treatment. In the complete process several steps are involved as shown in **Figure 1**. The tuning fork sensor is characterized by the length of carbon nanotube attached, radius of the ball capped on the top of MWCNT and orientation (angle with respect to the surface). All these parameters define the resolution, severity and image appearances of the tip.

2.1 Fabrication of tungsten tips

A 85% potassium hydroxide (KOH) was purchased from Sigma Aldrich® Company. A 2 molar concentration of KOH aqueous solution was prepared by dissolving KOH pallets in deionized water. After that, solution was agitated thoroughly to make it homogenous and left for 1 h. A gold wire with 200 μm diameter and 10 mm length was used as cathode. Tungsten (W) wire with diameter 50 and 25 mm length, with 99.95% purity, was purchased from Alfa Aesar® and used as anode. To improve the crystallinity, rapid thermal annealing (RTA) was done at 800°C for 40 s. DC voltages (typically 3–5 volts) are applied between gold and Tungsten

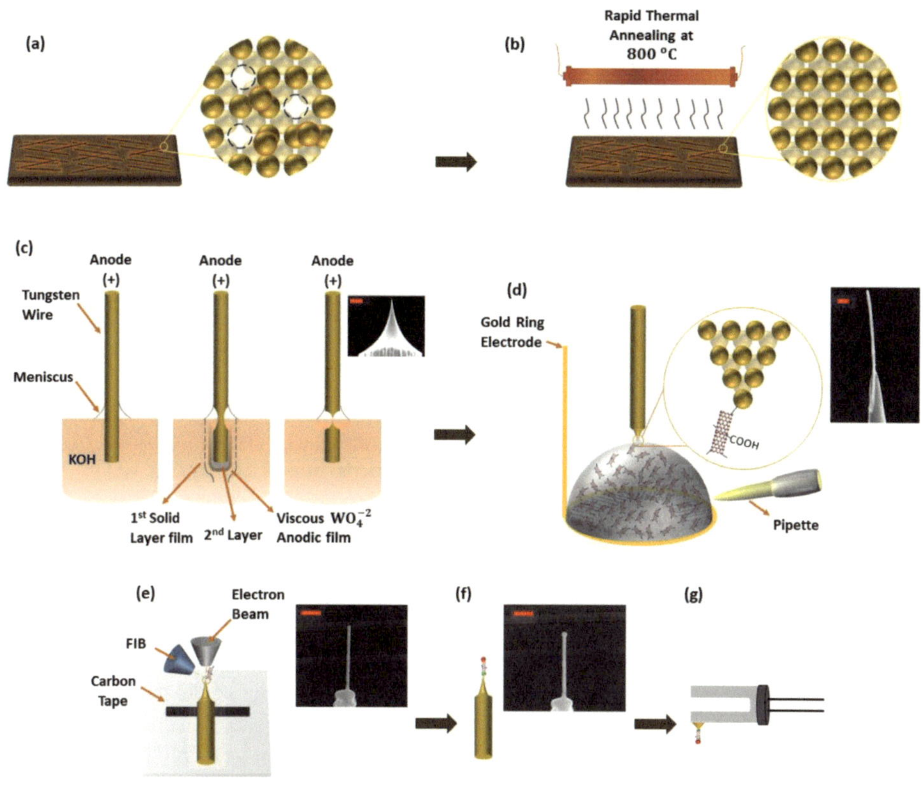

Figure 1.
Fabrication of tuning fork sensor followed by ball capped nanotube (a) Tungsten wire was annealed at 800°C for 40 s to improve the crystallinity. (b) Crystallinity of tungsten wire has been improved as the atoms are arranged at the lattice points. (c) Electrochemical etching using KOH produce very sharp tip of the order 65 nm. This diameter is ideal to attach all three types of CNTs (110–170 nm, 50–80 nm, 20 nm). (d) Small droplet of functionalized MWCNTTs is picked using micropipette and poured on gold ring electrode. Using Micromanipulator tip brings closer to droplet and meniscus is formed which covers the edge part of tungsten probe. AC voltages are applied across the electrodes (gold electrode and tungsten probe). SEM image of CNT attached at the apex of tungsten probe. Scale bar is 500 nm. (e) Tungsten probe followed by MWCNT is fixed on substrate using carbon tape and cutting, welding and capping was done using FIB. Carbon is deposited at the junction of MWCNT and tungsten so that it is firmly attached (welded) with tungsten. For stability purpose, excess length of CNT is cut-off. Gallium ions are used to cut the excess length of CNT. (f) Carbon atoms are deposited on the top of CNT to form a ball-capped shape to make it suitable for scanning of via holes. (g) Attachment of FIB treated tungsten probe with the quartz tuning fork.

electrodes using 2 M KOH aqueous solution. Keithley 32,220 function generator is used to apply DC voltage across the electrodes. By applying the DC voltage, redox reaction occurs at the electrodes and tungsten tip begins to narrow. The shape of the neck depends upon the meniscus formed at the air-liquid interface. The experimental setup for electrochemical etching is shown in **Figure A3**. Reaction takes place can be described by the following equation [26].

At anode

$$6H_2O + 6e^- \rightarrow 3H_2(g) + 6OH^-$$

At cathode

$$W(s) + 8OH^- \rightarrow WO_4^{2-} + 4H_2O + 6e^-$$

Overall, reaction can be summarized as:

$$W(s) + 2OH^- + 2H_2O \rightarrow WO_4^{2-} + 3H_2\ (g) \tag{1}$$

The hydrogen gas bubbles produced during the etching should be avoided as it disturbs the meniscus which affects the shape of probe. Due to this, glass slide is placed between the electrodes. During the etching process, variation in the thickness of tungsten wire and current was monitored using the CCD camera and oscilloscope. When the current drops below the set point, shut-off circuit trigger the solid state relay (shut-off time is 0.5 ns), sharp edge of the tungsten wire is formed and hence in this way over-etching does not happen. The home-built automatic cutoff circuit to avoid over-etching is used as shown in **Figure A1**. After the wire has been etched, it was washed with the hot distilled water (D.I) to reduce the oxide layer around the edge and examined under the scanning electron microscope. Tips with the sharp edges are reserved for next step (attachment of MWCNT with tungsten probe) as these are the best candidates used for D.E.P.

2.2 Functionalization and multiwall carbon nanotubes attachment with tungsten tip using dielectrophoresis

We have used commercially available MWCNTs (Sigma Aldrich). First two MWCNTs samples having 110–170 nm (10 μm in length) and 60 nm (15 μm in length) diameter are grown by chemical vapor deposition (CVD) and 20 nm (30 μm in length) diameter MWCNTs was grown by arc discharge method. The reason why we used the un-purified carbon nanotubes is to reveal that anyone can easily make the carbon nanotubes tips in one's laboratory, because MWCNTs made by arc discharge are very difficult to purify. To obtain a single MWCNT, fictionalization of MWCNTs is good choice. Other than that, it is difficult to attach single MWCNT on tungsten tip. Firstly, to functionalize the MWCNTs (with different diameter as 100–170 nm, 60 nm, 20 nm), 1 g of PVA was completely dissolved in 60 ml of water at room temperature. After that, 0.2 mg of MWCNTs was added in to the above-mentioned solution and mixture was stirred at 70°C for four 12 h. Secondly, centrifugation machine was used to remove the impurities or particles from the CNT-PVA solution. Centrifugation at 4000 RPM for 30 min was done several times unless the solution becomes clear and no CNT left in solution. Thirdly, to remove excess PVA from PVA-CNT solution, filter paper Anodisc 47 (Whatman) with the pore size 200 nm and diameter 47 mm is used. Filtered CNT-PVA are then collected and dried under vacuum.

Finally, CNT-PVA solution is sonicated in an ultrasonic bath for 6 h to uniformly disperse in solution. After attachment of carboxylic group, MWCNTs were investigated by Fourier Transform Infrared spectroscopy (FTIR). **Figure 2** shows the FTIR spectrum of PVA treated MWCNTs in which peaks appears at 3425 and 1390 cm^{-1} are identified with O-H bond bending and stretching. The frequency associated with these vibrations is due to hydrogen bonding linked with functional groups. The peak observed at 1260 cm^{-1} is due to C–O bond stretching vibrations in phenol and alcohol. Another peak appeared at 1663 cm^{-1} is due to C=O stretching of functional group (COOH) considered at the surface of MWCNTs. In addition, C–H stretch vibrations correspond to 2856 and 2924 cm^{-1} band. Thus, FTIR is a powerful tool to see the chemical changes in the carbon system. MWCNTs are attached to the tungsten tips not only to increase the spatial resolution but also to increase the aspect ratio as well. According the literature survey, many groups have used CNTs as a scanning probe. Some of them used direct grown of CNT on cantilever or by picking the single CNT using special system installed within the SEM.

However, these instruments are costly or require special techniques to attach single MWCNT on AFM probe. Here, we have used dielectrophoresis method which is not only the cost-effective process but also a time saving and a simple process to attach the CNT. To attach the single MWCNTs on chemically etched tungsten probe, sharpness of the tip, applied frequency and diameter of the tip are the key parameters. One electrode is the etched tungsten wire while other is gold ring electrode. Using micropipette small droplet of functionalized MWCNTs is dropped into the gold ring electrode and with the help of micro-manipulator, the tip is brought closer to MWCNTs solution droplet and form a meniscus which cover only the edge part of etched tungsten tip. Using this technique, possibility to attach the CNT at the edge is increased. Otherwise, CNTs may attach at the surface of tungsten tip. Alternating voltages (6 volts at 2.2 MHz) were applied to the electrodes causes to generate the electric filed and CNTs were began to align parallel and eventually attached to the sharp edge of tungsten tip. However, at this stage MWCNT cannot be used for scanning purpose because CNT is attached with tungsten probe using Van der Waals and not sufficient to hold it during scanning. The homemade 3D AFM setup along with all electronics components is shown in **Figure A2**.

2.3 MWCNTs welding and cutting using FIB

We have utilized a focused beam of gallium ions (FIB) to align and attach the CNT on tungsten tip. The FIB system used (QUNTA 3D FEG) is a dual beam type

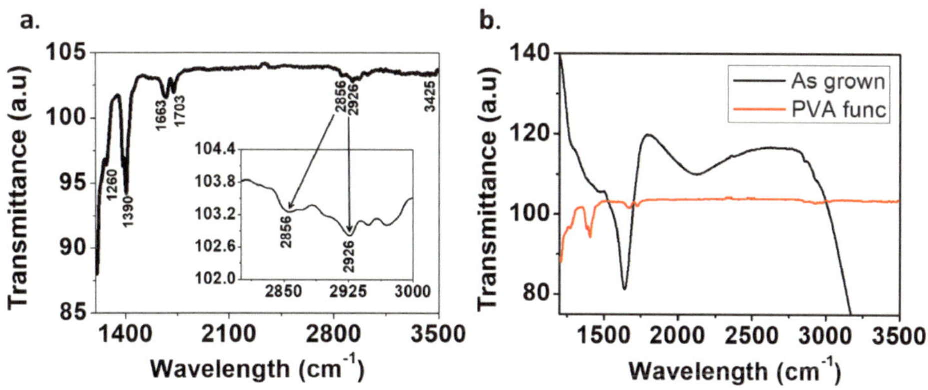

Figure 2.
Functionalization of MWCNT. (a) FTIR spectra depicts attachment of functional groups in CNTs: Peaks reveal 1200–3500 confirm the presence of functional group in purified CNTs. Inset image clearly shows the peaks at 2856 and 2926. (b) Comparison between the PVA-functionalized and as grown MWCNT.

where the electron and ion column are placed vertically (at 0°) and at an angle 52°, respectively. Alternatively, the same viewing angle of another gun can be obtained by tilting the sample back and forth by 52°.

Using the FIB process, CNT can be welded and shorten to use for AFM scanning. Our Fabrication of MWCNT based AFM probe comprises of four steps. Firstly, using electron beam induced deposition (EBID) hydrocarbon, CNT is welded on the etched part of tungsten tip (**Figure 3(b)**). The carbon atoms are deposited at the interface of tungsten which makes the CNT to attach firmly with tungsten. Secondly, CNT is aligned in the FIB system through the ion beam bending (IBB) [27] (**Figure 3(c)**). FIB is irradiated from the top. In the complete process of CNT alignment, acceleration voltage was set to 30 kV, resolution was 512 pixels × 442 pixels and FIB image magnification was 5000. Beam current as well as acceleration voltages can be increased to reduce the alignment time, however, should be optimized to avoid sample damage. We have used this method on 110–170, 50–80 and 20 nm CNT and it has been found that exposure time (for alignment process) to MWCNT is directly proportion to the diameter of MWCNT. For example, MWCNT with the diameter 20 nm takes less exposure time as compared to 50–80 or 110–170 nm MWCNT. The variables that affect the alignment process includes FIB magnification, exposure time, beam current and accelerating voltage. To produce good welding, many combinations can be made that produce good results. On the

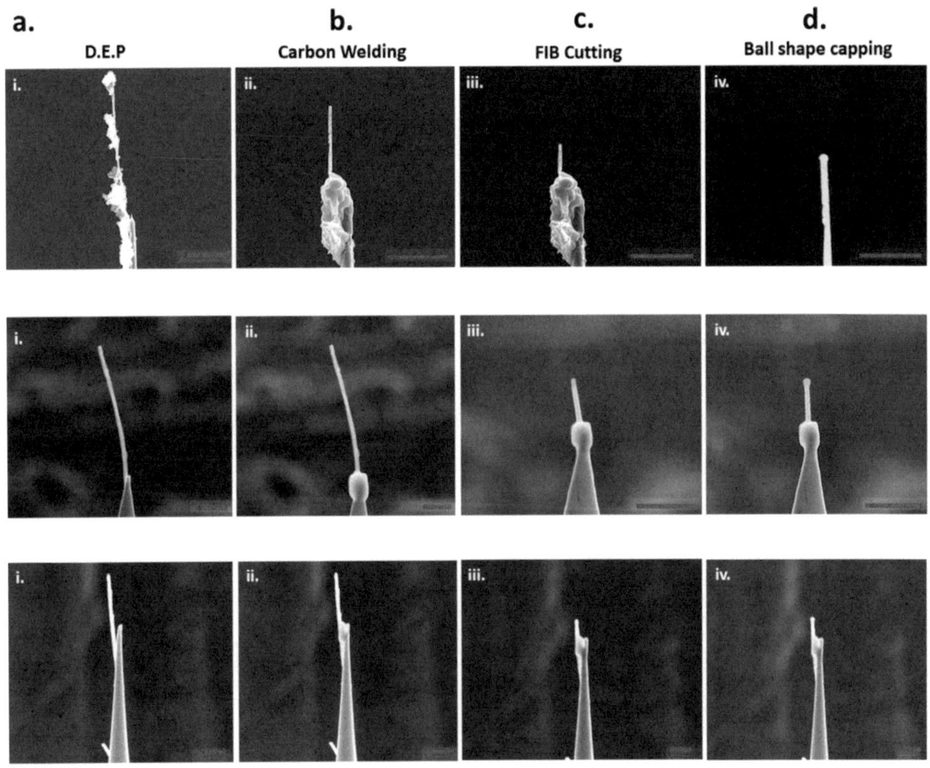

Figure 3.
Fabrication of MWCNT AFM probe. (a) As attached using DEP, some impurities are also attached with MWCNT which can be removed using lower current ion beam. (b) After hydrocarbon deposition (welding) on the interface of MWCNT and Tungsten probe. (c) Alignment of MWCNT using IBB process (d) using gallium ion bombardment extra length of CNT is cut off. To give some offset from the sidewall (in case of AAO sample or samples having vertical side walls), ball shape (using hydrocarbon) was deposited on the top of MWCNT. Diameter of MWCNT is 20, 60 and 120 nm. Scale bar is 1 μm.

other hand, all these parameters are carefully chosen while considering the CNT diameter under investigation, length and initial orientation. Thirdly, gallium ions are used to shorten the CNT (**Figure 3(c)**). After the selection of suitable length of CNT, i.e. we choose 300, 500, 700 nm and 1μm length, extra length was cut-off using gallium ions (**Figure 3(d)**). **Figure 3** shows the welding, alignment, cutting and ball shape capping of CNT on tungsten probe.

After that, the tip is shorten to less than 0.5 mm and attached to one end of quartz tuning fork (QTF) using the Torr Seal®. Quality factor, Q, of 45,000±500 in air was routinely obtained. Q factor of TF should be as higher as possible to obtain the highest sensitivity and resonance frequency, f, was shifted from 32,768 Hz to 32,310 Hz. QTF was excited at 15 mV which corresponds to 2.85 nm free oscillation amplitude.

3. Results and discussion

Herein, we have validated the working of the algorithm to measure the 3D profile of different samples (AAO and silicon pillars). Proposed algorithm enjoys all the advantages of conventional AFM. In addition to this, it has the capability to the measure the high aspect ratio features present above or below the sample. Here, we have used AAO sample to measure the internal morphology of the holes and 3D scanning of silicon pillars. To measure the internal morphology of the hole, firstly, the raster scan is performed using the MWCNTs based AFM probe: MWCNTs probe is attached using the FIB process on the Tungsten (W) apex which is further attached with the one prong of quartz tuning fork. Once raster scan is completed, feedback loop is turned off to start the 3D scanning of the hole. Positions of the holes and protrusion can be clearly seen from the raster scan image. Secondly, the algorithm is executed to scan the internal topography of via hole. Proposed algorithm can be summed up in four steps: Finding the location of hole(s), moving the tip to the bottom of the hole to be scanned internally, finding the first sidewall along the lateral direction, making the motion corresponding to the sidewall detected.

3.1 Finding the location of holes

Location of the hole can be founded in various ways. One of the easiest ways is do the raster scan over certain area at high resolution. However, due to the high aspect ratio of the AAO holes, the tip wear might occur which is not suitable to do the 3D scanning of the hole. For 3D scanning of the hole, the tip should have perfect symmetry so that the accurate imaging can be done inside of the hole. Other method is to do the force-distance curve or mapping, amplitude-distance curves or mapping over a certain area i.e. doing the amplitude-distance curve at each pixel and recording the coordinates and Z position. Lowest Z scanner value can be marked by red color and highest value with blue. Each individual pixel locally quantifies physical properties and interactions. By this, lot of information (elasticity, adhesion) can be mapped directly of the sample internal topography. In the **Figure 4**, we have done the mapping of certain area of AAO in which location of the can be clearly seen with different color contrast. In **Figure 4(i)**, FD curve is done on solid surface which has sharp change at 450 nm and on contrary side, **Figure 4(ii)** has rapid decrease in amplitude at 201 nm which shows which shows the tip can penetrate up to 249 nm. However, at this stage, tip is restricted not to go beyond the set point level. Later on, once the location of the hole is determined, tip can be pushed deep inside to measure the exact depth of the hole.

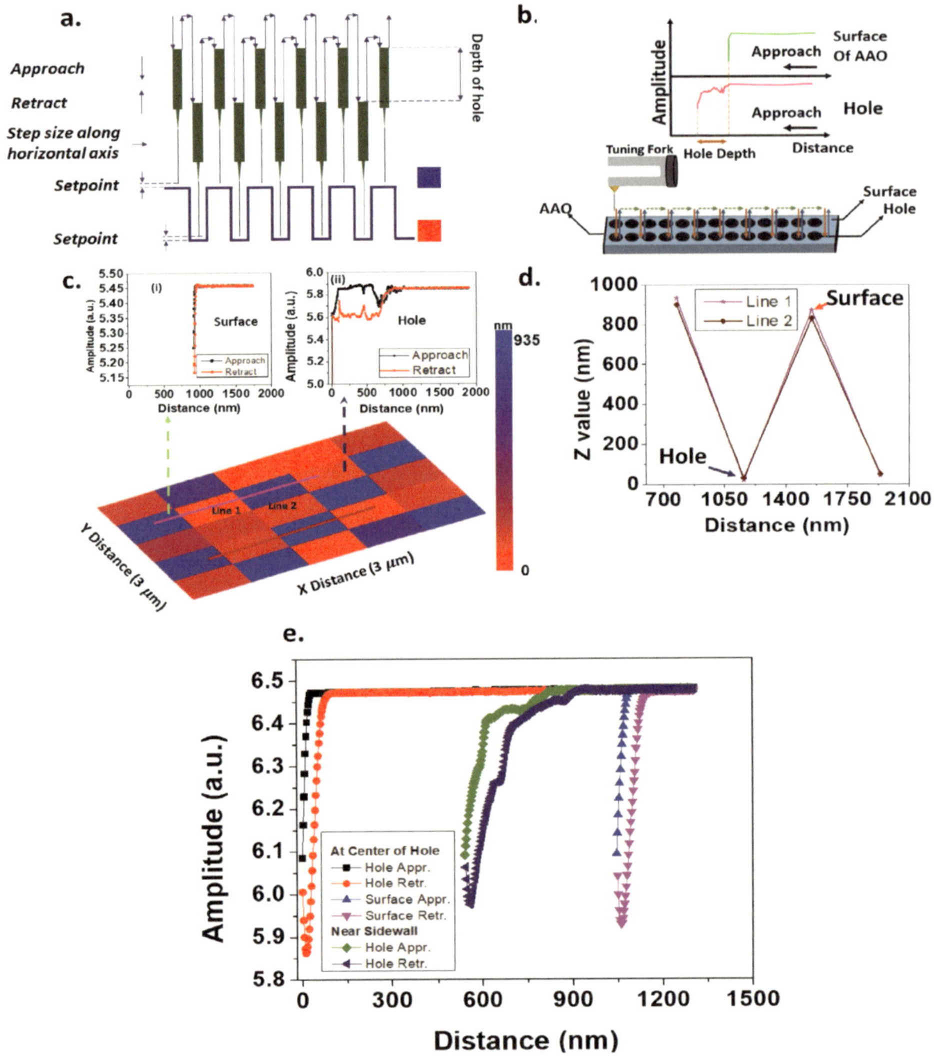

Figure 4.
Finding the location of holes by mapping. (a) FD mapping scheme in which the highest Z scanner value (surface or protrusion) is marked by blue color and lowest scanner value (hole) is marked by red color. (b) Using QTF, followed by MWCNT and tungsten probe, amplitude-displacement curve was measured at different position of AAO sample. Rapid decrease in amplitude (without wrinkle-top graph in green) shows the surface and amplitude-distance measurement with wrinkles shows the location of hole (lower graph in pink). Difference in both points (where the probe touches the surface and bottom) gives the depth of hole. (c) Mapping was done using 20 nm MWCNT on AAO surface over 3 micron area. From the graph (i) and (ii), it is clear that when Amplitude-displacement curve done on surface it does not have any hump while measured on hole, it shows some humps which corresponds to the Van der Waals interaction of probe with the inner sidewall of hole. CNT goes 935 nm gives the depth of the hole (d) Line 1 and line 2 shows the hole profile or depth profile of hole. From this data depth of the hole can be estimated which is 480 nm (Sample AAO – Actual depth—500 nm as purchased). (e) FD curve taken at surface, at the centre of hole, near the side wall of AAO hole.

3.2 Finding the bottom and depth measurement of via hole

First, for scanning the internal topography of the hole, one can click at the center of the hole and scanner will move at the center of the hole. Algorithm is designed in such a way that mouse click will bring the scanner to the required coordinates (at the center of the hole). Then the tip will move along $-Z$ axis to reach to the bottom of the hole. If the hole is not uniform, then the possibility still exist that the tip might not touch to the bottom due to the many local minima present inside. As the tip pierces in to the hole, amplitudes remains constant. However, if the tip is not at the center or non-uniform hole, then it might experience van der Waals interaction between MWCNT and sidewall of AAO sample as shown in **Figures 4(c)—(ii)** and **5**. Zongwei et al. [28] have used the DVD surface for force curve measurement at the pit edge as well as on plane surface and they have found that the ripple and wavelike wrinkles can be easily occurred if the FD measurement was done at the pit's edge. However, this van der Waals forces might restrict the tip to reach at the bottom. To resolve this, the FD is measured away from the sidewalls so that van der Waals are less likely to appear when the tip is at the center of the hole as shown in **Figure 4(e)**.

Once can interpret the interaction of MWCNT to the sidewall by observing the FD curve profile as shown in **Figure 5**. At some point, amplitude tends to decrease due to interaction of AFM probe with sidewalls. As the interaction dominates the setpoint, i.e., amplitude decreased more than the setpoint, tip is retracted. **Figure 4** shows a good method to find the exact location of the hole. However, to measure the actual depth of the hole, this is not the point where the tip has to stop. We try to push the tip inside of the hole by tracking the amplitude change and it has been found that when the amplitude drops more than 60%, then we can say that it might be the bottom of the hole or in worst case the tip has been stuck to the hole. However, this problem can be overcome by measuring the FD curve in near surrounding of hole and recording the XY-coordinates with scanner value. Coordinate with the lowest scanner value might be bottom of the hole. After that, the scanner will move to that coordinate having minimum $-Z$ value corresponds to the minimum position of the hole. We have used the AAO sample with pore diameter 350 nm and different depth (300 nm, 500 nm, 1 μm with tolerance±5%). FD curve is performed on different which gives the real depth of the sample. With 120 nm sample, depth of 1st AAO sample was measured 915 nm, 485 and 285 nm for second sample which corresponds to the vendor supplied values. In addition to this, we have done FD curves at the same position many times to check the reproducibility of the data as well as tip wear was less likely to found (Appendix). More than 10 times FD curves are performed and the difference between the first and final

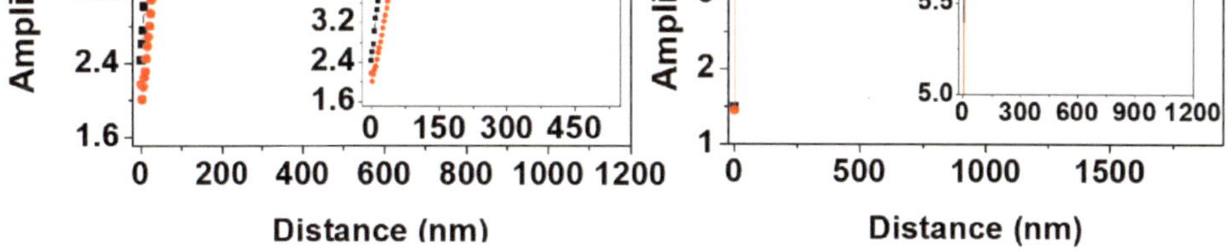

Figure 5.
MWCNT interaction with sidewall of via holes. (a) Amplitude-displacement curve was performed on AAO sample with depth 300 nm with pore diameter 350 nm. In 300 nm sample, first drop in amplitude shows that the tip entered in to the hole and it might experience the sidewall or AD curve is done near the sidewall. As tip moves towards the bottom of the hole, at some point, rapid drop in amplitude occurs which shows the tip touched to the bottom. (b) FD curve on AAO sample with depth 1 μm and pore diameter was 350 nm. The depth was scanned approximately 930 nm. In both case, 120 nm MWCNT was used.

depth was only 10 nm. However, the Van der Waals forces may vary measurement-to-measurement as the tip goes deep inside the hole. Moreover, there was no significant damage to the apex of the tip. In **Figure 5(c)**, 120 nm shows the best result in stability as well as depth measurement and 3D scanning of deep holes. On the contrary side, 20 nm is suitable for 3D scanning of shallow holes as well as protrusion.

3.3 Finding the first sidewall of hole

As tip find the real bottom of the hole, it will start moving in either direction (along +X axis) to find the first sidewall. Starting from the center (bottom) of the hole, tip start piercing the direction of sidewall by moving forward and backward along all the axis individually as shown in **Figure 6**. At each step, either the sidewall is present or not, its location is stored in variable. This is important because at some point the tip might not found the sidewall, then the direction of motion of probe will be decided by 'location of previous sidewall' and it can only be done storing the direction of sidewall at each step. In addition to this, if the tip has lost the sidewall (due to drift), then tip will move to the nearest sidewall present by tracking the history of motion that probe has moved.

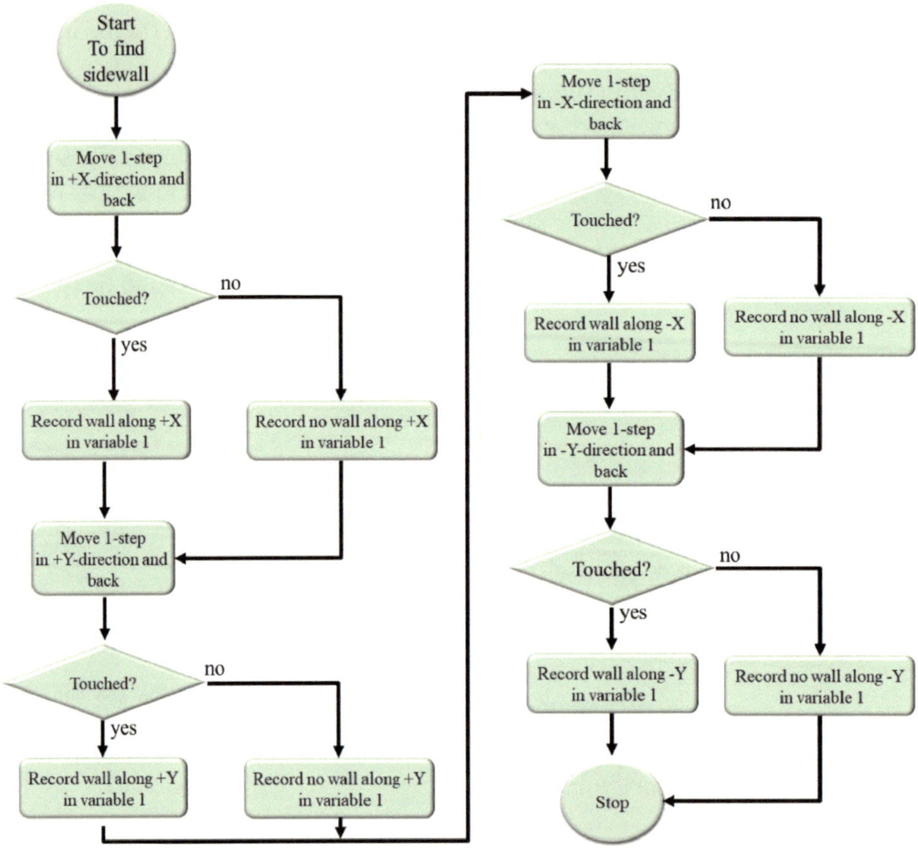

Figure 6.
Algorithm to find the direction of sidewall. As the tip reaches at the bottom of the hole, it starts to find the sidewall in all direction by moving forward-backward motion and tracking the amplitude. Any encounter with the sidewall causes to decrease in amplitude. Direction of the sidewall is stored in the variable1 to make the movement accordingly.

For example, tip move one step in +X direction and move back by tracking the amplitude. If the amplitude remains the same, it means there is no sidewall along +X direction. Similarly, the same phenomena is repeated by scanner for other axes (+Y, −X, −Y). The direction of sidewall is checked at every step. If the sidewall is detected in any direction, the corresponding motion will be performed by the scanner (Appendix).

3.4 Making the motion corresponding to the sidewall detected

As the tip is scanning to find the sidewall along all the direction, at some point, the sidewall will be detected along +X axis and according the algorithm the tip will move along +Y direction. After that, the direction of sidewall will be checked again. If the sidewall is along +Y direction, then the motion will be along -X axis. Similarly, following the counterclockwise scheme, motion will be along +X direction if the sidewall will be in -X direction. If at next step, tip found no sidewall then the scanner will move in the direction where the last sidewall was. This technique helps the tip to strictly follow the boundary of the hole and with the motion decided by the algorithm, the perfect boundary of the feature can be determined sophisticatedly. The situation is more interesting when the tip experience two or three sidewalls and motion of the scanner will be followed as described in **Table A1** (Appendix). However, if all sidewall direction is found then tip is assumed to be trapped and should be retracted along +Z direction for its safety. **Figure 6** shows the working strategy to find the side wall of algorithm to track the boundary of the hole and tracking the amplitude as shown in **Figure 7**. To know the exact shape of the boundary, X and Y step size can be varied. However, there are two types of steps, i.e., jump step and motion step. The first one is jump step along any axis to check the direction of sidewall, which is usually greater than the motion step size. This is necessary to make the tip at a safe distance so that the tip not wear or to minimize the van der Waals forces between the tip and the sidewall. The motion step can be decreased to few nanometers for high resolution and to measure the exact shape of feature. After that, arithmetic is done by the algorithm, by considering the jump step, to calculate the diameter of each complete rim scan by adding the jump step to diameter obtained from 3D scanning.

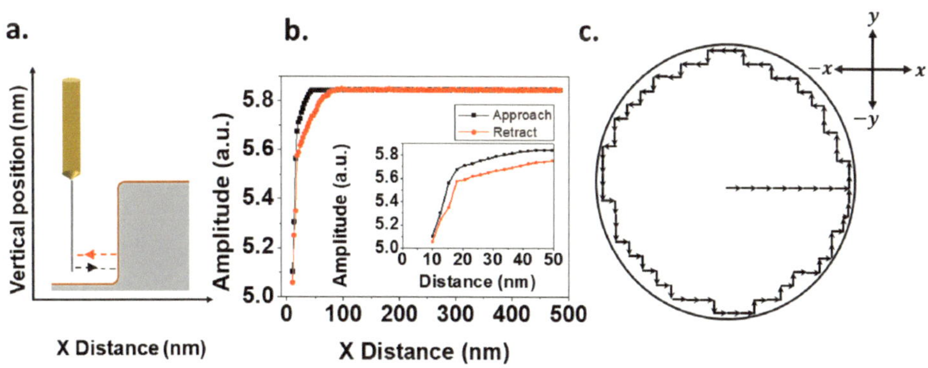

Figure 7.
Working strategy of Algorithm to find the sidewall and scan the via holes. (a) Algorithm record the amplitude before and after moving towards the position of sidewall. If there is sidewall, rapid decrease in amplitude is detected as can be seen in (b). (c) Motion of the tip to follow the boundary of the via hole. The step size can be varied for high resolution. Tip start from the centre of the hole and start moving along +X direction to find sidewall. As it finds the sidewall, it starts following the boundary of hole in xy plane in counter clockwise.

3.5 3D scanning of hole AAO

To test the feasibility of the proposed three-dimensional imaging algorithm of via holes as well as protrusion, we first done the simulation in the LABVIEW software (Appendix). The versatility of the software can be seen in such a way that any irregular shape of hole as well as protrusion can imaged using this algorithm.

The above-mentioned algorithm is implemented in LabVIEW software and tested on Anodic Aluminum Oxide (AAO) sample bought from Vida Biotechnology CO. LTD. We have three samples with 300 nm, 500 nm and 1 μm depth. Diameter of the holes were 350 nm. For protrusion, silicon pillars bought from NT-MDT having the approx. Height of 500 nm and 3 μm periodicity along lateral and 2.1 μm along diagonal direction. Dimension of the sample was as: 2.5 × 2.5.

Figure 8(b) shows the 3D image of AAO hole having the depth of 650 nm when the algorithm is applied to scan the internal topography of hole. Z step is set to 34 nm, however for higher resolution it can be decreased. Moreover, **Figure 8(c)** is the AFM image obtained using conventional AFM which cannot go deep more than 530 nm due to low aspect ratio. This shows our 3D AFM algorithm has strong potential to scan the features having high aspect ratio. As the tip goes deep inside the hole more than a micron, Van der Waals forces between the tip and sidewall also increases which may lead to the instability of the CNT. However, this problem can be overcome by using increasing number of layers of CNT so that it will be rigid, less flexible and stable for 3D scanning.

As the location of the holes (coordinates) is evident from the FD mapping (Z scanning) as shown in **Figure 4(c)**, next holes can be easily scanned. For example,

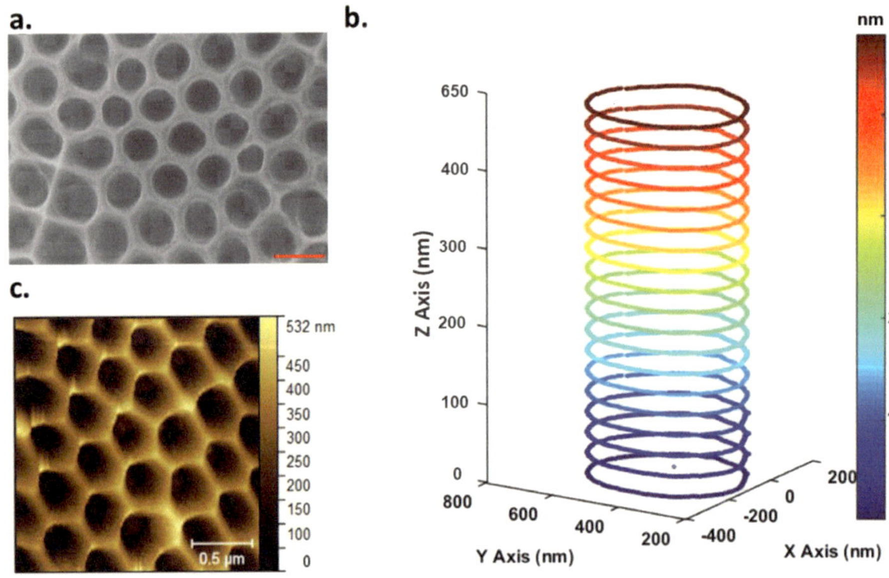

Figure 8.
AAO image with the depth 1um and pore diameter 350 nm. (a) SEM image of AAO. (b) 3D image of inside of hole which shows that the tip scanned 650 nm depth. Tip goes deep inside of hole and find the bottom where it has to start the 3D scanning as marked by the dot in figure. Then it moves in either direction (+X) to find the sidewall. Once the sidewall is found, tip start to follow the boundary. After completing the one rotation (360°), scanner retracted along +Z direction to start new rim scan. When the straight line is found during scanning, it means that the tip is escaped from the hole or tip finished scanning the hole (straight line not shown in figure). (c) Raster image obtained from conventional AFM used to find the location of hole. Due to the low aspect ratio of the conventional AFM, it cannot go deep and limited to 532 nm, however, the original sample depth is 1 μm. Scale bar is 500 nm.

once the 3D scan of first hole is completed and tip is escaped from the hole (in line profile straight line shows that the tip is escaped from the hole), using the coordinates of next hole, tip moves along –Z direction and reaches at the bottom of the hole. Once it touches to the bottom, 3D scanning is done in the same fashion as of first hole.

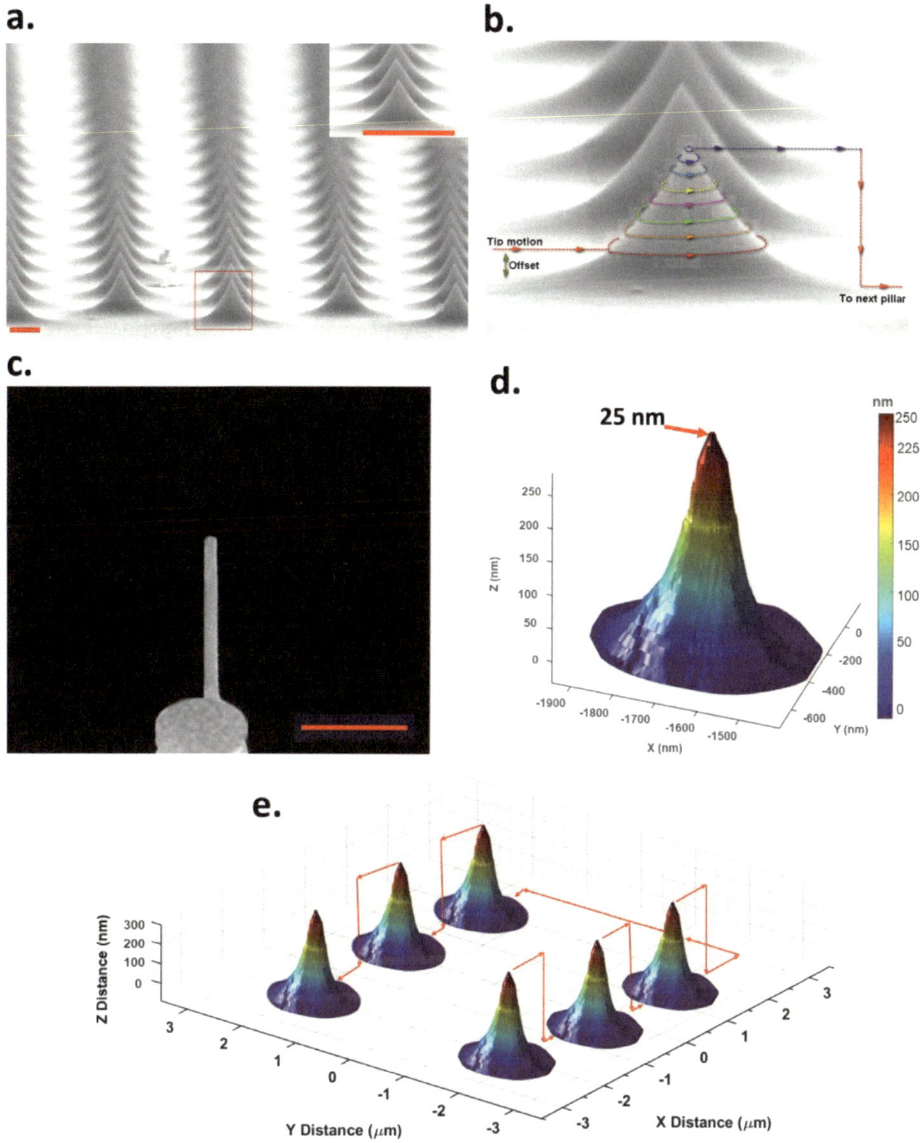

Figure 9.
Image of silicon pillars. (a) SEM image of silicon pillars. (b) Lines showing the scan direction of tip. As the tip comes near the surface it is retracted (offset) few hundreds of nm to avoid any interface of surface with the tip. As the tip is moving in +X direction to find the sidewall, at some point it find the sidewall of silicon pillar and according to algorithm, it start to find the boundary of the pillar. As once rotation completed, tip retracted along +Z direction to scan next rim. If the straight line is found in the scan, as shown in figure, it means that the tip has completed scanning the pillar and similarly next pillar can be scanned using the same scheme. (c) FIB treated 20 nm MWCNT (without capping) used to obtain the 3D profile of silicon pillar (d) 3D image of silicon pillar obtained from the 3D scanning algorithm with the step size 19 nm along X, Y direction and 20 nm along +Z direction. However, the step size along all axis can be even smaller for high resolution. The scale bar is 500 nm. The edge of the tip is limited to the radius of the probe. In our case, tip radius of arc grown CNT is 20 nm. (e) 3D image of silicon pillars: After completing one pillar, tip is brought near to the surface again (moved along –Z direction) to scan the next pillar. Motion of the probe is shown by the red arrow in figure.

In addition to this, algorithm is applied to obtain the 3D topography of the silicon pillars. Tip start to scan the pillar in either direction. Once it encounter the sidewall of pillar, it start to follow the boundary of the silicon pillar and on completion of one rotation, scanner retracted along +Z direction to complete the next rim scan. The process continuous unless it complete scanning the pillar. Resolution of 3D image depends upon the thickness of AFM probe. We have used 20 nm tip to scan the pillar which means the last (top edge) rim scan should be approximately equals to the diameter of the CNT as shown in **Figure 9**.

In addition to this, as the tip completes once pillar, the tip can be pushed deliberately near to the surface and moved in +X direction to find the second pillar and so on.

4. Conclusion

MWCNT tips are one of the best candidates to further investigate different materials in AFM imaging world, enhancing tip lifetime, eliminating the artifacts and increasing the resolution by decreasing the tip-surface forces. Many algorithms has already been proposed and implemented to scan the deep trenches, sidewall roughness and to measure other 3D parameters. However, none of them can able to 3D scan the deep via holes, which has the vertical sidewalls and real time 3D imaging protrusions features like silicon pillars. As TSVs have potential application in NAND flash memories, 3D meteorology, and 3D system integration. So, exact location of via holes, depth and internal roughness of sidewall should be measured accurately. In this work we have proposed an algorithm which has ability to scan the features above and below the surface such as TSV and silicon pillars. We have tested the algorithm on AAO in which location of holes, depth of holes as well as internal roughness of sidewalls are measured. In addition to this, real time 3D scanning has been performed by the algorithm on silicon pillars (protrusion) and it has been found that it is a best choice to measure the features having high aspect ratio. We believe that impact of this algorithm will not be limited to only the semiconductor field. In general, getting the 3D topography is a crucial issue for measuring the parameters like surface roughness, hole depth and locations etc. So far, several algorithms has been proposed which employs special type of probes, but these are specially designed to measure the structures like trenches as discussed earlier. The new method described here is a unique since it can image any kind of feature (present above or below the surface).

A. Appendix

A.1 Functionalization of MWCNT

To functionalize the MWCNTs, 1 g of PVA was completely dissolved in 60 ml of water at room temperature. After that, 0.2 mg of MWCNTs was added into the above-mentioned solution and mixture was stirred at 70°C for 4–12 h. A centrifugal separator was used to remove the impurities or particles from the CNT-PVA solution. Centrifugation at 4000 RPM for 30 min was done several times unless the solution becomes clear and no agglomerated CNTs left in solution. To remove excess PVA from PVA-CNT solution, filter paper (Anodisc 47, Whatman) with the pore size 200 nm and diameter 47 mm was used. Filtered CNTs-PVA are then

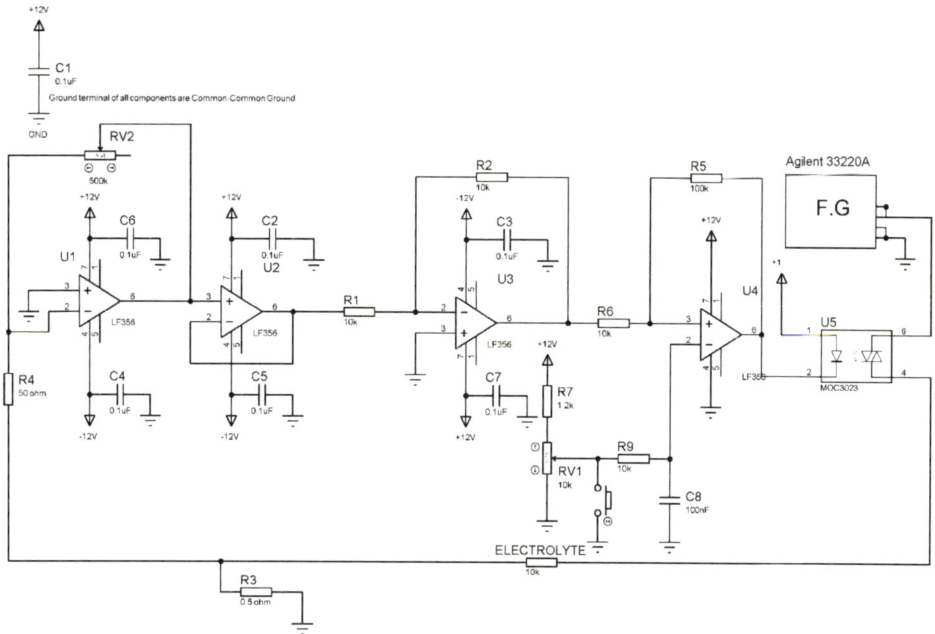

Figure A1.
Automatic shutoff circuit for electrochemical etching of tungsten wire. Etching circuit is divided in to two resistors, i.e., 0.5 and 50 Ω (Connected to real and virtual ground, i.e., input to the current amplifier). Second stage amplifier is a voltage follower which makes the signal stable during the etching. Third stage amplifier is an inverting amplifier which inverts the signal from the first stage so that it can be compared to the set point. At the very last moment of etching, when lower part of the tip is about the drop, there might be some fluctuation in the signal and tip can be blunt or uniformity of the tip can be disturbed. To avoid this, Schmitt trigger is used at the comparator which makes the signal stable at the last moment of etching. Solid state relay is used to reduce the shut-off time. This automatic cut-off circuit has the shut-off time 0.5 μs.

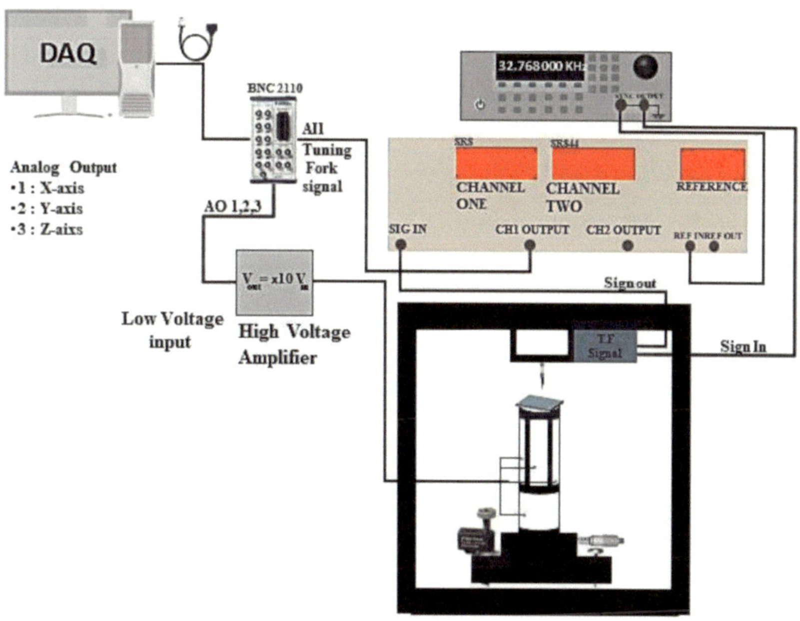

Figure A2.
Homemade setup for 3D AFM, composed of tuning fork sensor, lock-in amplifier, PC based digital controller, and a tube scanner.

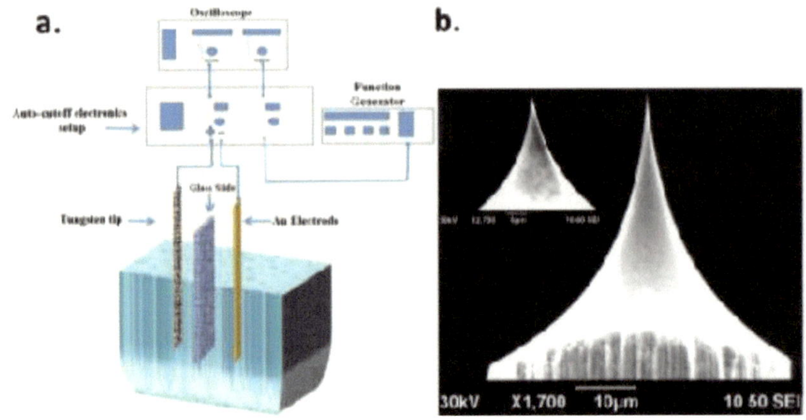

Figure A3.
Experimental setup for tungsten wire etching. (a) Gold wire is used as anode and tungsten wire acts as cathode. As the etching begins, bubbles travel from anode to cathode which may deform the meniscus shape. A glass slide is placed between the electrodes to avoid this problem. (b) SEM image of tungsten wire after it has been etched using KOH solution. Once the wire has been etched, it is washed with hot D.I water to reduce the oxide layer around the etched part. We have used 2 M concentration of KOH solution. Etching takes approximately 70 s to complete the etching. After etching 5–6 tips, the solution needs to be replaced. Otherwise, etched shape may be not uniform or the surface of the tungsten becomes rough. Almost all of tips are produced uniformly and the average tip radius is found to be 80 nm.

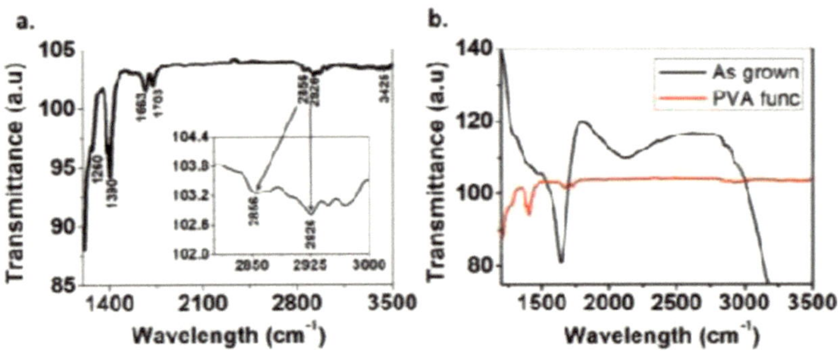

Figure A4.
FTIR spectra for functionalized MWCNT. (a) FTIR spectrum depicts attachment of functional groups in CNTs: Peaks in the range of 1200–3500 confirm the presence of functional group in purified CNTs. Inset image clearly shows the peaks at 2856 and 2926. (b) Comparison between the PVA-functionalized and as grown MWCNT.

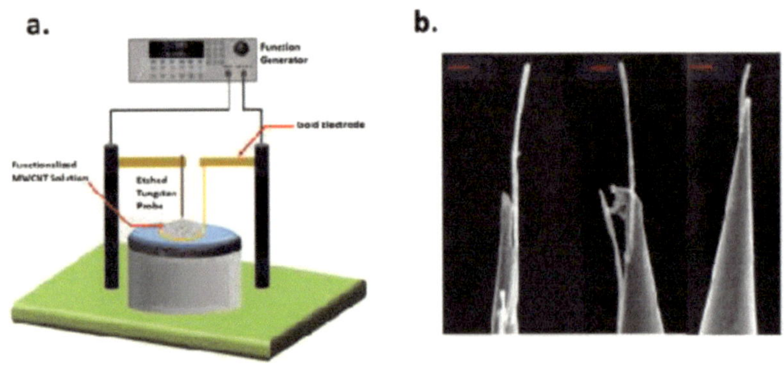

Figure A5.
Dielectrophoresis setup for attachment of multiwall carbon nanotube on chemically etched tungsten probe. (a) Function generator is used to apply AC voltage which generates electric field and consequently attracts the CNT towards the tungsten probe. (b) SEM image shows a single MWCNT attached to the tungsten probe via van der Waals forces. Scale bar is 500 nm.

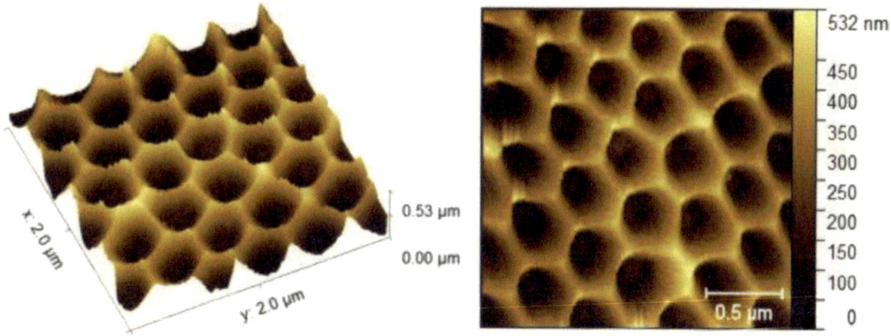

Figure A6.
AFM image of AAO sample obtained from home built AFM using the MWCNT. Raster scan mode was applied to find the location of holes.

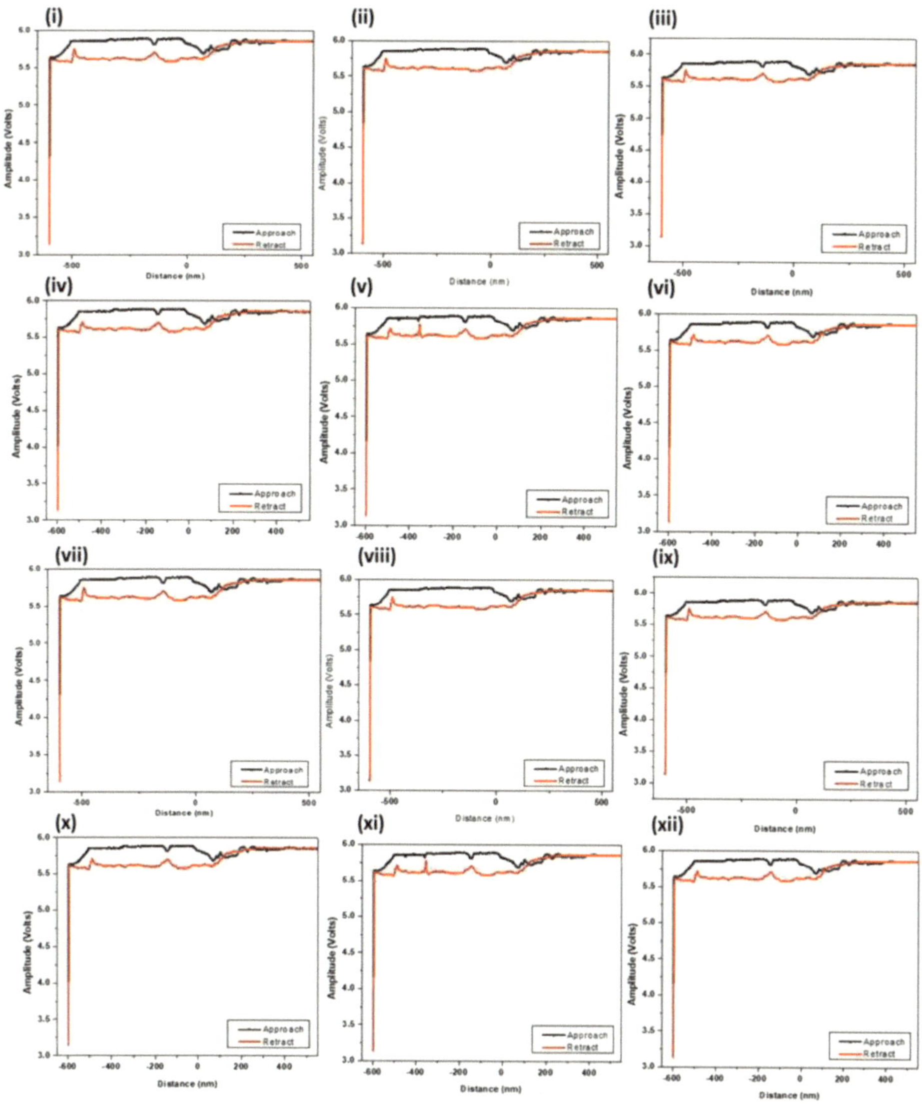

Figure A7.
FD curves were measured at the different holes in AAO sample having the depth of 1 μm. Sharp decreases imply that the tip touches to the bottom of the hole, and the depths are estimated as ∼935 nm similar from all data.

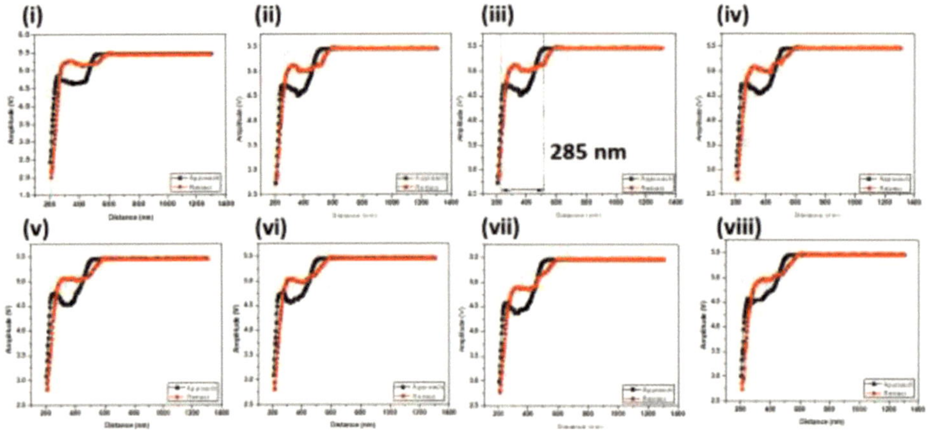

Figure A8.
FD curves are measured at different holes on AAO sample having the depth of ~300 nm. Measured data show that the depth is approximately 285 nm. Sharp increases indicate that the tip touches to the bottom of the hole.

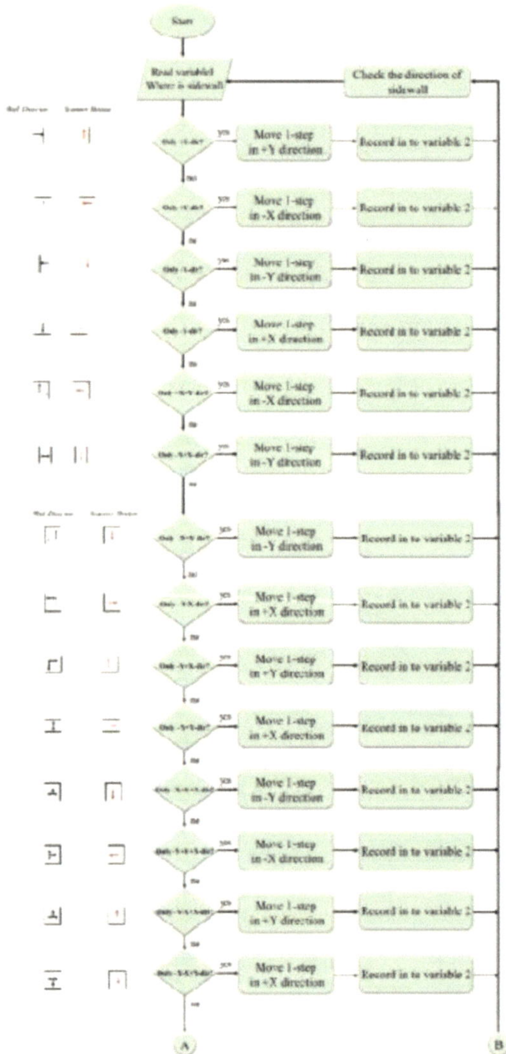

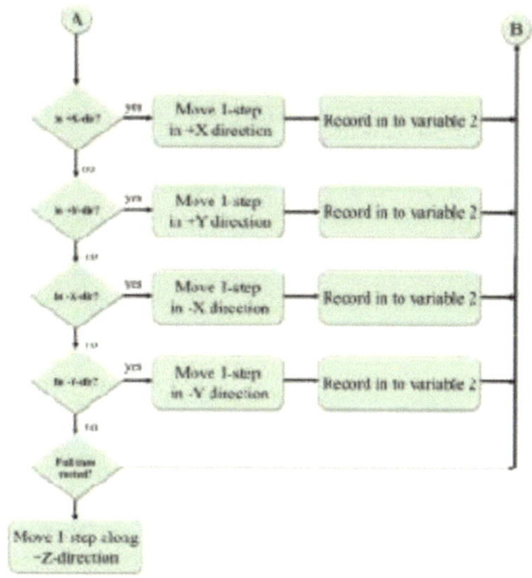

Figure A9.
*After the wall is encountered by the probe by tracking the difference in amplitude, motion of the scanner is determined according to **Table A1**. However, if the probe is unable to find the sidewall, it will move along the direction where the previous sidewall was. This gives the appropriate correction of the sidewall direction to follow the exact shape of the boundary. By this strategy, any shape of either hole or protrusion is scanned perfectly.*

S.#	Sidewall direction				Scanner motion
	+X	+Y	−X	−Y	
1	—	—	—	—	Previous direction of sidewall
2	—	—	—	SW	Along +X
3	—	—	SW	—	Along −Y
4	—	—	SW	SW	Along +X
5	—	SW	—	—	Along −X
6	—	SW	—	SW	Along −X
7	—	SW	SW	—	Along −Y
8	—	SW	SW	SW	Along +X
9	SW	—	—	—	Along +Y
10	SW	—	—	SW	Along +Y
11	SW	—	SW	—	Along −Y
12	SW	—	SW	SW	Along +Y
13	SW	SW	—	—	Along −Y
14	SW	SW	—	SW	Along −X
15	SW	SW	SW	—	Along −Y
16	SW	SW	SW	SW	Along +Z

SW: sidewall detected, "—": not detected.

Table A1.

collected and dried under vacuum. Finally, CNT-PVA was dissolved into DI water and the solution is sonicated in an ultrasonic bath for 6 h to uniformly disperse in solution. After attachment of carboxylic group, MWCNTs were investigated by Fourier transform Infrared spectroscopy (FTIR), as shown in **Figure A4**. The peaks observed at 1390 and 3425 cm^{-1} are identified with O–H bond bending and stretching, respectively [29]. The peaks observed at 1260 and 1663 cm^{-1} are due to C–O and C=O stretching of functional groups, respectively [29]. In addition, C–H stretching vibrations correspond to 2856 and 2924 cm^{-1} peaks [30].

A.2 Attachment of MWCNT by dielectrophoresis

MWCNTs are attached to the tungsten tips not only to increase the spatial resolution but also to increase the aspect ratio as well. To attach the single MWCNT on chemically etched tungsten probe, sharpness of the tip, applied frequency, and field intensity are the key parameters. One electrode is the etched tungsten wire,

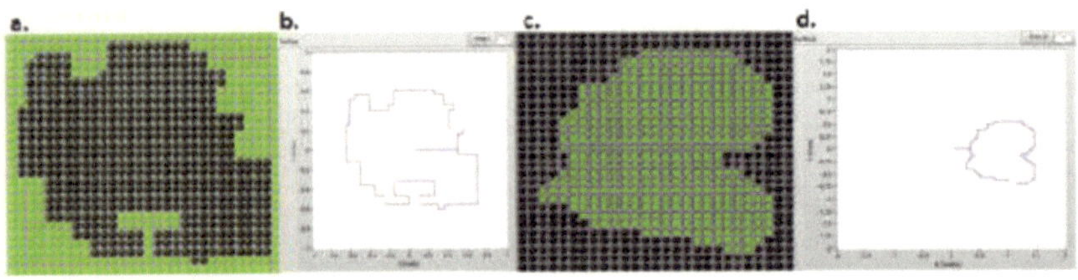

Figure A10.
Simulation results of the algorithm for complicated hole and protrusion. (a) A random shape hole was drawn to check the versatility of the algorithm, where dark portion indicates the empty hole. (b) Line profile shows a simulation result of the tip trance. The tip starts scanning from the centre and move in along +X axis direction to find the side wall. Once the sidewall is detected, tip start to follow the boundary of the hole. Afterwards, one complete rotation is completed. (c) A random shape protrusion was drawn, where green portion represents the protrusion. (d) The simulation result of scanning shows a line profile of protrusion. Working methodology to scan the protrusion is the same as the hole.

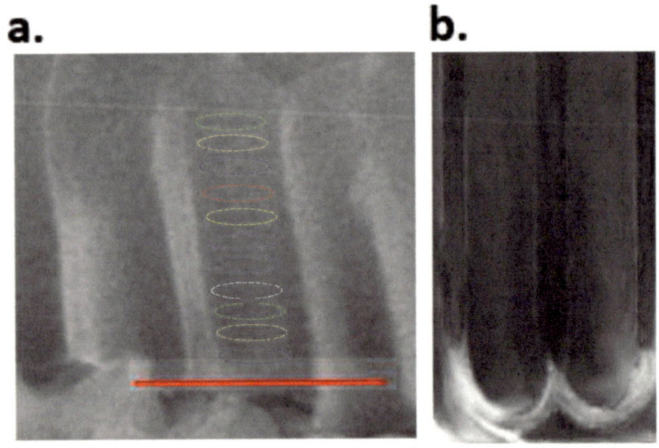

Figure A11.
Cross section views of fractured AAO samples (a) in the middle part (b) near the bottom. Circles show the expected 3D scanning traces. Tip starts scanning from the bottom and raised a step along +Z direction as one rotation completed. Scale bar is 1 μm.

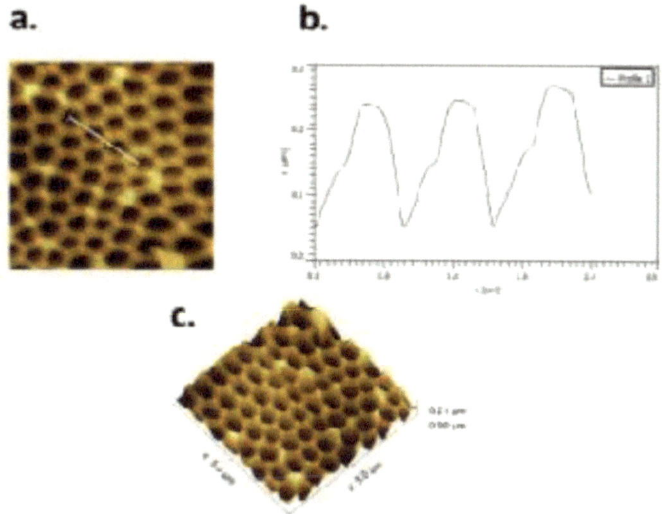

Figure A12.
Raster scan image of AAO sample using commercially available AFM with silicon cantilever. (a) Raster scan image with silicon cantilever showing that the AFM probe cannot go beyond 210 nm. (b) Line profile of the hole showing the depth of the hole is only 100 nm. However, original sample has depth around 1 μm. (c) 3D view of AAO sample.

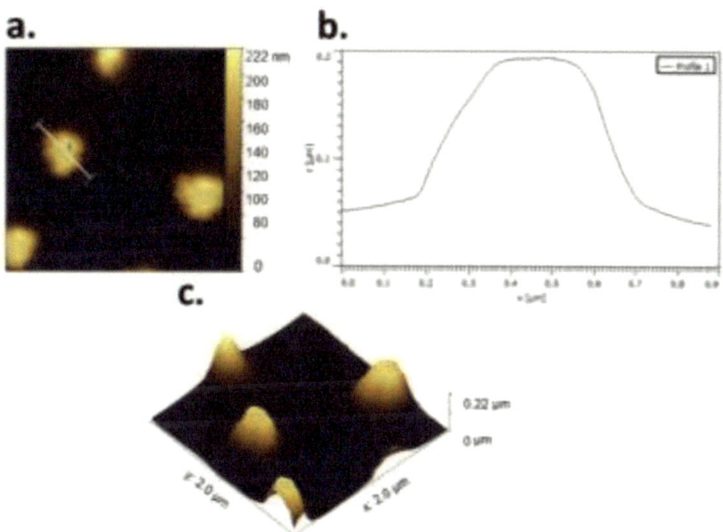

Figure A13.
Raster scan image of silicon pillars using commercially available AFM using silicon cantilever. (a) Raster scan image with silicon cantilever showing that the AFM probe cannot go beyond 222 nm due to low aspect ratio. (b) Line profile of the silicon pillar which shows the height of one pillar is 120 nm as well as broad peak unlike the real sample. However, original sample has height around 500 nm*. (c) 3D view of silicon pillars.*

while other electrode is a gold ring. AC voltages (6 V at 2.2 MHz) are applied to the electrodes, and CNTs are attached to the sharp edge of the tungsten tip. At this stage, van der Waals forces hold the CNTs, which are not strong enough to be used for scanning purpose.

Author details

Imtisal Akhtar*, Malik Abdul Rehman and Yongho Seo
Department of Nanotechnology and Advanced Materials Engineering, Sejong University, Seoul, South Korea

*Address all correspondence to: writetoimtisal@hotmail.com

References

[1] Benkart P, Kaiser A, Munding A, Bschorr M, Pfleiderer HJ, Kohn E, et al. 3D chip stack technology using through-chip interconnects. IEEE Design and Test of Computers. 21 November 2005; **22**(6):512–518

[2] Topol AW, La Tulipe DC, Shi L, Frank DJ, Bernstein K, Steen SE, et al. Three-dimensional integrated circuits. IBM Journal of Research and Development. July 2006;**50**(4.5):491–506

[3] Cho S-J, Ahn BW, Kim J, Lee JM, Hua Y, Yoo YK, et al. Three-dimensional imaging of undercut and sidewall structures by atomic force microscopy. Review of Scientific Instruments. 24 February 2011;**82**(2): 023707

[4] Miyashita H, Kanada H, Ogawa T. Three dimensional shape measurement of via-holes for the production of multi-layers circuit boards. Journal of The Japan Institute of Electronics Packaging. 1 July 2004;**7**(4):333–338

[5] Akhtar I, Rehman MA, Choi W, Kumar S, Lee N, Cho SJ, et al. Three-dimensional atomic force microscopy for ultra-high-aspect-ratio imaging. Applied Surface Science. 1 March 2019; **469**:582-592

[6] Nakamura M, Kitada H, Sakuyama S. Direct depth measurement tool of high aspect ratio via-hole for three-dimensional stacked devices. Journal of Surface Analysis. 2014;**20**(3): 182–186

[7] Fischer AC, Forsberg F, Lapisa M, Bleiker SJ, Stemme G, Roxhed N, et al. Integrating MEMS and ICs. Microsystems & Nanoengineering. 28 May 2015;**1**:1-6

[8] Wood J. The top ten advances in materials science. Materials Today. 2008;**11**:40–45

[9] Kim D, Sahin O. Imaging and three-dimensional reconstruction of chemical groups inside a protein complex using atomic force microscopy. Nature Nanotechnology. 2015;**10**:264–269

[10] Dufrêne YF et al. Imaging modes of atomic force microscopy for application in molecular and cell biology. Nature Nanotechnology. 2017;**12**:295–307

[11] Dai H, Hafner JH, Rinzler AG, Colbert DT, Smalley RE. Nanotubes as nanoprobes in scanning probe microscopy. Nature. 1996;**384**:147–150

[12] Wiesendanger R, Mulvey T. Scanning probe microscopy and spectroscopy. Measurement Science and Technology. 1995;**6**:600

[13] Wilson NR, Macpherson JV. Carbon nanotube tips for atomic force microscopy. Nature Nanotechnology. 2009;**4**:483–491

[14] Astrova E, Fedulova G, Zharova YA, Gushchina E. Side-wall roughness of deep trenches in 1D and 2D periodic silicon structures fabricated by photoelectrochemical etching. Physica Status Solidi C. 2011;**8**:1936–1940

[15] Hua Y, Buenviaje-Coggins C, Lee YH, Lee JM, Ryang KD, Park SI. New three-dimensional AFM for CD measurement and sidewall characterization. In: Metrology, Inspection, and Process Control for Microlithography XXV. Vol. 7971. International Society for Optics and Photonics. 20 April 2011. p. 797118

[16] Dai G, Wolff H, Pohlenz F, Danzebrink H-U, Wilkening G. Atomic force probe for sidewall scanning of nano-and microstructures. Applied Physics Letters. 2006;**88**:171908

[17] Ku Y-S, Huang KC, Hsu W. Characterization of high density

through silicon vias with spectral reflectometry. Optics Express. 2011;**19**: 5993–6006

[18] Jin J, Kim JW, Kang C-S, Kim J-A, Lee S. Precision depth measurement of through silicon vias (TSVs) on 3D semiconductor packaging process. Optics Express. 2012;**20**:5011–5016

[19] Lee JS et al. Multifunctional hydrogel nano-probes for atomic force microscopy. Nature Communications. 2016;**7**:11566

[20] Gu JM, Hong SJ. In: 2015 IEEE Electrical Design of Advanced Packaging and Systems Symposium (EDAPS). IEEE; 2015. pp. 27-30

[21] Orji NG et al. Tip characterization method using multi-feature characterizer for CD-AFM. Ultramicroscopy. 2016;**162**:25–34

[22] Liu H-C et al. Proceedings of SPIE: 69222J

[23] Choi J et al. Evaluation of carbon nanotube probes in critical dimension atomic force microscopes. Journal of Micro/Nanolithography, MEMS and MOEMS. 2016;**15**:034005

[24] Kneedler R et al. Proceedings of SPIE:905–910

[25] Seo Y, Park J-Y, Kim K, Lee N. General algorithm and method for scanning a via hole by using critical-dimension atomic force microscopy. Journal of the Korean Physical Society. 2014;**64**:1643–1647

[26] Kulakov M, Luzinov I, Kornev K. Capillary and surface effects in the formation of nanosharp tungsten tips by electropolishing. Langmuir. 2009;**25**: 4462–4468

[27] Park BC, Jung KY, Song WY, Ahn SJ. Bending of a carbon nanotube in vacuum using a focused ion beam. Advanced Materials. 2006;**18**:95–98

[28] Xu Z, Fang F, Dong S. Electronic Properties of Carbon Nanotubes. Rijeka, Croatia: InTech; 2011

[29] Kim UJ, Furtado CA, Liu XM, Chen GG, Eklund PC. Raman and IR spectroscopy of chemically processed single-walled carbon nanotubes. Journal of the American Chemical Society. 2005;**127**(44):15437–15445

[30] Kazmi SJ, Shehzad MA, Mehmood S, Yasar M, Naeem A, Bhatti AS. Effect of varied Ag nanoparticles functionalized CNTs on its anti-bacterial activity against *E. coli*. Sensors and Actuators, A: Physical. 2014;**216**:287–294